Lecture Notes in Physics

Springer-Verlag Berlin Heidelberg GmbH

Physics and Astronomy ONLINE LIBRARY

http://www.springer.de/phys/

The Editorial Policy for Proceedings

The series Lecture Notes in Physics reports new developments in physical research and teaching – quickly, informally, and at a high level. The proceedings to be considered for publication in this series should be limited to only a few areas of research, and these should be closely related to each other. The contributions should be of a high standard and should avoid lengthy redraftings of papers already published or about to be published elsewhere. As a whole, the proceedings should aim for a balanced presentation of the theme of the conference including a description of the techniques used and enough motivation for a broad readership. It should not be assumed that the published proceedings must reflect the conference in its entirety. (A listing or abstracts of papers presented at the meeting but not included in the proceedings could be added as an appendix.)
When applying for publication in the series Lecture Notes in Physics the volume's editor(s) should submit sufficient material to enable the series editors and their referees to make a fairly accurate evaluation (e.g. a complete list of speakers and titles of papers to be presented and abstracts). If, based on this information, the proceedings are (tentatively) accepted, the volume's editor(s), whose name(s) will appear on the title pages, should select the papers suitable for publication and have them refereed (as for a journal) when appropriate. As a rule discussions will not be accepted. The series editors and Springer-Verlag will normally not interfere with the detailed editing except in fairly obvious cases or on technical matters.
Final acceptance is expressed by the series editor in charge, in consultation with Springer-Verlag only after receiving the complete manuscript. It might help to send a copy of the authors' manuscripts in advance to the editor in charge to discuss possible revisions with him. As a general rule, the series editor will confirm his tentative acceptance if the final manuscript corresponds to the original concept discussed, if the quality of the contribution meets the requirements of the series, and if the final size of the manuscript does not greatly exceed the number of pages originally agreed upon. The manuscript should be forwarded to Springer-Verlag shortly after the meeting. In cases of extreme delay (more than six months after the conference) the series editors will check once more the timeliness of the papers. Therefore, the volume's editor(s) should establish strict deadlines, or collect the articles during the conference and have them revised on the spot. If a delay is unavoidable, one should encourage the authors to update their contributions if appropriate. The editors of proceedings are strongly advised to inform contributors about these points at an early stage.
The final manuscript should contain a table of contents and an informative introduction accessible also to readers not particularly familiar with the topic of the conference. The contributions should be in English. The volume's editor(s) should check the contributions for the correct use of language. At Springer-Verlag only the prefaces will be checked by a copy-editor for language and style. Grave linguistic or technical shortcomings may lead to the rejection of contributions by the series editors. A conference report should not exceed a total of 500 pages. Keeping the size within this bound should be achieved by a stricter selection of articles and not by imposing an upper limit to the length of the individual papers. Editors receive jointly 30 complimentary copies of their book. They are entitled to purchase further copies of their book at a reduced rate. As a rule no reprints of individual contributions can be supplied. No royalty is paid on Lecture Notes in Physics volumes. Commitment to publish is made by letter of interest rather than by signing a formal contract. Springer-Verlag secures the copyright for each volume.

The Production Process

The books are hardbound, and the publisher will select quality paper appropriate to the needs of the author(s). Publication time is about ten weeks. More than twenty years of experience guarantee authors the best possible service. To reach the goal of rapid publication at a low price the technique of photographic reproduction from a camera-ready manuscript was chosen. This process shifts the main responsibility for the technical quality considerably from the publisher to the authors. We therefore urge all authors and editors of proceedings to observe very carefully the essentials for the preparation of camera-ready manuscripts, which we will supply on request. This applies especially to the quality of figures and halftones submitted for publication. In addition, it might be useful to look at some of the volumes already published. As a special service, we offer free of charge LaTeX and TeX macro packages to format the text according to Springer-Verlag's quality requirements. We strongly recommend that you make use of this offer, since the result will be a book of considerably improved technical quality. To avoid mistakes and time-consuming correspondence during the production period the conference editors should request special instructions from the publisher well before the beginning of the conference. Manuscripts not meeting the technical standard of the series will have to be returned for improvement.
For further information please contact Springer-Verlag, Physics Editorial Department II, Tiergartenstrasse 17, D-69121 Heidelberg, Germany

Series homepage – http://www.springer.de/phys/books/lnpp

A. Borowiec W. Cegła
B. Jancewicz W. Karwowski (Eds.)

Theoretical Physics Fin de Siècle

Proceedings of the XII Max Born Symposium
Held in Wrocław, Poland, 23-26 September 1998

Springer

Editors

Andrzej Borowiec
Wojciech Cegła
Bernard Jancewicz
Witold Karwowski
Institute of Theoretical Physics
University of Wrocław
pl. Maksa Borna 9
50-204 Wrocław, Poland

Library of Congress Cataloging-in-Publication Data applied for

Die Deutsche Bibliothek - CIP-Einheitsaufnahme

Theoretical physics, fin de siècle : proceedings of the XII Max Born Symposium held in Wroclaw, Poland, 23 - 26 September 1998 / A. Borowiec ... (ed.).
(Lecture notes in physics ; Vol. 539)
ISBN 978-3-662-14291-2 ISBN 978-3-540-46700-7 (eBook)
DOI 10.1007/978-3-540-46700-7

ISSN 0075-8450
ISBN 978-3-662-14291-2

Originally published by Springer-Verlag Berlin Heidelberg New York in 2000
Softcover reprint of the hardcover 1st edition 2000

Typesetting: Camera-ready by the authors/editors
Cover design: *design & production*, Heidelberg

Printed on acid-free paper
SPIN: 10720686 55/3144/du - 5 4 3 2 1 0

XII Max Born Symposium
Wrocław 1998
Theoretical Physics
Fin dé Siecle
September 23 - 26
1998
Wrocław

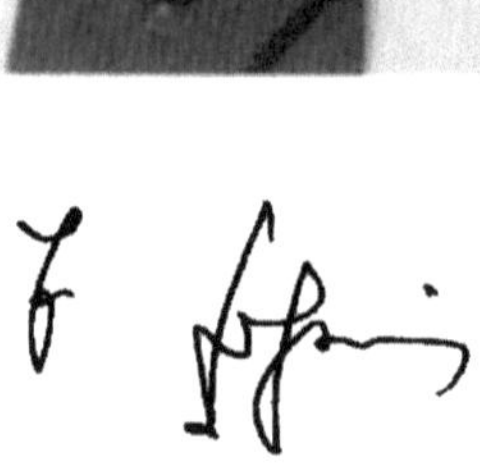

FOREWORD

The XII Max Born Symposium has a special character. It was held in honour of Jan Łopuszański on the occasion of his 75th birthday.

As a rule the Max Born Symposia organized by the Institute of Theoretical Physics at the University of Wrocław were devoted to well-defined subjects of contemporary interest. This time, however, the organizers decided to make an exception.

Łopuszański's influence on and contribution to the development of theoretical physics at Wrocław University is highly appreciable. His personality and scientific achievements gave him authority which he used to the best advantage of the Institute. In fact we still profit from his knowledge, experience and judgment. Łopuszański's scientific activity extended over about half a century. He successfully participated in research on the most important and fascinating issues of theoretical physics. During his scientific career he met and made friends with many outstanding physicists who shaped theoretical physics to the present form.

For this reason, as well as the coincidence of the approaching end of the century, we thought that it would be interesting and instructive to give the symposium a retrospective character. We decided to trust the speakers' judgment and intuition for the choice of subjects for their talks. We just asked them to give the audience the important message based on their knowledge and experience.

The beginning of the XII Max Born Symposium had a particularly solemn character. It took place in Aula Leopoldina, the beautiful baroque hall in the main building of our University. In the audience were present the participants and invited guests. Seven speeches were delivered in honour of Professor Jan Łopuszański. Professors from Wrocław, Z. Bubnicki, Z. Latajka and J. Ziółkowski, spoke on the academic career of Jan Łopuszański and his activity in the Wrocław division of the Polish Academy of Sciences. Professor J. Lukierski, as a director of our Institute, welcomed all the guests and, as a friend of Jan Łopuszański, gave a very personal history of Jan's life, showing also some photos starting form his childhood up to recent days.

Professor K. Zalewski from Cracow still remembers Łopuszański's PhD defense at the Jagiellonian University where he was present in the audience as a young student. Professor R. Haag recalled some humorous stories of his early meetings with Łopuszański. He underlined Łopuszański's honesty and sincerity in scientific research. It was Łopuszański who introduced him to supersymmetry , which resulted in a very influential paper by Haag, Łopuszański and Sohnius.

Among the guests of honour there was also Dr. Roland Kliesow, Consul General of the Federal Republic of Germany. He spoke of Łopuszański's contribution to German–Polish understanding. He considered Łopuszański as a man of deep knowledge of the German language, history and culture. At the time , when the political circumstances were unfavorable for German–Polish relations he co-worked with German scientists and helped to develop personal contacts and collaboration between German and Polish colleagues.

The opening session ended with a short piano recital given by the young pianist Michał Ferber.

The organizing committee takes the opportunity to thank warmly the sponsors:

University of Wrocław
Stiftung für deutsch–polnische Zusammenarbeit
The British Council
Ministry of National Education
Polish Academy of Sciences

Their financial help made the organization of the Symposium possible. Moreover, the Stiftug für deutsch–polnische Zusammenarbeit financially supported the publishing of the proceedings.

The organizing committee

Andrzej Borowiec
Wojciech Cegła
Bernard Jancewicz
Witold Karwowski

Jan Łopuszański – the Man and His Achievements

During the opening session of XII Max Born Symposium I had the honour and pleasure to present the life of Jan Łopuszański from his pre-scientific period in Lvov. Let me therefore first recall these first twenty two years of his life.

Jan Łopuszański was born on 21st October 1923, in Lvov, as the only child of Janina Łopuszańska, de domo Kuźmicz. His father, Władysław Łopuszański was, until Piłsudski's coup d'état in 1926, in governmental service, but after these events he left the state post, became the Head of the Local Landowners Association, and further the Director of the Insurance Company "Floryanka". The most well-known in the Łopuszański family were Professor Jan Łopuszański, Jan's uncle, who was the Head of the Ministry of Public Works in the 1920s, and also the Rector of the Lvov Institute of Technology, and Tadeusz Łopuszański, the Head of the Ministry of Religious Confessions and Education in the first years of independence. From the early years of Jan's life he had impeccable knowledge of the German language; the primary school education he mastered while being tutored by his German private teacher, Fräulein Henriette. The family of Jan Łopuszański belonged definitely to the upper class of Lvov's social circles. As a youngster he was neither interested nor involved in politics. Only from the perspective of many years, after the Second World War, did he recall complex and not always socially just relations between Polish, Ukrainian and Jewish communities. His traveling – a part of his duties as an international scientist – began quite early. For example in 1938 his summer vacation was spent in Italy, on the beach near Ancona.

In 1939 the Second World War started and Lvov was incorporated into the Ukrainian Soviet Republic. Jan attended the last classes of Soviet elementary school, the so-called "desjatiletka". However, he did not finish it. Under the accusation of participating in a subversive pupil's organization he was arrested and sentenced to 10 years of prison camp in Siberia. He was still in Lvov prison for the German offensive in June 1941. Only because of great luck and his very alert attitude, was he able to avoid being shot by escaping Soviet security forces. He escaped from prison a few moments before the beginning of the extermination of all prisoners. He confessed later that this was the most dramatic, and the most fortunate, moment of his life, which left a trauma for the rest of his life.

During the German occupation of Lvov (1941–44) Jan finished the clandestine high school and passed maturity exams together with the well-known Polish poet Zbigniew Herbert. He also worked for his living in the research institute for epidemic diseases and provided his blood by feeding lice needed for medical experiments. This permitted him to avoid the exportation to forced labour in Germany. After the second arrival of the Soviets in Lvov, in 1944, Jan started to attend the Polytechnical Institute. After the death of his father he decided to move with his mother to Wrocław.

The second part of his life and his whole scientific career was linked to Wrocław University.

Already at the beginning of his studies in Lvov he realized that his interests and research activities were linked more to pure science; his choice was the field of theoretical physics. After his arrival in Wrocław in 1945 he immatriculated as a student of physics at Wrocław University. At that time there were in Wrocław only three lecturers of physics all three from Lvov: Professors Stanisław Loria and Jan Nikliborc, and Roman Ingarden, the son of a famous philosopher, who became Loria's assistant. Jan Łopuszański obtained his master degree in 1950, and in 1952 he became a lecturer. His scientific career developed quickly; after defending his Ph.D. thesis in Cracow in 1955 Jan obtained the position of Docent and finally, in 1959, was nominated to the post of Professor in Physics.

The first eight years of the scientific career of Jan Łopuszański was devoted to the problems of statistical physics. He studied the statistical models of cosmic rays and cosmic cascades. By applying the theory of stochastic equations he obtained concrete solutions, providing good hints on how to compare the theory with experiment in cosmic rays physics.

1958 began a new period in the scientific career of Jan, related to three one-year research visits abroad: Utrecht University (1958), New York University (1961/62) and the Institute for Advanced Study in Princeton (1964/65). His new scientific passion was quantum field theory. In Utrecht he studied soluble field-theoretic models; two years later, in New York he became involved in the mathematical foundations of quantum field theory. In Princeton, Jan, together with Helmut Reeh, started the main scientific subject of his life: the problem of symmetries in classical and quantum physics. In particular, in 1965 with H. Reeh, Jan obtained important results concerning so-called spontaneous symmetry breaking in quantum models, which is related to the famous Goldstone theorem and the existence of degenerate physical vacua. Further, during his visit to Stony Brook in 1970/71, Jan studied the mathematical properties of generators in axiomatic field theory, and obtained the classification of all possible generators of internal symmetries. Unfortunately by introducing too restrictive assumptions he discarded the possibility of a new symmetry – supersymmetry. However, when, in the early 1970s, supersymmetry appeared as a new idea, transforming bosonic into fermionic fields, Jan was very well prepared to consider the classification theorem for all physically allowed supersymmetry generators. In 1975 his most famous paper appeared, written with R. Haag and M. Sohnius during his stay at Karslsruhe University and CERN, entitled "All possible generators of supersymmetries of the S-matrix" (Nucl. Phys. B **88**, 257 (1975)). In this paper appeared the first classification of four-dimensional supersymmetry algebras which are permitted by the axioms of local quantum field theory and the relativistic scattering theory described by the so-called S-matrix. This paper is at present the most well known single publication in the domain of theoretical physics from Wrocław after the Second World War – at present it has over 300 citations by other authors.

Now the scientific recognition of Jan's outstanding research results is complete. In 1976 Jan Łopuszański became the corresponding member of the Polish Academy of Sciences. He continued his research, in particular by considering further the notion of central charges, the mathematical object in supersymmetry scheme introduced by him earlier. He collaborated with Polish (M. Wolf) as well as foreign (D. Buchholz) specialists in algebraic methods, and visited several times the Max Planck Institute in Munich and the universities in Göttingen and Bielefeld. In particular Łopuszański obtained the rigorous definition of nonlocal symmetry charges as well as the definition of generators in the presence of massless excitations.

In the early days of his employment at Wrocław University Jan Łopuszański was already involved in administrative duties. He was elected in 1957 the Deputy Dean of the Faculty of Mathematics, Physics and Chemistry of Wrocław University, and in the period 1962–1968 a Dean of the Faculty. In the period 1954–1968 Jan Łopuszański worked also in the branch of the Polish Academy of Sciences in Wrocław. The most essential period, however, for theoretical physics at Wrocław University is the period 1970–84 when Jan Łopuszański was the Director of the Institute for Theoretical Physics. On one side he promoted new research domains (supersymmetry, quantum field theory) which engaged theoretical physics in Wrocław in front-line research in the world. Another important side of Jan's activities as director of the institute was very just handling of personal matters, with a unique and proper blend of tolerance and firmness. One can call the years 1970–1984 the golden period of theoretical physics at Wrocław University, characterized by the lot of contact with research centers abroad and quick development of new branches of research. From this period I would like only to mention the contacts with Stony Brook University and the head of theoretical physics there, Prof. C.N. Yang, Nobel Prize winner in 1957. In the late 1970s at least half of the members of the Institute for Theoretical Physics at Wrocław University visited Stony Brook, and obtained important scientific results in the framework of this scientific collaboration.

In the period 1984–1994 until retirement, Jan Łopuszański was an unquestionable moral authority, not only among his colleagues at the Institute, but also at the University of Wrocław as well as in the community of physicists in Poland. In 1986 he became a real member of the Polish Academy of Sciences, and in 1996 was nominated, as the only physicist from Wrocław, the real member of the Polish Academy of Arts and Science in Cracow. In that decade he publishes two books on "Spinorial Calculus" (PWN Wrocław, 1984) and "An Introduction to Symmetry and Supersymmetry in Quantum Field Theory" (World Scientific, Singapore, 1991). The last book also contains collected results from Łopuszański's research papers during 25 years on the subject of symmetry and supersymmetry.

The academic year 1993/94 was the last before Jan's retirement. He was not happy with his new situation after leaving university without any didactic and academic duties. Since 1996 he has again been employed at the institute, with a part-time contract, and every semester presents a monographic lecture on recent scientific developments. He is also scientifically active and in 1998, began preparing his new book about the research results obtained in the collaboration with P. Stichel and J. Cisło.

Now Jan approaches 75 years. He is quite often present in our institute, and very much interested in all scientific and human developments. His ability to give much advice on all important matters was always very essential for me personally. I hope that we shall be able to enjoy Jan's presence among us and his warm and friendly personality for many years in the Institute for Theoretical Physics.

Jerzy Lukierski

Contents

About the Volume

The editors 3

Max Born and Molecular Theory
H. Rechenberg 7

Part I Quantum Physics: Basic Problems

Trying to Divide the Universe
R. Haag 23

Modal Interpretation of Quantum Mechanics and Classical Physical Theories
R.S. Ingarden 32

On Localisation in Relativistic Quantum Mechanics
Th.W. Ruijgrok 52

Dynamical Equivalence, Commutation Relations and Noncommutative Geometry
P.C. Stichel 75

Quantum Channels of the Einstein–Podolsky–Rosen Kind
A. Uhlmann 93

Part II Quantum Physics: Different Approach

Quantum Dynamics in the Proper Time
Z. Haba 109

Is It Possible to Construct Exactly Solvable Models?
O. Haschke and W. Rühl 118

Names of the lecturers are underlined.

Summational Invariants
H. Reeh 141

Relative Entropy Estimates in Statistical Mechanics and Field Theory
B. Zegarliński 149

Partial *-Algebras: A Retrospective
J.-P. Antoine 161

Banach Partial *-Algebras and Quantum Models
C. Trapani 180

Part III Supersymmetry and Quantum Deformations

New Superparticle Models Outside the HLS Supersymmetry Scheme
I. Bandos and J. Lukierski 195

On the Implementation of Supersymmetry
D. Buchholz 211

On Generalized Quantum Statistics
W. Marcinek 221

q-Deformed Minkowski Algebra and Its Space-Time Lattice
J. Wess 227

Part IV Particles, Fields and Geometry

The New Inverse Problem of the Newton Law
M. Gusiew-Czudżak 237

Particle in Curved 2d Space–Time
G. Jorjadze and W. Piechocki 245

Are There Two Sterile Neutrinos Cooscillating with ν_e and ν_μ?
W. Królikowski 251

Magnetic Monopoles and Gravity
D. Maison 263

Critical Behavior of Classical Spin Models and Local Cohomology
A. Patrascioiu and E. Seiler 279

Bose–Einstein Correlations in High Energy Multiple Particle Production Processes
K. Zalewski 291

Reduction of Couplings in Massive Models of Quantum Field Theory
W. Zimmermann 304

Abstracts

Particle–Hole Asymmetry in the BCS Thermodynamics
J. Czerwonko 317

On Generalizations of the Gravitational Interaction
L. Halpern 318

List of Participants

Jean-Pierre ANTOINE e-mail: antoine@fyma.ucl.ac.be
Institut de Physique Théorique et Mathématique , Université Catholique de Louvain, Chemin du Cyclotron 2 , B-1348 Louvain-la-Neuve, BELGIUM ,,

Bogdan BANASIEWICZ
Electric Engineering Department , Technical University of Lublin , ul. Nadbystrzycka 38 , 20-618 Lublin , POLAND,,

Andrzej BOROWIEC e-mail: borow@ift.uni.wroc.pl
Institute of Theoretical Physics, University of Wrocław , pl. Maksa Borna 9 , PL-50-204 Wrocław, POLAND ,,

Wojciech CEGŁA e-mail: cegla@ift.uni.wroc.pl
Institute of Theoretical Physics , University of Wrocław , pl. Maksa Borna 9 , PL-50-204 Wrocław, POLAND ,,

Jerzy CISŁO e-mail: cislo@ift.uni.wroc.pl
Institute of Theoretical Physics , University of Wrocław , pl. Maksa Borna 9 , PL-50-204 Wrocław, POLAND ,,

Jerzy CZERWONKO
Institute of Physics , Wrocław University of Technology , Wybrzeże Wyspiańskiego 27 , PL-50-370 Wrocław, POLAND ,,

Andrzej FRYDRYSZAK e-mail: amfry@ift.uni.wroc.pl
Institute of Theoretical Physics , University of Wrocław , pl. Maksa Borna 9 , PL-50-204 Wrocław, POLAND ,,

Roman GIELERAK e-mail: gielerak@ift.uni.wroc.pl
Institute of Theoretical Physics , University of Wrocław , pl. Maksa Borna 9 , PL-50-204 Wrocław, POLAND ,,

Dariusz GRECH e-mail: dgrech@ift.uni.wroc.pl
Institute of Theoretical Physics , University of Wrocław , pl. Maksa Borna 9, PL-50-204 Wrocław, POLAND,,

Magdalena GUSIEW-CZUDŻAK e-mail: gusiew@ift.uni.wroc.pl
Institute of Theoretical Physics, University of Wrocław, pl. Maksa Borna 9, PL-50-204 Wrocław, POLAND,,

Rudolf HAAG
Waldschmidtstr. 4b, D-87327 Schliersee-Neuhaus, GERMANY ,,

Zbigniew HABA e-mail: zhab@ift.uni.wroc.pl
Institute of Theoretical Physics, University of Wrocław, pl. Maksa Borna 9, PL-50-204 Wrocław, POLAND,,

Leopold HALPERN
Department of Physics, Florida State University, Tallahassee, Florida 32306-3016, USA,,

Roman S. INGARDEN e-mail: romp@phys.uni.torun.pl
Institute of Physics, Nicholas Copernicus University, ul. Grudziądzka 5, PL-87-100 Toruń, POLAND,,

Lech JAKÓBCZYK e-mail: ljak@ift.uni.wroc.pl
Institute of Theoretical Physics, University of Wrocław, pl. Maksa Borna 9, PL-50-204 Wrocław, POLAND,,

Bernard JANCEWICZ e-mail: bjan@ift.uni.wroc.pl
Institute of Theoretical Physics, University of Wrocław, pl. Maksa Borna 9, PL-50-204 Wrocław, POLAND,,

Janusz JĘDRZEJEWSKI e-mail: jjed@ift.uni.wroc.pl
Institute of Theoretical Physics, University of Wrocław, pl. Maksa Borna 9, PL-50-204 Wrocław, POLAND,,

Cezary JUSZCZAK e-mail: cjusz@ift.uni.wroc.pl
Institute of Theoretical Physics, University of Wrocław, pl. Maksa Borna 9, PL-50-204 Wrocław, POLAND,,

Bartłomiej JURKIEWICZ e-mail: bajur@ift.uni.wroc.pl
Institute of Theoretical Physics, University of Wrocław, pl. Maksa Borna 9, PL-50-204 Wrocław, POLAND,,

Witold KARWOWSKI e-mail: wkar@ift.uni.wroc.pl
Institute of Theoretical Physics, University of Wrocław, pl. Maksa Borna 9, PL-50-204 Wrocław, POLAND,,

Małgorzata KLIMEK e-mail: klimek@matinf.pcz.czest.pl
Institute of Mathematics, Technical University of Czstochowa, pl. Dąbrowskiego 73, PL-42-200 Czstochowa, POLAND,,

Michael KNYAZEV e-mail: knyazev@iaph.bas-net.by
Instute of Applied Physics, National Academy of Sciences of Belarus, Academicheskaya ul. 16, BY-220072 Minsk, BELARUS,,

Sylwia KONDEJ e-mail: kondej@ift.uni.wroc.pl
Institute of Theoretical Physics, University of Wrocław, pl. Maksa Borna 9, PL-50-204 Wrocław, POLAND,,

Jan ŁOPUSZAŃSKI e-mail: lopus@ift.uni.wroc.pl
Institute of Theoretical Physics, University of Wrocław, pl. Maksa Borna 9, PL-50-204 Wrocław, POLAND,,

Piotr ŁUGIEWICZ e-mail: piotrlug@ift.uni.wroc.pl
Institute of Theoretical Physics, University of Wrocław, pl. Maksa Borna 9, PL-50-204 Wrocław, POLAND,,

Jerzy LUKIERSKI e-mail: lukier@ift.uni.wroc.pl
Institute of Theoretical Physics, University of Wrocław, pl. Maksa Borna 9, PL-50-204 Wrocław, POLAND,,

Dieter MAISON dim@mppmu.mpg.de
Max-Planck Institut für Physik, Föhringer Ring 6, D-80805 München, GERMANY,,

Władysław Adam MAJEWSKI e-mail: fizwam@univ.gda.pl
Institute of Theoretical Physics and Astrophysics, Gdańsk University, ul. Wita Stwosza 57, PL-80-952 Gdańsk, POLAND,,

Władysław MARCINEK e-mail: wmar@ift.uni.wroc.pl
Institute of Theoretical Physics, University of Wrocław, pl. Maksa Borna 9, PL-50-204 Wrocław, POLAND,,

Paweł MAŚLANKA e-mail: pmaslan@krysia.uni.lodz.pl
Department of Theoretical Physics II, Łódź University, ul. Pomorska 149/153, PL-90236 Łódź, POLAND,,

Anatol NOWICKI e-mail: anowicki@omega.im.wsp.zgora.pl
Institute of Physics, Pedagogical University, pl. Słowiański 6, PL-65-069 Zielona Góra, POLAND,,

Zbigniew OZIEWICZ e-mail: oziewicz@ift.uni.wroc.pl
Institute of Theoretical Physics, University of Wrocław, pl. Maksa Borna 9, PL-50-204 Wrocław, POLAND,,

Włodzimierz PIECHOCKI e-mail: wlodzimierz.piechocki@fuw.edu.pl
Theory Division, Institute for Nuclear Studies, ul. Hoża 69, PL-00-681 Warszawa, POLAND,,

Ziemowit POPOWICZ e-mail: ziemek@ift.uni.wroc.pl
Institute of Theoretical Physics, University of Wrocław, pl. Maksa Borna 9, PL-50-204 Wrocław, POLAND,,

Jerzy PRZYSTAWA e-mail: przystaw@ift.uni.wroc.pl
Institute of Theoretical Physics, University of Wrocław, pl. Maksa Borna 9, PL-50-204 Wrocław, POLAND,,

Szymon RABSZTYN
Institute of Mathematics, Silesian Technical University, ul. Zwycistwa 42, PL-44-100 Gliwice, POLAND,,

Helmut RECHENBERG e-mail: her@mppmu.mpg.de
Max-Planck Institut für Physik und Astrophysik, Föhringer Ring 6, Postfach 40-12-12, D-8000 München, GERMANY,,

Wolfgang RÜHL e-mail: ruehl@physik.uni-kl.de
Universität Kaiserslautern, P.O.Box 3049 D-67653 Kaiserslautern, GERMANY,,

Theodor W. RUIJGROK e-mail: t.w.ruijgrok@phys.uu.nl
Institute of Theoretical Physics, University of Utrecht, Princetonplein 5, NL-3508 TA Utrecht, THE NETHERLANDS,,

Erhard SEILER e-mail: ehs@mppmu.mpg.de
Max-Planck Institut für Physik, Föhringer Ring 6, D-80805 München, GERMANY,,
Jan SOBCZYK e-mail: jsobczyk@ift.uni.wroc.pl
Institute of Theoretical Physics, University of Wrocław, pl. Maksa Borna 9, PL-50-204 Wrocław, POLAND,,
Peter STICHEL e-mail: hanne@physik.uni-bielefeld.de
An der Krebskuhle 21, D-33614 Bielefeld, GERMANY,,
Camillo TRAPANI e-mail: trapani@unipa.it
Dipartimento di Science Fisiche e Astronomiche, Università di Palermo, Via Archirafi 56, I-90123 Palermo, ITALY,,
Ludwik TURKO e-mail: lturko@ift.uni.wroc.pl
Institute of Theoretical Physics, University of Wrocław, pl. Maksa Borna 9, PL-50-204 Wrocław, POLAND,,
Armin UHLMANN e-mail: armin.uhlmann@itp.uni-leipzig.de
Institute of Theoretical Physics, Universität Leipzig, Augustusplatz 10/11, D-04109 Leipzig, GERMANY,,
Julius WESS e-mail: julius.wess@physik.uni-muenchen.de
Max-Planck Institut für Physik, Föhringer Ring 6, D-80805 München, GERMANY,,
Uzal N. ZAKIROV e-mail: zakirov@sci.kcn.ru
Department of General Relativity and Gravitation, Kazan State University, Kremljevskaja ul. 18, RU-420008 Kazan, RUSSIA,,
Kacper ZALEWSKI e-mail: zalewski@chall.ifj.edu.pl
Institute of Physics, Jagellonian University, ul. Reymonta 4, PL-30-059 Kraków, POLAND,,
Bugusław ZEGARLIŃSKI e-mail: b.zegarlinski@ic.ac.uk
Department of Mathematics, Imperial College, 180 Queens Gate, London SW7 2BZ, UNITED KINGDOM,,
Wolfhart ZIMMERMANN
Max-Planck Institut für Physik, Föhringer Ring 6, D-80805 München, GERMANY,

About the Volume

ABOUT THE VOLUME

As it was mentioned in the Foreword, we invited contributions from the scientists who have been in close relations with Jan Łopuszański. This restricted the scope of the volume to the subjects related to his research interests. Still these frames allowed for rather broad spectrum of topics. Most generally they can be characterized by the key words: Quanta, Relativity, Symmetry, Statistics.

The volume is organized as follows. After a historical article by Helmut Rechenberg the contributions are collected in four chapters, and in addition two abstracts are put at the end of the Proceedings.

As explained by Helmut Rechenberg, already in early stage of its development the quantum mechanics provided a framework for physical models of atoms and molecules. The article presents Max Born conceptual and calculational contributions to the theory of molecules.

In *Chapter 1* some fundamental questions about the understanding of quantum theory are discussed.

Rudolf Haag poses some questions and gives suggestions concerning physical observations made in an infinite environment, the notions of individual objects, events and their relations to the space-time.

Roman S. Ingarden expresses an opinion that the understanding of quantum mechanics requires that its very formulation is based on a logic which abandons the distributive law, and discusses a special version of the quantum modal logic.

It is well known that in both relativistic and nonrelativistic quantum mechanics the localized states spread over all space under the time translation. Theodor W. Ruijgrok argues that it should not be interpreted as noncausality. This is so because a nonlocality in the coordinate representation of the eigenstates of the Newton-Wigner position operator is of the order of the Compton wavelength of the particles involved. Thus it is not relevant experimentally.

In standard classical mechanics the dynamics is determined by the Hamiltonian. Peter Stichel discusses a fundamental question concerning the canonical description of physically equivalent systems: whether the interaction can be represented by a nonstardard choice of the symplectic form (classical mechanics) or equal time commutation relations (quantum mechanics). The author presents also interesting observations concerning quantum field theory.

The contribution by Armin Uhlmann presents a view of the quantum information theory that the Einstein, Podolsky and Rosen effect is not a paradox but a channel or a part of a protocol to transfer so called "quantum information" from one system to another. The author uses the von Neumann quantum theory of measurements, Tomita-Takesaki theory and copositive operators introduced by Woronowicz, and translates properties of the density operators into properties of quantum channel maps. Among others, he poses the question: why do we restrict ourselves to the Hermitian operators? and gives the answer: we can use normal operators, and their complex eigenvalues may be treated as point on a screen.

Chapter 2 consists of papers which put emphasis on mathematical aspects of more specific problems in the quantum theory.

Zbigniew Haba proposes an explicitly relativistically invariant quantization scheme which treats simultaneously particle's wave function and quantum field fluctuation.

Oliver Hashke and Werner Rühl develop an algorithm which permits to construct new exactly soluble models. There are constructed prepotentials, defining the ground state wave functions of the Schrödinger operator associated with the solvable model. These prepotentials are in a one- to-one relation with the orbits of the Coxeter group, corresponding to a given Lie algebra.

Helmut Reeh discusses representations for the summational invariants. In a particular case he expresses them as linear combinations of the Schrödinger, respectively Poincaré groups. The complexity of a more general situation is illustrated by examples.

Bogusław Zegarliński gives a careful review of application of relative entropy estimates in the statistical mechanics and field theory. All this is done by strict mathematical methods.

Jean-Pierre Antoine and Camillo Trapani devote their articles to partial *-algebras, i.e. the algebraic structures for which multiplication is not always defined. Antoine gives a review of general mathematical theory as developed during its 15 years old history. First article on this subject appeared in the volume devoted to Jan Łopuszański on his 60-th birthday. Trapani concentrates on a special class of the so called Banach partial *-algebras and discusses their physical relevance.

The materials in *Chapter 3* are based on modern concepts of supersymmetry and quantum deformations.

Igor Bandos and Jerzy Lukierski consider a new class of superparticle models invariant under extended class of supersymmetries, with tensorial central charges. In 1975 Haag, Łopuszański and Sohnius introduced the supersymmetry algebra with scalar central charges (HLS scheme). About twenty years later it was observed that it is desirable to extend the HLS scheme by introducing tensorial central charges. In particular, such an algebra with two-tensor and five-tensor central charges describes at present very fashionable symmetry of eleven-dimensional M-theory.

Detlev Buchholz considers implementations of supersymmetry transformations by Hilbert space operators in the framework of supersymmetric asymptotically abelian C^*-dynamical systems. He shows that the only states admitting such an implementation are pure supersymmetric ground states and mixtures or elementary excitations thereof.

Władysław Marcinek proposes a generalization of quantum statistics which includes the one-dimensional models. The essential structure for such a generalization is a cross symmetry instead of the braided one. The Fock space representation and existence of the well defined scalar product are discussed.

Julius Wess studies the question how the relativistic Heisenberg algebra is deformed if we assume its invariance under the action (or rather coaction using the language of quantum groups) of q-deformed Lorentz group. After discussing the Hermiticity of deformed relativistic phase space generators Wess discusses their real spectra. It appears that the time variable and three-dimensional radius are becoming commuting observables with discrete eigenvalues which are described by explicitly given functions of the deformation parameter q.

Chapter 4 contains the papers on particle theory and various aspects of symmetry and geometry.

Magdalena Gusiew-Czudżak applies the formalism of differential forms to consider the so called inverse problem i.e. finding a Lagrangian for a given equation of motion. Her approach can be treated as the higher grade generalization of the Hamilton-Jacobi equation.

G. Jorjadze and Włodzimierz Piechocki present in two-dimensional spacetime a specific metric with nonzero curvature which exhibits already on classical level a kind of annihilation and creation of particle trajectories. Quantization in this scheme means finding appropriate representations of $sl(2, \mathbf{R})$ algebra on a set of trajectories.

Wojciech Królikowski investigates consequences of his model of "fermion texture". He finds that the model implies existence of two sterile neutrinos, i.e. those interacting only gravitationally. The mixing mechanism with the conventional neutrinos is proposed which may explain deficits of solar and atmospheric neutrinos.

The article of Dieter Maison recalls the development of last 20 years in the theory of fundamental interactions. The author finds a "half- supersymmetric" solution of the Einstein-Yang-Mills theory which means that it is annihilated by one half of the supersymmetric generators. The paper connects the standard model with supersymmetry and supergravity and finds gravitational confinement of nonabelian monopoles.

In the paper by A. Patrascioiu and Erhard Seiler the reflection positivity of Osterwalder and Schrader is used to establish a connection between the existence of a critical point in the classical spin models and the triviality of a certain cohomology class in the continuum limit. The rigorous result by Fröhlich and Spencer for $N = 2$ shows the Kosterlitz-Thouless transition. The standard wisdom is that for $N > 2$ the model does not become critical at any finite β. The authors, however, present arguments according to which all the $O(N)$ models have the transition to a spin phase.

Kacper Zalewski presents from diverse points of view a thorough analysis of the multiple production of bosons. He arrives at the conclusion that the momentum distribution, the particle cumulant and p-th correlation function can be expressed in terms of one function of two single particle momenta. He also explains under which conditions Einstein's correlation occurs.

Wolfhart Zimmermann discusses problems related to formulating method for reduction of the number of coupling constants in the quantum theory of massless fields. He shows that the principle of reduction is independent of renormalization scheme used. A possibility of eliminating the mass parameters is also discussed.

The Editors

Max Born and Molecular Theory

H. Rechenberg

Max-Planck-Institut für Physik,
Münich (Germany)

Introduction

While the 20th century is approaching its conclusion, the historian may look back and assemble the essential scientific fruits of the this period. Nearly fifty years ago, Werner Heisenberg stated in a lecture that in quantum or wave mechanics "a new, unified science of matter has arisen, where the separation between chemistry and physics essentially lost any meaning", because (Heisenberg 1953):

"The chemical properties of atoms have at least in principle become accessible to calculation, and already in the first years after the rise of quantum mechanics the simplest chemical binding, namely that of the two hydrogen atoms in the hydrogen molecule was calculated with the help of the new methods and was found in closest agreement with chemical experience. Thus the chemical valency-forces were explained on a physical basis, and the application of the new knowledge in industrial practices became only a matter of time."

Evidently, the new unification of physics and chemistry constitutes one of the most eminent results of 20th-century science, and to those who accomplished the great enterprize belonged, besides Heisenberg himself, Max Born and Friedrich Hund.

75 years ago, at about the time when Professor Łopuszański was born, Born began in Göttingen his course of 'Lectures on Atomic Mechanics'. These lectures summarized the theoretical foundations of atomic physics in 1923, notably the theory of multiply-periodic system, which was believed then to describe the detailed behavior of atoms and molecules. The author explained the goal of his course in the introduction of the later published book (Born 1925):

"The title 'atomic mechanics' has been shaped according to the notion of 'celestial mechanics', (and it) should express that here the facts of atomic physics are treated under the particular point of view of applying the mechanical principles. This implies that we are dealing with an attempt to treat atomic theory deductively ... a logical experiment, whose meaning is just to mark out the limitations up to which today's valid principles of atomic and quantum theory are substantiated, and to trace the paths leading beyond those limitations."

After having summarized the contents of the lectures, Born continued: "That I succeeded to edit these lectures as a book, I owe in first place to the devoted labor of my assistant Friedrich Hund. Large portions of the text, which I have only little worked over, are due to him" (Born 1925, p. VII).

For the Göttingen professor of theoretical physics the lectures on atomic mechanics in the winter term 1923/24 provided the structural basis of the later reformulation of the classical dynamics as quantum mechanics, which he himself helped to establish after in July 1925 the initial revolutionary step was taken by his other assistant, Werner Heisenberg. For Friedrich Hund, however, assisting Born in editing his lectures opened the door to a career in atomic and especially molecular theory; he indeed became the first pioneer of the wave-mechanical description of molecules.

In Max Born's scientific work the theory of molecules does not occupy a very prominent place. The list of major topics includes rather: relativity theory and kinetic theory of solids before 1920, general quantum theory and quantum mechanics in the twenties, and solid state theory, fluid mechanics and a little field theory and elementary particle theory afterwards. Still a sharper look at the bibliography reveals a number of papers devoted to molecular questions. These papers begin after 1915 with a couple of investigations on the dispersion of light in molecular gases and fluids. Upon the 1922 paper, entitled 'On the model of the hydrogen molecule', which marked Born's entry into the decisive period of atomic studies culminating with the discovery of quantum mechanics, there followed a set of significant molecular investigations, including papers written in collaboration with Erich Hückel, Heisenberg and James Franck, and finally the paper with with Robert Oppenheimer in 1927. On the other hand, besides Hund's pioneering work since 1926 on the quantum-mechanical theory of molecules, the investigations of Walter Heitler, Gerhard Herzberg and Eugene Wigner, also performed in Göttingen, established Born's institute as a center of molecular theory beyond national borders. The professor's extended review article on 'Chemical binding and quantum mechanics', published in the 1931 *Ergebnisse der exakten Naturwissenschaften* finished this enterprize.

Born's main interest in molecular theory was devoted (like in other topics of his scientific work) to *general principles*, while his associates worked out 'details'. His pushing forward the formalism often annoyed collaborators: thus Hückel did not enjoy "the tedious perturbation calculations" (Suchy 1980), and Heisenberg complained to Pauli on 7 December 1923: "The paper on molecules of Born and myself now is completed at last; it contains bracket symbols up to 8 indices and probably will be read by nobody" (Pauli 1979). Indeed, in Born's scientific work often formal, mathematical-technical aspects seem to suffocate the physical contents. However, his discipline and endurance in formalism paved the way again and again to important physical ideas and implications.

1 Molecules and Chemical Forces in the Old Quantum Theory (1920-1923)

As we have mentioned, Born approached topics of molecular physics only after more than a decade of scientific work had established his reputation in relativity and quantum theory. However, when he moved in 1919 to Frankfurt to assume his first independent and full professorship, he became quite interested in problems of of physical chemistry, like the hydratization heat of ions or the electric affinity of oxygen and sulfur atoms. In those days, he had to deliver a series of semi-popular lectures in order to collect money for the experimental investigations of his assistant Otto Stern and his student Elisabeth Bormann; and he published some of such accounts in the journal *Naturwissenschaften*, which addressed quite general questions in science. Thus he wrote an essay on 'The bridge between chemistry and physics' for the issue of 14 May 1920; there he wrote, after having reported on the results of previous generations (Marcellin Berthelot, Hendricus Van't Hoff and Walther Nernst) in explaining chemical affinity by physical, thermodynamical concepts (Born 1920):

"And still that task has not been solved by these, the task which must be present as the ideal in the mind of the physicists, namely, the proof of unity of all physical and chemical forces, and the reduction of those to the interaction between elementary constituents, the electrons and atoms or atomic nuclei, respectively. Today's physics already possesses pictures that surely approach reality to a certain degree, and it can with the help of those explain numerous mechanical, electrical, magnetic and optical properties of substances. It cannot stop in front of the chemical properties, and it must try to reduce them to atomic forces as well".

Especially the binding forces of diatomic molecules, Born argued further in 1920, seemed to come out – at least in the case of the so-called 'heteropolar' or ionic type – fairly accurately from the well-known model of atomic constitution, as developed since 1913 by Niels Bohr and Arnold Sommerfeld.

Two years later, after his return to Göttingen as professor of theoretical physics, Born penetrated more deeply into the recent atomic physics and began a systematic exploration of its theoretical basis. We should recall at that point that, as a former PhD student of the mathematician David Hilbert, Born had already in the previous Göttingen period (i.e., from 1908-1915) carried out his teacher's program of 'axiomatization of physics', which meant in particular the strict, methodical and mathematical formulation of the kinetic theory in the domain of solid state physics (Born and others called it 'crystal dynamics'). Now in the twenties, with Hilbert's interest in physics still continuing, he undertook a quite similar 'axiomatization', or perhaps we should say 'mathematical penetration', of the so far rather intuitively presented, refined atomic models of Bohr, Sommerfeld, Alfred Landé and others. In the short note of July 1922 already mentioned, Born selected the example of the hydrogen molecule in order to indicate the direction of how to approach the problem of atomic structure with the help of the classical perturbation scheme for multiply-periodic many-body systems. Together with his young, extremely talented personal assistants Wolfgang Pauli and Werner Heisenberg, he developed the quantum-theoretical adaptation of this scheme. "I am now working with Born to improve and refine the Born-Pauli method (of perturbation theory); with its help, for instance, one can prove that the quantum theory demands phase relations between the electrons of an atomic system. This method essentially comes from (Henri) Poincaré (notably, the latter's book on celestial mechanics)", Heisenberg wrote to Sommerfeld on 4 December 1922" (Sommerfed, Letter Archiv.).

In a paper entitled 'On the phase relations in Bohr's models of atoms and molecules', Born and Heisenberg worked out the perturbation-theoretical method for multiply-periodic systems having degenerate degrees of freedom, and they derived strict phase relations between the motions of electrons in atoms or molecules (Born and Heisenberg 1923a). With these subtle methods the authors then attacked the problem of the helium atom, but they found that the excited states thus calculated (with the proper quantum conditions employed) did not describe the observed spectra (Born and Heisenberg 1923b). Previously, Born and Erich Hückel – since 1921 *physikalischer Hilfsassistent* of David Hilbert and until September 1922 also associated with Born – had studied explicitly the quantum theory of polyatomic molecules in a detailed paper submitted in November 1922 (Born and Hückel 1923). Especially, they had worked out the coupling relations between the oscillational and rotational motions in these molecules on the basis of the Born-Pauli formalism (i.e., without any degeneracy). The tedious but straightforward formalism annoyed (as we have mentioned above) Hückel, who soon joined Peter Debye in Zurich to work on the theory of strong electrolytes. Heisenberg, on the other hand, did not give up easily; after having completed his doctorate with Sommerfeld in July 1923 at the Munich University, he again came to Born and got involved in the next paper 'On the quantum theory of molecules', in which both described systematically what they called 'Entwicklungsstufen (steps of development)' of molecular quantum theory, namely (Born and Heisenberg 1924):

(i) the simplest rotator (or spinning top for polyatomic molecules);
(ii) the consideration of harmonic nuclear oscillations;
(iii) the interactions between rotations and nuclear oscillations;
(iv) the inclusion of electronic and nuclear angular momentum;
(v) the full treatment of the nuclear and electronic structure of molecules.

To achieve this purpose, the authors wrote down a perturbation scheme, in which they expanded the energy of the states in powers of the square root of the ratio mass of the electron over mass of the nuclei, $\lambda = \sqrt{m/M}$, that is,

$$H = H_0 + \lambda^2 H_2 + \lambda^3 H_3 + \lambda^4 H_4 ,$$

with no first-order term, since for nuclei at rest the term linear in λ drops out.

In a general review lecture on 'The chemical binding as a dynamical problem', presented in September 1924 at the *Naturforscherversammlung* in Innsbruck, Born discussed some of the consequences from the Göttingen work on molecular theory for the understanding of the problem of chemical binding. At the end of his lecture, he emphasized that "every new result of the atomic theory will also throw light on the theory of molecules" (Born 1924), a quite adequate remark in the light of the many discrepancies of the existing quantum theory in those days. Indeed, since 1923 the good old Bohr-Sommerfeld theory of atomic structure had fallen into discredit: i. e., after the calculation of the helium states by Born and Heisenberg had failed, a series of other defects had emerged, reducing increasingly the range of application of the whole theory that had started out so successfully with Bohr explaining the hydrogen spectrum in 1913. The most active theoreticians, including Hendrik Kramers, Pauli, Heisenberg and Born himself, now suggested several mathematical tricks, such as those involved in the dispersion-theoretical approach, in order to cure the worst discrepancies. Only after a *new property of the electron*, its proper angular momentum or spin, had been discovered and a *new, quantum-mechanical formalism* based on non-commuting variables had been established, a consistent atomic theory resulted whose principles could be applied successfully to explain the structure and all properties of molecules (Mehra and Rechenberg 1982a). In these efforts towards the final solution of the problem, Born's students and associates played a principal role, and also the professor himself contributed importantly by own considerations, including those on molecular physics – here he collaborated with his experimental colleague and friend James Franck, whose presence and deep interest in molecular questions helped much to create an expertise of the Göttingen physicists in this field.

2 Göttingen, a Center of the Quantum Mechanics of Molecules (1925-1930)

In the period between 1925 and 1929 a new theoretical description of molecules containing also a physical explanation of chemical forces – Born's ideal of 1920! – arose, to which the Göttingen school of Max Born played a quite decisive part.. While Born himself participated mainly in the general problems of creating the appropriate quantum-mechanical formalism and its physical interpretation, he left the detailed investigation and the discovery of crucial effects and concepts mostly to his former students and collaborators, who now worked at Göttingen and few other places (like Copenhagen and Rostock).

The miraculous year 1925 opened in January with Pauli's discovery of the 'fourth quantum number' of the electron, which was associated in October of that year by George Uhlenbeck and Samuel Goudsmit in Leyden – with the proper angular momentum (or 'spin') of the subatomic particle. In July 1925 Heisenberg handed his paper, entitled 'On the quantum-theoretical re-interpretation of kinematic and mechanical relations', over to Born, who recognized in the formal relations given there the mathematical calculus of matrices and worked out, in the Heisenberg's absence with Pascual Jordan, the so-called 'matrix mechanics' (September 1925) (Heisenberg 1925; Born and Jordan 1925; Mehra and Rechenberg 1982b). Then the team Born, Heisenberg and Jordan developed the full matrix theory of quantum mechanics in a great memoir of 60 pages, submitted in November 1925 (Born et al. 1926). Amongst other items it contained a complete quantum-mechanical reformulation of the classical (celestial) perturbation scheme, where the dynamical variables and the action function were expressed as Hermitean matrices.

Soon afterwards, in the early months of 1926, Erwin Schrödinger of Zurich published his version of modern atomic theory, wave mechanics, in which a corresponding perturbation theory could be written down for operators expressing dynamical variables. Both the Heisenberg-Born-Jordan and the Schrödinger theory could be immediately applied to molecular systems – see, e.g., the paper of Erwin Fues, 'The spectrum of eigenoscillations of diatomic molecules according to undulatory mechanics', received on 21 April 1926 by the Annalen der Physik (Schrödinger 1926; Fues 1926; Mehra and Rechenberg 1987). Still an essential ingredient to the molecular theory was missing, until Heisenberg came up in July 1926 with his quantum-mechanical treatment of many-body systems; notably, he proposed to consider two or more coupled identical systems as one resonating system, and introduced the so-called 'exchange integral' to describe the situation mathematically (Heisenberg 1926a). He applied this idea immediately to solve the helium problem, which he had tried unsuccessfully back in 1923 with Born (Heisenberg 1926b). We may add perhaps at this point that already in 1922 Heisenberg had suggested an exchange of two 'equivalent' electrons in the helium ground state, thus anticipating somehow the new resonance phenomenon in quantum mechanics (Mehra and Rechenberg 1982b).

Having quantum mechanics at hands, no further obstacle stood in the way of the physicists in developing a new molecular theory. According to his own taste, Born composed in August 1927, together with the American guest Robert Oppenheimer, a systematic paper entitled 'On the quantum mechanics of molecules' (Born and Oppenheimer 1927). Here the previous classical perturbation treatment for molecular systems (as given in 1922/23 by Born, Pauli, Hückel and Heisenberg) now was reformulated consistently in the language of wave mechanics. The authors refined the old approach by using as a parameter in the perturbation expansion the fourth root of the ratio electron mass over mass of the nuclei, $\lambda = (m/M)^{1/4}$ – instead of the previous $(m/M)^{1/2}$ – and found in particular :

1. Nuclear oscillations correspond to terms of second order;

2. rotations correspond to terms of fourth order;
3. the first- and third-order terms disappear.

Also, to determine the eigenfunctions of the molecule describing its electronic motions in zeroth order, one needs the energy expression up to fourth order, and (Born and Oppenheimer 1927, p. 460):

"Higher than fourth-order approximations are not treated in the paper; they would correspond to couplings between the three principal types of motion. A calculation of these effects would make sense only if simultaneously all degeneracies of the electron motion for nuclei at rest are taken into account, in particular Heisenberg's resonance degeneracy due to equivalent electrons (perhaps also of some equivalent nuclei!) and, for the diatomic molecules, the degeneracy of the proper rotation about the axis connecting the nuclei".

While Born and Oppenheimer provided the general line of the procedure, others began to deal with a more detailed theory of molecules. The first of them was Born's former assistant Friedrich Hund, who had thought – as an entry in his scientific diary of November 1923, stating 'molecular spectra?', shows – already in fall of 1923 about the problem. In February 1924 he had studied eagerly diatomic molecules, e.g., H_2 and HCl , and in fall of the same year he had suggested (again in his diary) a systematic program of investigating diatomic and polyatomic molecules, including H_2O, CO_2, NH_3 and CH_4 (Hund 1924). He then had noticed, in a paper on 'The shape of polyatomic polar molecules' which he submitted in December 1924 for publication, especially (Hund 1925):

"The calculation of the structure of molecules is not possible today on the basis of quantum- theoretical principles. However, we can reduce the constitution of *polar molecules*, i.e., of those in which the single constituents (ions) are contained relatively unaltered, to the properties of the constituents themselves. ... The formation of a polar molecule rests on the electric forces between positive and negative ions; in their calculation one can neglect quantum theory".

At the time when quantum mechanics was established by Heisenberg and other Göttingen colleagues, Hund occupied himself with working on a detailed, 'phenomenological' description of complex atoms. He returned in fall of 1925 to study molecular problems and submitted since March 1926 continuously papers on the subject for publication, which would distinguish him eventually as a great pioneer of the new molecular theory (Hund 1926–27).

In particular, Hund turned to a wave-mechanical treatment of the molecular constitution while staying, without teaching duties, from October 1926 to March 1927 on a fellowship of the International Education Board with Bohr in Copenhagen. Starting from the two-center problem – i.e., the case of diatomic molecules – which had been considered earlier (say, in Born's lectures) with the help of the classical perturbation theory, he arrived in three papers submitted to *Zeitschrift für Physik* between December 1926 and May 1927 at a quantum-mechanical interpretation of the molecular energy terms, as had been obtained recently from experiment by the Americans Robert Mulliken and Raymond Birge (Mulliken 1926, Birge 1926). In the last of this series of three publications (entitled 'On the interpretation of molecular spectra' close of citation), which dealt with polyatomic molecules, Hund discovered a new effect that became later known as 'tunnel effect'. This effect connected, in particular, the states of optically isomeric molecules, in which the electrons can penetrate quantum-mechanically (but not classically) through a potential barrier; the transition from one state of the isomeric molecule to the other occurred in times of the order of many thousand years if the proper values for potential barriers and excitation energies were used (Hund 1926–27). In the following investigations, Hund applied the symmetry-group approach, which had just been pioneered by Heisenberg and Wigner in atomic theory, to molecular problems (Heisenberg 1927; Wigner 1927). Thus he quickly advanced to Germany's top expert in this field and was invited as such to present a review at the *Deutsche Physikertag*, held in September 1927 at Bad Kissingen (Hund 1927). There he mentioned also a work of Walter Heitler and Fritz London from Zurich, entitled 'Interaction of neutral atoms and covalent binding on account of quantum mechanics' (Heitler and London 1927). These authors had developed recently a method of molecular theory competing with the scheme, which Hund and Robert Mulliken had derived together for explaining molecular constitution and the chemical consequences (Mulliken 1927).

The American John Slater, himself an expert on molecular theory, later called Heitler and London's approach – which was explicitly based on an extension of Heisenberg's exchange-integral method to the two-center problem of the hydrogen molecule "a great sensation among physicists", because: "Here was the first suggestion of a theory that could explain the covalent bond, that between two neutral atoms which did not form a ionic molecule" (Slater 1975). After Schrödinger left Zurich in fall of 1927 (to take over Planck's chair at the University of Berlin), Heitler came to Göttingen to replace Hund (who had been called as an extraordinary professor of theoretical physics to the University of Rostock, see Rechenberg 1996) as assistant to Born. Heitler continued to work in Born's institute on the Heitler-London theory, for which Born had immediately shown great interest; especially, he used the then fashionable group-theoretical method (which he had begun to study already in Zurich) and treated the formation of molecules from atoms having single valency-electrons (Sugiura 1927; Heitler 1927, 1928). In Göttingen at that time also Eugene Wigner and the American guest E.E.Witmer joined their forces to derive rules for the electronic terms of molecules from group-theoretical considerations (Wigner and Witmer 1928). Finally another guest arrived at Born's institute, Gerhard Herzberg. He had obtained his PhD at Darmstadt, with an experimental investigation on the properties of nitrogen-band spectra. Now he joined Heitler in substantiating the Heitler-London theory of the covalent binding (Heitler and Herzberg 1929).

Meanwhile Hund and Mulliken had perfected their alternative theory, which they later named 'the method of molecular orbitals'. The central idea of the Hund-Mulliken scheme was that an electron in a molecule should, just like one in a many-electron atom, move in the field of nuclei and the other electrons, while in the Heitler-London method the electron would not belong to a single center. In 1928 and 1929 (and also later) a fight arose between the two camps as to which method described the properties and data of molecules better: it seemed that concerning the data description, the Hund-Mulliken scheme had to be preferred, while concerning the explanation of chemical binding the Heitler-London method was superior. However, in summer 1929 Herzberg completed in Göttingen his fundamental paper 'On the constitution of diatomic molecules' (Herzberg 1929). "We have not invented quantum chemistry", Hund admitted 40 years later, and continued (Hund 1968):

"Heitler and London began this discipline. Within the framework of molecular orbitals, it was Gerhard Herzberg who explained chemical binding in a simple and convincing manner with bonding and antibonding electrons, where an antibonding electron could counterbalance a bonding electron".

In 1971 Herzberg would receive, partly for his work in Göttingen, the Nobel Prize in chemistry.

3 Born and the Theory of Chemical Binding (1930-1933)

For several further years Göttingen remained a center of molecular research. Besides Heitler (who got his *Habilitation* in 1929 and stayed until 1933), the mathematician (and since spring of 1930 Hilbert's successor) Hermann Weyl got involved and considered in two notes the calculation of the molecular binding energies as a problem of mathematical group theory (Weyl 1930). The work of Heitler and Weyl in turn stimulated Max Born to write himself a series of papers dealing with the quantum theory of chemical forces. In the first of these, submitted on 29 July 1930 to *Zeitschrift für Physik*, he showed how to obtain Heitler's formulae for the chemical binding forces of two unequal atoms without the help of group theory, i.e., by applying rather John Slater's determinant method – by which the American had defeated what he called the '*Gruppenpest* (group pestilence)' in the theory of atomic spectra (Slater 1929). Born suggested especially a simple vector-diagram method using the total spin of the (valency-) electrons in the molecules (Born 1930). Heitler and George Rumer then applied the Born-Slater method to investigate the complex situation of polyatomic molecules (Heitler and Rumer 1931).

Finally, Born summarized the quantum-mechanical foundation of chemistry in his review article for the *Ergebnisse der exakten Naturwissenschaften*, entitled 'Chemical binding and quantum mechanics'. In the introduction, he pointed out the the problems that had been solved since 1920 (Born 1931):

"The task of reducing chemical valencies to physical forces meets the difficulty that within chemistry itself the concept of valency loses its originally clear determination. Initially, it was doubtlessly meant in the way that the picture of valency-dashes or, mechanically stated, of pair-like saturation forces was taken seriously and applied in every case. Already the existence of compounds, in which the same atom assumes variable valuedness, shook this orthodox conception. Then came, in addition, the discovery of the different nature of polar and non-polar bindings, the meaning of the coordination number, and finally the recognition that all these new concepts may, taken in the strict sense, only be applied to ideal limiting cases, while in the general case they overlap and cross over. Thus in anorganic chemistry hardly anything has remained of the original valency idea. However, in organic chemistry it has been kept, and still constitutes today an important tool of research, though even there everywhere defects and gaps have shown up. One may say that the valency schemes represent the bone-skeleton, which the living structure of facts have covered totally".

Born discussed all available methods, besides those of Heitler and London and Hund and Mulliken also the attempt of Slater and himself to avoid the group theory altogether, an attempt which he admitted had failed in some sense, because: "It seems that the binary invariants due to Weyl turn out to be the adequate tool for a computational treatment of valences according to the Heitler-London scheme" (Born 1932). Still he presented a short account of the Slater-Born scheme to explain 'the saturation of valencies' in a contribution, which he read at the Centenary (1931) Meeting of the British Association for the Advancement of Science in London (Born 1932, p. 249). While for diatomic molecules, he argued there, the original principle of Heitler and London – namely", that every valency of an atom corresponds to an electron, the spin of which is not compensated by that of another; and the saturation of two valencies of two atoms corresponds to the mutual interaction of two such valency electrons (one for each atom), the spins being antiparallel to each other – was "entirely serviceable", already in the case of triatomic molecules combinatory difficulties arose, as "the number of valency-bond schemes of the three atoms is not equal to the number of possible spin configurations" (Born 1932, p. 252).

The author then represented the possibilities for the free valencies v of these atoms and noticed in a particular case of three atoms A, B, C, having valencies respectively $a = 3$, $b = 2$, $c = 1$. If arranged according to finally free valencies, there are four types available:

$v = 6$: one case, where all atoms are separate and exhibit their 6 free valencies;

$v = 4$: three cases, where in case (i) A and B are coupled and C is separate, in (ii) A and C are coupled and B is separate, and in (iii) A is separate and B and C are coupled;

$v = 2$: three cases, where in two cases all A, B and C are coupled, and in the third A and B are coupled and C is separate;

$v = 0$: one case, where all A, B and C are coupled thus that no valency is free.

Now considering the cases $v = 2$, in particular, Born found that to *three possible valency-bond schemes* only *two different spin configurations* corresponded, which seemed to represent a quite unsatisfactory situation. However, due to Weyl's investigation one should assign to every atom one letter – say x to A, y to B, z to C – and to every valency occurring between the atoms a square bracket – say – $[x, y]$ to a valency between A and B, with $[x, l]$ denoting a free valency of the atom A. Born then found for $v = 2$ exactly three bond schemes , namely:

$$\phi_1 = [x,l]^2[x,y][x,z]; \quad \phi_2 = [x,l][x,y]^2[z,l]; \quad \phi = [x,l][x,y][y,l][z,x]\,,$$

and asked the question as to whether on could "assign a real physical significance to the symbolic quantities ϕ_1, ϕ_2, ϕ_3." This was indeed the case, and "the ϕ products already described are simply all the independent spin-invariants, which can be formed from the spin vectors x, y, z of the atoms and the 'void vector' l in such a way that they are of the order a (i.e., 3) in $x, \ldots$ and finally of the order v $(= 2)$ in l," and (Born 1932 p. 252).

"Accordingly, every valency-bond scheme corresponds to a 'pure valency state', characterized by a definite spin invariant ϕ, and the theory of combination of valencies is completely identical with the theory of combination of invariants of two-dimensional vectors or 'binary forms'. It is very interesting to find that this formal connection was noticed already more than fifty years ago by some mathematicians, Cayley, Sylverster and others."

Now a discrepancy arose between the number of spin configurations and the number of valency schemes, "because the vector addition of spins does not include all invariants of ϕ, but just only the linearly independent ones". As a rule, there are identities between the $\phi's$, like $(\phi_1 + \phi_2 + \phi_3 = 0$. Further, "it can be seen that the pure valency states , which are assigned to the valency-bond schemes, do *not* in general correspond to states of definite energy," Born continued, or "the idea of a 'state of definite energy' is wider than that of a 'chemical molecule' ", because: "Whereas the former refers to arbitrary positions of of the atomic nuclei, in a chemical molecule, the nuclei have definite positions in which there is a stable equilibrium, i.e., minima of the curve showing the energy as a function of nuclear separation." On the other hand, several minima or nuclear types can belong to 'state of definite energy'. Still the author finished his talk (Born 1932, p. 253 and p. 254):

"Notwithstanding, I should like to express the opinion that this extension of Heitler and London's theory is of great value. For it shows why throughout all changes in modern views on valency the old scheme of valency bonds still stands out, clearly visible amongst the maze of facts involved."

The presentations and discussions at the 1931 British Association Meeting provided proper contributions also to the Faraday jubilee – 100 years had passed since the discovery of the electric-voltage induction by a changing magnetic field – celebrated on the same occasion. Besides Born, there spoke in the same session on molecular structure Peter Debye, John-Edward Lennard-Jones, Ralph Howard Fowler, Victor Henri, Werner Heisenberg and William Lawrence Bragg. Born certainly succeeded to demonstrate to a very distinguished international circle the importance of his Göttingen school also in the field of molecular theory. Afterwards he contributed only twice to molecular theory: in a paper with Siegfried Flügge on a specific point in the description of diatomic molecules, published in 1933 shortly before he had to leave his Göttingen position, and in a later note – a little addendum to the 1927 paper of Born and Oppenheimer – which he contributed to the Göttingen Academy in November 1951, about two years before he returned to Germany again (after the retirement from the Edinburgh professorship) (Born and Flügge 1933; Born 1951). In spite of the comparatively short period of a little more than a decade, which Born devoted to molecular problems in a scientific carrier extending over nearly sixty years, the contributions discussed here were quite respectable. They again confirm the quality of this great scholar and teacher, *who combined in an extraordinary manner the spirit of mathematics and physics, thus continuing the brilliant Göttingen tradition of Gauss, Riemann and Hilbert.*

References

Birge, R.T., (1926): The structure of molecules, Nature **117**, 300–302

Born, M., (1920): Die Brücke zwischen Chemie und Physik, Naturwiss **8**, 373–382, esp. pp. 373–374

Born, M., Heisenberg, W., (1923a): Über Phasenbeziehungen bei den Bohrschen Modellen von Atomen und Molekeln. Z. Phys. **14**, 44–58

Born, M., Heisenberg, W., (1923b): Die Elektronenbahnen im angeregten Heliumatom, Z. Phys. **16**, 229–243

Born, M., Hückel, E., (1923): Zur Quantentheorie mehratomiger Molekeln, Phys. Z. **24**, 1–12

Born, M., Heisenberg, W., (1924): Zur Quantentheorie der Molekeln, Ann. d. Phys. (4) **74**, 1–31

Born, M., (1924): Die chemische Bindung als dynamisches Problem, Naturwiss **12**, 1199–1207

Born, M., (1925): Vorlesungen über Atommechanik, J. Springer, Berlin, p. V.

Born, M., and Jordan, P., (1925): Zur Quantenmechanik, Z. Phys. **34**, 858–888;

Born, M., Heisenberg, W., Jordan, P., (1926): Zur Quantenmechanik. II, Z. Phys. **35**, 557–615

Born, M., Oppenheimer, R., (1927): Zur Quantentheorie der Molekeln, Ann. d. Phys. (4) **84**, 457–484

Born, M., (1930): Zur Quantentheorie der chemischen Kräfte. Z. Phys. **64**, 729–740

Born, M., (1931): Chemische Bindung und Quantenmechanik, Ergeb. exakt. Naturwiss. **10**, 387–444

Born, M., (1932): The application of the theory of homopolar valency to polyatomic molecules, In Chemistry at the Centenary (1931) Meeting of the British Association for the Advancement of Science, W. Heffter & Sons, Cambridge pp. 249–254

Born, M., (1933): Flügge, S.,: Zur Quantenmechanik des Zweiatomsystems, Ann. d. Phys. (5) 16, 768–780;

Born, M., (1951): Kopplung der Elektronen- und Kernbewegung in Molekeln und Kristallen. Nachr. Akad. Wiss. Göttingen, Math.-Phys. Kl. (IIa) 1951, No. **6**, 1–3

Fues, E., (1926): Das Eigenschwingungsspektrum zweiatomiger Moleküle, Ann. d. Phys. (4) **80**, 367–396

Heisenberg, W., (1925): über quantenmechanische Umdeutung kinematischer und mechanischer Beziehungen, Z. Phys. **33**, 879–888;

Heisenberg, W., (1926a): Mehrkörperproblem und Resonanz in der Quantenmechanik, Z. Phys. **38**, 411–426

Heisenberg, W., (1926b): über die Spektra von Atomsystemen mit zwei Elektronen, Z. Phys. **39**, 499–518

Heisenberg, W., (1927): Mehrkörperprobleme und Resonanz in der Quantenmechanik II, Z. Phys. **41**;

Heisenberg, W., (1953): Die Beziehungen zwischen Physik und Chemie in den letzten 75 Jahren. Naturwiss. Rundschau **6**, 1–7, esp. p. 4

Heitler, W., London, F., (1927): Wechselwirkung neutraler Atome und homöopolare Bindung nach der Quantenmechanik, Z. Phys. **44**, 455–472

Heitler, W., (1927): Störungstheorie und Austausch bei Mehrkörperproblemen, Z. Phys. **46**, 47–92

Heitler, W., (1928): Zur Gruppentheorie der homöopolaren Bindung; Z. Phys. **47**, 835–858

Heitler, W., Herzberg, G., (1929): Eine spektroskopische Bestätigung der quantenmechanischen Theorie der homöopolaren Bindung. Z. Phys. **53**, 52–56

Heitler, W., Rumer, G., (1931): Quantentheorie der chemischen Bindung für mehratomige Moleküle, Z. Phys. **68**, 12–41

Herzberg, G., (1929): Zum Aufbau der zweiatomigen Moleküle. Z. Phys. **57**, 601–630

Hund, F., (1924): Wissenschaftliches Tagebuch, entries of November 1923, February 1924 and 9 October

Hund, F., (1925): Die Gestalt mehratomiger Molekeln. Z. Phys. **31**, 81–106, esp. p. 81

Hund, F., (1926): Zur Deutung einiger Erscheinungen in den Molekelspektren, Z. Phys. **36**, 657–674; Hund, F., (1927): Zur Deutung der Molekelspektren I, II, III; Z. Phys. **40**, 742–764; **42**, 93–120; **43**, 805–826

Hund, F., (1927): Fortschritte in der Systematik und Theorie der Molekelspektren, Phys. Z. **28**, 779–784

Hund, F., (1968): After Dinner Speech for Robert Mulliken Memorial Lectures, Chicago

Mehra, J., Rechenberg, H., (1982): The Historical Development of Quantum Theory, Vol 1, Springer-Verlag, New York, Heidelberg, Berlin, Chapters III to VI.

Mehra, J., Rechenberg, H., (1982): The Historical Development of Quantum Theory, Vol. 2 and Vol. 3, Springer-Verlag, New York, etc.

Mehra, J., and Rechenberg, H., (1987): The Historical Development of Quantum Theory, Vol. 5. Springer-Verlag, New York, etc., Chapters III and IV,

Mulliken, R.S., (1926): Systematic relation between electronic structure and band-spectrum structure in diatomic molecules, Proc. Natl. Acad. USA **12**, 144–151; 151–158;

Mulliken, R.S., (1927): Electronic states and band-spectrum structure in diatomic molecules IV, Hund's theory; second positive nitrogen and Swan bands; alternating intensities. Phys. Rev. **29**, 630–649

Pauli, W., (1979): Wissenschaftlicher Briefwechsel, Vol. I: 1919–1929, Springer-Verlag, New York, Heidelberg, Berlin, p. 132

Rechenberg, H., (1996): Die erste Göttinger Zeit und die Rostocker Professur (1919–1929), in: M. Schroeder, ed.: Hundert Jahre Friedrich Hund (Nachr. Akad. Wiss. Göttingen II, Math.-naturwiss. Klasse). Vandenhoeck & Rupprecht, Göttingen, pp. 4–32

Schrödinger, E., (1926): Quantisierung als Eigenwertproblem I, II, Ann. d. Phys. (4) **79**, 361–376; 489–527;

Slater, J., (1929): The Theory of complex spectra, Phys. Rev. **34**, 1293–1322

Slater, J., (1975): Solid State and Molecular Theory, A Scientific Autobiography, John Wiley, New York, London, p. 93

Suchy, K., (1980): Erich Hückel zum Gedenken, Phys. Bl. **36**, 248–249, esp. p. 249

Sugiura, Y., (1927): über die Eigenschaften des Wasserstoffmoleküls, Z. Phys. **45**, 484–492;

Weyl, H., (1930): Zur quantenmechanischen Berechnung der molekularen Bindungsenergien I, II, Nachr. Ges. Wiss. Göttingen, pp. 285–294; ibid 1931, 33–39

Wigner, E., (1927): Die nichtkombinierenden Terme in der neuen Quantenmechanik I, II. Z. Phys. **40**, 493–500; 883–892

Wigner, E., Witmer, E.E., (1928): über die Struktur der zweiatomigen Molekelspektren nach der Quantenmechanik. Z. Phys. **51**, 859–886

Part I

Quantum Physics: Basic Problems

Trying to Divide the Universe

R. Haag

Waldschmidtstr. 4[b]
D-83727, Schliersee-Neuhaus
Germany

Dedicated to Jan Łopuszański
on the occasion of his 75th birthday

Abstract. Questions and suggestions concerning the notions of individual objects, events and their relation to space-time.

1 Introduction

In recent years I attended several conferences centered around spectacular experimental progress in atomic physics, quantum optics and their implications for our understanding of Quantum Theory. One of them, entitled "Quantum Future" was the Max Born Symposium preceding our present gathering; another, entitled "Mysteries, Puzzles and Paradoxes in Quantum Mechanics" took place three weeks ago at Lake Garda. Both were excellent conferences. We heard authoritative reports on the breeding of Schrödinger kittens, EPR-type experiments, quantum erasers, non- demolition experiments, teleportation and much more. Among the predominant mysteries were the "extremely non-local aspects of Quantum Mechanics" and questions about "reality". When I told my young neighbor at dinner that I had written a book entitled "Local Quantum Physics" he asked me: "when did you write that book?". Well; it was not long ago and I do not want to apologize for it. There is obviously a question of language. What do we mean by "Locality" or "Reality"? I think that some of the paradoxes loose their bite if one uses concepts and a language which differ somewhat from the standard ones. But there remain problems, mysteries if you wish, which in my opinion cannot be clearly resolved within the scope of the presently existing theory.

I have chosen the somewhat peculiar title for my talk in order to stress a point whose significance is not duly appreciated in many discussions on mysteries. I shall then address two other controversial issues: indeterminism and realism and briefly state my position on them. This leads to a rough conceptual picture which, in my judgment, incorporates in a natural way essential lessons of Quantum Physics but also raises new questions.
The division problem. Every scientific endeavor begins by dividing complex situations into individual elements. This is sometimes called "reductionism" in contrast to "holism" which stresses that everything hangs together. Now we have painful experiences with the fact that any kind of "ism" is dangerous.

Of course the holistic criticism of reductionism, whether it concerns medicine or even physics, is justified. But holism does not provide a scientific method. We have to divide and, as it turns out, we are highly successful in doing so up to some point. Let us look at various approaches to the division problem in physics.

The first is the division into individual objects, beginning from bulk material and proceeding with its subdivision into molecules, atoms and ultimately "elementary particles". The reduction of physics to the study of interaction between such objects is the point of view of mechanics. Quantum mechanics takes over this vocabulary but shows that the notion of individual objects can have a precise meaning only under very special circumstances. One requirement is isolation. An isolated, stable object has some distinctive attributes: mass, magnitude of spin, an array of charge quantum numbers and an internal structure. Thus every carbon atom in the ground state has the same internal structure, described by an internal wave function (after separating off the part that refers to the object as a whole). It is one of the triumphs of Quantum Mechanics to achieve a classification of stable objects and to describe their distinctive attributes. But isolation can never be perfectly realized. It is an asymptotic notion. Furthermore it does not even suffice to allow us to speak of an individual object. There may be entanglement[1] with other far separated objects. The Pauli principle tells us that all electrons in the world are entangled. What do we mean if we speak of an individual electron or of an individual "subsystem" in the world?

Another approach to the division problem is provided by the notion of space-time. In all successful theories till now space-time is considered as a 4-dimensional continuum which provides the arena in which physics can play. By definition it is infinitely divisible. But an empty arena is not very interesting. We need the players. If we disqualify objects what remains?

Indeterminism and Reality. Before addressing this question let us look at another aspect of Quantum Theory: *indeterminacy.* The theory does not provide us with the ability to predict future phenomena from the knowledge of the past with certainty. Instead it gives us probability assignments for the occurrence of one phenomenon among a number of alternative possibilities. The majority of physicists accepts indeterminism as an intrinsic feature of the laws of nature but there is a very outspoken minority trying to restore determinism by looking for hidden variables or regarding the wave function as an attribute of an individual object. In my opinion such attempts have

[1] "Entanglement" is the translation of the term "Verschränkung" introduced by Schrödinger 1935. It means that there are pure states of a compound system which yield stronger correlations in the joint probability distribution of measuring results on the subsystems than those which can arise from correlations between individual states of the subsystems. A quantitative criterion distinguishing entanglement from any interpretation in terms of hidden variables has been given by Bell 1964. It also excludes an interpretation in terms of any assignment of individual states to the subsystems (Clauser and Horne 1974).

up to now not met with much success. In particular I do not understand the enthusiasm with which some colleagues propagate "Bohmian Theory". It seems hard to envisage how such an essentially mechanistic picture could cope with basic phenomena in quantum optics (e.g., Rabi oscillations between atomic levels induced by laser radiation) or produce a natural alternative to relativistic quantum field theory.

So I believe in the intrinsic indeterminacy of the laws of nature. But this means that we have to distinguish between *possibilities* which are encoded in the notion of *state* (if you wish the state of the universe) and *facts* which involve a decision between different alternatives. What can we say about such decisions? In orthodox quantum language the facts are cautiously called "results of observation". This emphasis on "observation" has raised another controversy: the meaning of *reality.* An observation needs an observer and this is usually a human being. Does Quantum Theory mean that the establishment of a fact is ultimately tied to the realm of the mind? This is a position shared in various degrees of conviction and clarity by many. Most decisively and clearly it has been advocated by Wigner. In his "Remarks on the Mind-Body Problem" (1963) he writes: "If one formulates the laws of Quantum Mechanics in terms of probabilities of impressions these are *ipso facto* the primary concepts with which one deals.... The principal argument is that thought processes and consciousness are the primary concepts, that our knowledge of the external world is the content of our consciousness". In spite of my great admiration for Eugene Wigner to whose thinking I owe so much it seems to me that the empirical material of Quantum Physics does not relate at all to the mind-body problem and that it is not helpful to invoke consciousness for its interpretation. If the decision as to whether a detector has clicked or a dot on a photographic plate has been produced depended on the consciousness of some spectator it would be far too unreliable to serve the purposes of physics. The scope of physics is limited to the study of relations between phenomena which are reproducible at different places, different times and recognized by different people. It is the independence of the physically relevant phenomena from the state of mind of any individual human observer which is meant if we speak of an "exterior (physical) world" called nature.

There is an early controversy (1927) about the "decision" involved in the realization of a particular result in a measurement. Dirac called it "a decision by nature". Heisenberg wanted to reserve the decision making to the observer. Bohr, in later years used Dirac's formulation but warned that this should not be understood as implying any personification of nature. If we consider measuring results as facts, not relying on conscious perception, and in addition believe in intrinsic indeterminism then we can only attribute a limited amount of decision making to the observer. He may limit the range of possibilities by the choice of his measuring device but he certainly has no power to decide on the realization of a particular outcome. Thus we come back essentially to Dirac's formulation, though, heeding the warning by Bohr we might better say: it is a decision in nature instead of by nature. In order to rule out clearly the mental interpretation I propose to generalize the term "result of observation" to the term "event", a concept not tied to any passage into consciousness or to the performance of an experiment. It is important, however, that the notion of *event* is independent of and sharply distinguished from the notion of state. The latter refers to possibilities the former to the establishment of a fact. What is an event? Of course if we think of the example of a click in a detector this is a very coarse event. In standard language it is produced by the interaction of an "atomic object" with a macroscopic device. Again there is the problem of subdivision. Can we define *elementary events?* Under what circumstances can we regard the ionization of an atom by a photon as a closed event? Or the death of a Schrödinger kitten? The formalism of present day Quantum Theory suggests that there is no sharp boundary. While the definition of individual objects is limited by entanglement the definition of individual events is limited by coherence. Both have their root in the assumed unrestricted validity of the superposition principle. On the experimental side entanglement is demonstrated in EPR-type experiments, coherence in interference experiments. Nevertheless I believe that a theory based on the notion of real events can be developed in a way which is not at variance with established experimental results. I shall give a few arguments for this and for my motivation in proposing the ensuing picture.

Motivation. Bohr's epistemological arguments, leading to the need for a cut between the "system" in which we are interested and the "observer" with his equipment is an excellent description of what is done and what can be learned in an experiment. But experiments have always been regarded as a means to penetrate the "mysteries of nature", not as a purpose in itself. This is mirrored by the gap between empirical evidence and a coherent theory. There is a discontinuous step, involving, in Einsteins words, "free inventions of the mind", introducing concepts and mathematical structures which have no direct counterpart on the empirical side and thus transcend what we can possibly know. A theory aims at providing an ontological model which should be in harmony with known experimental facts but is not synonymous with them. As a model it has a limited range of applicability. I do not believe in the possibility of a "theory of everything" achieving a mental isomorphy with the world of appearances. In Quantum Theory we have on the one side a very sophisticated mathematical formalism allowing "in principle" an enormous variety of observables whereas in practice we are restricted by the need to use objects subjected to the laws of nature and to control their placement in space-time. We should not overrate the range of applicability of the formalism. The perspective changes already when we pass from Quantum Mechanics to Quantum Field Theory. The characterization of observables and states is completely different. In Quantum Mechanics the observables are built from positions and momenta of particles. In Quantum Field Theory the basic observables relate to space-time regions. Though Quantum Field

Theory is the more encompassing theory from which Quantum Mechanics should result in a non relativistic approximation, it is clear that in it the relation between observables and space-time is over idealized. By appealing to the Bohr-Heisenberg cut (which in Quantum Field Theory is only needed if we think of detectors as the prototypes of observables and useful within the limits in which their placement in space-time can be controlled) we avoid the formulation of a self-consistent theory of nature. This is adequate for the recognition of attributes of particles and of reaction cross sections within the theory (at least in principle). But the shifting of the cut to small space-time regions of, say, one Fermi in diameter becomes completely unrealistic. If in high energy physics one talks about distances of 10^{-16} cm this cannot be interpreted in terms of sharply localized observables (in the sense of the cut) but rather as the intrinsic extension of an event. Restrictions in the realizability of hypothetical observables lead to limitations of coherence. This is exemplified by the study of "effective decoherence" due to "practical impossibility" of observation. One way of expressing restrictions of observability in the existing formalism is provided by the notion of "superselection rules". We know strict ones, relating to charge quantum numbers. They result from the requirement of global gauge invariance of the theory. Since they concern global properties they are not relevant in our context without the possibility of a clear-cut subdivision of the charges in the universe. But we also know that *local* gauge invariance plays an important role in Quantum Field Theory. It means that the transport of (generalized) phases from one point to another is a highly nontrivial matter.- "Effective" (approximate) superselection rules also result from large numbers. They become sharp in the thermodynamic limit. Examples are temperature, "long range order" but also the center of mass position of a heavy body. This leads to the remark that the possibility of observation depends on the state. If, by a "self consistent theory" we mean an ontological model of the physical universe then the prevailing state (of the universe) should restrict (though not determine) the possibilities for the occurrence of subsequent events. In other words: superselection rules do not need the existence of a center in an algebra which is defined without reference to the state. In the standard formalism the idea of "consistent histories" forwarded by Griffiths, by Gell-Mann and Hartle and by Omnés, is a step towards the recognition of this problem. It is, however, not the answer.

Considering all this I feel that it is rather safe to predict that the future development of the theory will lead to strong limitations of the superposition principle, irrespective of the problems of Quantum Gravity and relevant much before the Planck scale. They may allow to give a clear meaning to the notion of an "elementary event" and thereby provide the physical counterpart to the divisibility of space-time.

On the other side of the coin we have experiments demonstrating an amazing degree of coherence. In particular, coherence is not destroyed in the interaction of an "atomic object" with a system encountering negligible back reaction (typically any system that can be idealized as an "external field"). A striking example is provided by atomic interference experiments where a laser beam produces Rabi oscillations between internal states of the atom and may be used as a beam splitter for the atomic beam. Here the laser beam, being a coherent state of many photons, acts like an external field. In this context it should be remarked that photons play a very special role, not only due to their vanishing rest mass but because they are the only particles which carry no charge quantum number of any kind, a feature essential for the existence of the coherent states with undefined photon number. Coherence is apparently not lost in the down conversion of a photon to two photons of lower frequency by a non linear crystal. Thus this process may not be regarded as an event. The crystal plays the role of an external field. There are other experiments whose significance in posing limits to the notion of individual events I have not yet understood. A prime example is the intensity interferometry of Hanbury Brown and Twiss. In any case there is a fascinating area where both theory and experiment must contribute to clarification.

2 Proposal of a Picture and Terminology

We distinguish sharply between facts and possibilities as it is demanded if we believe in intrinsic indeterminism. An "event" is regarded as the realization of an individual fact (and thus an "element of reality") whereas a "state" subsumes the probability assignment for the realization of a pattern of future events. It is not an element of reality. This implies, however, that the arrow of time must have fundamental significance. A fact is created; it did not exist prior to some time and it is irrevocable. Thus the picture describes an evolving universe with a growing pattern of events representing the respective past as opposed to an open future. The pattern of realized events determines the probability assignment for subsequent growth of the pattern (the state).

In addition to the events we must consider *causal links* between them. A link is a messenger connecting a source event with a target event. In the simplest case (low density situation) links correspond to the aforementioned objects (stable particles). It must, however, be borne in mind that the evolutionary picture demands that a link becomes established (an "element of reality", if one wishes) only when the target event is concluded. In general the source event does not even determine the nature of the potential links originating from it. Potential links belong to the realm of possibilities, not facts. This illustrates the sense in which we can speak of individual objects. For example we can speak of the individual electron which was ejected from a metal surface by a radiation pulse and subsequently caused a click in a detector. But it makes no sense to talk about an individual electron without reference to specific events.

Events and (established) links are the needed players in the arena of space-time. Specifically, an event has the attribute of localization in space-time, the sharpness of which depends on the nature of the event. It will be very diffuse in low energy events and rather sharp in the case of high energy-momentum transfer. No independent localization property can be assigned to causal links. This corresponds to the orthodox statement that "a particle does not have a position at any given time unless it is measured". In our words: a causal link does not have space-time attributes (apart from those due to the events it connects). It is the event, the interaction process, to which the attribute of localization can be assigned.

The impossibility of assigning space-time attributes to "atomic objects" is one of the earliest, most striking lessons of Quantum Physics. It is usually called the "wave aspect" of matter. Consider for instance the radioactive decay of a nucleus. The emitted α–particle is described prior to its detection by a spherical wave whose origin marks roughly the position and time of the source event (here the decay process) and which subsequently sweeps over wide regions of space. It will ultimately cause an event which selects again a reasonably sharp localization among many alternative possibilities. In standard language this is called the "particle aspect". In the terminology corresponding to our picture the "wave-particle duality" is replaced by an object-event duality (object = potential causal link). The dual aspects refer to different things and the relation to space-time is carried only by one of them (the events). In the case of a macroscopic object the assignment of a trajectory in space-time results from the involvement of the object in many minor events which do not change its internal structure. In this view the "spontaneous localization" postulated by Ghirardi, Rimini and Weber is replaced by the realization of events and the occurrence of many unnoticed events corresponds to the discussion of decoherence due to background by E. Joos and H.D. Zeh.

The picture emphasizes the importance of distinguishing between individual cases and statistical statements, between causal relations linking individual events and statistical correlations. Thus, the "non locality" in EPR-type experiments does not concern causal effects travelling faster than light but statistical correlations in the joint probabilities for the occurrence of several far separated events. Note that the probability for a specific single event depends only on the events in its causal past and is not altered by anything which may be done in space-like distances. The perseverance of the correlations between events over large distances is in itself no paradox. What appears paradoxical and differs from the case of classical correlations is only that the correlations are of such a nature that they cannot be attributed to correlations between assumed "states of subsystems" but only as correlations between the events themselves. This corresponds in our picture to the statement that unaccomplished (potential) links belong to the realm of possibilities, not facts. They have neither a well defined individuality nor any independent localization properties.

Assessment. The picture provides some model for a divisibility of the universe into individual elements which are considered as elements of reality. It injects probably the maximal amount of realism which can be tolerated by existing experimental evidence and at the same time the minimum needed in a reductionist approach.

Since all "observation results" are coarse events, about whose real existence and finality one need not quibble (at least not from a practical point of view) there can arise no contradiction between our picture and experimental findings as long as one focuses only on such coarse events and their correlations. But this does not provide a basis for a self consistent theory. The central problem which a theory of events has to face is the question of subdivision of coarse events into finer ones. As Bohr has pointed out it is the indivisibility of a "quantum process". Can we single out individual quantum processes and call them events; where are the limits of divisibility? What is an irreducible, an elementary event? As mentioned, one prerequisite for this is a limitation of coherence. I have given some reasons for the expectation that such limitations will be recognized in the future development. This would allow a self consistent theory. More precisely: an ontological model along the lines of the sketched picture. As any model it must have a limited range of validity and such limits are provided by holistic features which are neglected in a reductionist approach.

Some remarks concerning space-time should be added. Since we consider a single universe which is realized by the (growing) pattern of events and causal links it is this pattern itself which constitutes the realized aspect of space-time, the events corresponding to points, the links to connecting lines. In the definition of a geometry of such a pattern the rest masses of existing particles play an essential role. Since the laws of nature allow many different patterns there is also the aspect of potentiality i.e., the set of all possible patterns. It is not likely that they can be embedded in a 4-dimensional continuum but the set of possibilities itself may be regarded as replacing the aforementioned arena.- Events are not points in a pregiven continuum and our knowledge about their localizability in classical space-time is pitifully small. In high energy physics where we may expect a sharpness of events of the order of 10^{-16} cm the experimental accuracy in determining the vertex of a reaction is of the order of 10^{-4} cm. Also the establishment of a space-time reference system has limitations which are far more serious than the existence of the Planck length. There are no ideal rigid bodies, a photon does not define a straight line etc. All this must be taken into consideration if one aims at a self consistent theory in which we cannot retire behind the protection by the Bohr-Heisenberg cut. This list of ignorances could be continued. But it may be taken as a positive message. There is much to do and not all is restricted to fashionable areas of speculation.

References to previous published work by myself on the topics addressed here as well as to related ideas by others may be found in chapter 7 of the second edition of my book "Local Quantum Physics", Springer Verlag 1996, and in the later articles "An evolutionary picture for quantum physics" Com-

mun. Math Phys. **180**, 733 (1996) and "Objects, Events and Localization", Proc. XI Max-Born Symposium 1997.
References to other work alluded to in the text can be found in the book by R. Omnès: "The interpretation of Quantum Mechanics", Princeton University Press 1994 and the book by L. Mandel and E. Wolf: "Optical Coherence and Quantum Optics", Cambridge University Press 1995.

Modal Interpretation of Quantum Mechanics and Classical Physical Theories

R.S. Ingarden

Institute of Physics, N. Copernicus University,
Grudziądzka 5, 87–100 Toruń, Poland

Dedicated to Jan Łopuszański
with cordial wishes
on the occasion of his 75-th birthday

Abstract. In 1990, Bas C. van Fraassen defined the modal interpretation of quantum mechanics as the consideration of it as "*a pure theory of the possible*, with testable, empirical implications for what actually happens". This is a narrow, traditional understanding of modality, only in the sense of the concept of *possibility* (usually denoted in logic by the C. I. Lewis's symbol $\Diamond$) and the concept of *necessity* $\Box$ defined by means of $\Diamond$. In modern logic, however, modality is understood in a much wider sense as any intensional functor (i.e. non-extensional or determined not only by the truth value of a sentence). In the recent (independent of van Fraassen) publications of the author (1997), an attempt was made to apply this wider understanding of modality to interpretation of classical and quantum physics. In the present lecture, these problems are discussed on the background of a brief review of the logical approch to quantum mechanics in the recent 7 decades. In this discussion, the new concepts of sub-modality and super-modality of many orders are used.

1 Introduction. Many–Valued Logic and Nondistributive Logic

If we look at the XX-th century now, at *fin de siécle*, we see that in physics quantum mechanics and relativity theory were the most important achievements. Both theories appeared as a final result of long pressures of unexpected experiments, after many unsuccessful attempts to explain them theoretically on the base of the then accepted principles of physics. Both required a new paradigm of physical and not only physical concepts. And both were very successful theories, up to now they stand up to all the consecutive trials of falsification by more and more precise measurements. But at the same time they both presented painful challenges to human comprehension, even to classical logic and the most deeply rooted principles of common sense and physical intuition.

At first, relativity theory caused protests in some places, rarely among physicists, but more frequently among philosophers and intellectuals. Niels Bohr, one of main founders of quantum mechanics, said that if somebody can think about quantum mechanics without being a little dizzy, he certainly does not understand it. Actually, even Niels Bohr resigned to understand it fully in the usual sense of Western logic and philosophy. When he accepted his principle of complementarity, he appealed to Chinese philosophy for help. When he invented quantum jumps in his first model of the hydrogen atom, he actually "jumped" over his physical and logical conscience against the Latin proverb *Natura non facit saltus* (Nature does not make jumps) and the classical principle of differential continuity. When later on Max Born and Werner Heisenberg used algebras of infinite matrices with complex coefficients, they understood clearly the mathematical side of their theory, but not their physical interpretation. The principle of uncertainty formulated by Heisenberg a few years later, and the statistical interpretation of quantum mechanics given by Born at the same time, were a conscious resignation from the full description and understanding of physical processes, in a similar sense as the famous *hypotheses non fingo* of Newton who understood the universal gravity only through the picture of a falling apple, but not through a physical mechanism. Also Einstein understood special and general relativity on the level of his famous *Gedanken* experiments. There is an anecdote that he understood the relativistic concept of simultaneness when he looked at the reflections of the hands of the town-hall clock in the driver's mirror of a streetcar at the Kramgasse in Bern (Davies 1998, p. 909). Only Hermann Minkowski gave later on some generalized, but clear geometrical pictures to understand special relativity intuitively. But all the difficulties of understanding relativity, special and general, are nothing when we compare them with those connected with quantum mechanics. Therefore, since longer time relativity is treated as a part of classical physics, in contradistinction to quantum physics considered as a non-classical field of considerable philosophical difficulties. Even competent mathematical physicists and philosophers of physics, as Paul Busch, Pekka Lahti and Peter Mittelstaedt, said in 1991 at the beginning of their monograph on quantum theory of measurement (Busch et al. 1991, p. 1): "An understanding of quantum mechanics in the sense of a generally accepted interpretation has not yet been attained. The ultimate reason of this difficulty must be seen in the irreducibly probabilistic structure of quantum mechanics which is rooted in the nonclassical character of its language."

Max Jammer, in his excellent book *The Philosophy of Quantum Mechanics* said (Jammer 1977, p. 341):

> The decomposition of a physical theory T [...] into a mathematical formalism F, a set of epistemic relations R, and a physical picture M implied that an interpretation of T should concentrate on one or more of these components. All the interpretetions of quantum mechanics described so far were based on these assumptions. Certain developments in mathematics and philosophy, however, have led to the idea that the alternatives discussed so far were not exhaustive and that a fourth component, so to say, of the most general nature — which

> for this very reason had been altogether ignored — could also be an object of inquiry in the search for an interpretation: the formal structure of the deductive reasoning applied in formulating T. If a certain theory T leads to an impasse, it was claimed, it is not necessarily its mathematical formalism as such nor the meaning of its extralogical concepts that may have to be modified; it may equally well be logic underlying the formulation of T which has to be revised. A search for an interpretation of quantum mechanics along these lines is usually called a quantum logical approach.

It is well-known that there were so far two such attempts. The first one consisted in formulating a so-called non-Chrysippean logic (sometimes also called, historically less correctly, non-Aristotelian logic), namely, a many-valued logic. The second is an application of a nondistributive logic based on the algebraic concept of a lattice.

In the beginning of the XX-th century there appeared several independent attempts to formulate a three-valued (later many-valued) logic, namely by Hugh MacColl (1832–1909) in London 1906, Charles Sanders Peirce (1839–1914) in New York 1909, Nikolai Aleksandrovich Vasil'ev (1880–1940) in Kazan 1910, Jan Łukasiewicz (1878–1956) in Warsaw 1918, and Emil Leon Post (1897–1954), born in Poland, in New York 1920. Only the latter two logicians formulated their theories mathematically, while the three former ones expressed their ideas by the names: "logic of three dimensions", "trichotomic mathematics" and "imaginary non-Aristotelian logic", without any interpretation. Only Łukasiewicz interpreted the third logical value as 'possibility' (for example, of a future event) and connected it with indeterminism in physics. (In 1969 his pupil Jerzy Słupecki interpreted the third value as 'change', so his values are: non-exitence 0, change 1/2, existence 1.) Łukasiewicz also "envisaged the possibility of generalizing his system in an infinitely-many-valued logic" (Jammer 1977, p. 344), in a way similar to probability calculus. An application of the three-valued logic to quantum mechanics was first proposed by Zygmunt Zawirski (1882–1948) in Poznań 1931, then independently by Hans Reichenbach (1891–1953) in Berlin 1932 and Fritz Zwicky in Pasadena 1933. The first one used Łukasiewicz's three-valued logic with the third value as 'uncertainty', while all of them interpreted probability by infinitely-many-valued logic with a continuous scale of truth values. This point of view was criticized by Henry Margenau (1934) saying that the truth of a physical law has only two values and should be distinguished from the truth of an experiment which is "in a state of flux". I think that this is the question of type or order of proposition, i.e., whether we speak in the concrete (objective) language about elements of reality or in the meta-language about sentences or sets of real elements. Physicists are interested mainly in the the former language, philosophers in the latter, but both languages are equally admissible in science. The interpretation of Zawirski–Reichenbach–Zwicky was meant as the physical one, i.e., formulated in the objective, but many-valued language, while the meta-language can be two-valued. We shall come back to this problem later on from a more general point of view.

The mentioned many-valued interpretations of quantum mechanics remained, as Jammer said, "virtually unknown in the world of physics" (Zawirski wrote in Polish and French, Reichenbach in German, Zwicky in English), they were only noticed by few philosophers interested in modern physics. Jammer remarked that such new basic ideas require, as a rule, a very long time to become rape for wider comprehension and application, as was the case with the non-Euclidean Riemannian geometry applied to physics by Einstein almost one century after the first appearance of this idea in mathematics. But, as Jammer said, *loc. cit.* p. 346

> In fact, the first serious breakthrough of nonclassical logic in quantum mechanics was made in 1936. Not the law of bivalence, but rather the distributive law of classical logic was the major target of this assault.

This was done in the famous paper by Garrett Birkhoff and John von Neumann "The logic of quantum mechanics" published in 1936. This idea has now an extensive literature and is well-known among mathematical physicists. Instead of the phase space Γ of a classical mechanical system, Birkhoff and von Neumann used a complex separable Hilbert space $\mathcal{H}$ of a quantum system. They replaced the subsets of Γ by closed linear subspaces of $\mathcal{H}$ as representing quantum events or quantum propositions occurring in experiments. Then the truth value of a quantum proposition a is the eigenvalue, 1 or 0, of the projection operator on a. Quantum logic is an orthocomplemented modular lattice of closed subspaces of $\mathcal{H}$ or of projection operators on these subspaces. As is well-known, a general lattice is defined as a partially ordered set with order relation $\leq$, in $\mathcal{H}$ represented by inclusion $\subseteq$. The concept of lattice was introduced by Ernst Schröder in 1890 and was called a dual group by Richard Dedekind 1897, since a lattice has dual operations: join $\cup$ and meet $\cap$ represented in $\mathcal{H}$ by the linear sum and the intersection of the subspaces, respectively. In quantum logic join is interpreted as alternative, and meet as conjunction. Orthocomplementation is defined in $\mathcal{H}$ as the passage from a closed subspace a to the subspace orthogonal to it, and logically means negation. The relation of inclusion of subspaces is interpreted in logic as the relation of implication. Abstractly, in a lattice with the zero element O and the unit element I (in $\mathcal{H}$, the zero vector and the whole $\mathcal{H}$, respectively), the *orthocomplement* $\neg a$ of a is defined as a dual automorphism of a, i.e., such that

$$\neg(a \cap b) = \neg a \cup \neg b, \quad \neg(a \cup b) = \neg a \cap \neg b \tag{1}$$

fulfilling additionally the two conditions:

$$\neg\neg a = a, \tag{2}$$

and

$$a \cap \neg a = O, \quad a \cup \neg a = I. \tag{3}$$

In general, *complementarity* is a wider property than orthocompelmentarity since it is defined only by condition (3), and is not necessarily unique, not an automorphism. Of course, each orthocomplement is a complement, but not necessarily conversely. But in a distributive lattice, as a Boolean lattice, the complement is unique and is also the orthocomplement. In Hilbert space $\mathcal{H}$ the orthocomplement as a subspace, being a complement in the lattice sense, is not a complement in the set-theoretical (i.e., Boolean) sense. Physically, this fact, a linear, not set-theoretical, addition of events, is connected with the phenomenon of interference, or the principle of superposition, of the so-called material waves or probability amplitudes. This experimental fact actually destroys the classical concept of a particle as a physical object. In consequence, also the philosophical concept of an object as an individual element loses its sense in quantum physics. So in quantum domain there is no more "reification" in the strict sense, no more "objectivity". We have only a weaker concept of "reality" as a 'propensity' or a possibility of a particle, or a potential, virtual, not actual thing. Does it mean that this reality is "subjective"? Yes and no. In the wider sense 'yes' since the interference is connected with the preparation measuring devices which are prepared by the observer, but 'no' in the narrower sense, since the phenomenon of interference is independent of the will of the observer when the devices are fixed. Somebody can say that *res*, an object, a particle appears again in such phenomena as photoeffect, scintillation etc. No, because electrons, and other quantum particles as photons, mesons etc., are only quanta of quantized fields and have no individuality, they do not behave according to the classical, Boltzmann statistics. Therefore, speaking about individual interpretation of quantum mechanics, as is frequently done, is misleading, better to speak about realistic interpretation. We have only real phenomena, events, physical intersubjective and repeatable facts, but not "hard substances" or "individual things" in the classical sense, not the indestructible atoms of Demokritos. This point of view is in agreement with the irreducible character of quantum probability and the refutation of hidden variables by many "no–go–theorems" of Gleason 1957, Bell 1966, Kochen and Specker 1967 etc.

The idea that the world is composed of facts, not of things, was formulated in philosophy before quantum mechanics. Namely, in 1918 Ludwig Wittgenstein wrote his famous theses: "1 The world is all that is the case. 1.1 The world is the totality of facts, not of things." (Wittgenstein 1961, p. 7), see also (Wolniewicz 1968), (Suszko 1968). Actually, in quantum physics we have something more than only facts: not substances, but propensities (pure and mixed states: probability amplitudes or probabilities), in addition to observables as properties (facts). Therefore, from the Aristotelian concepts potentialities and actualities (properties) remained, but the substances vanished.

In a quantum lattice modularity is, on the other hand, a weaker form of the distributivity in classical, Boolean logic. Namely, it is defined by the condition

$$\text{if } a \subseteq b, \text{ then } a \cup (b \cap c) = (a \cup b) \cap c. \tag{4}$$

As already mentioned, classical Boolean lattice is a complemented distributive lattice, but because of the mentioned theorem it can be also defined as an orthocomplemented distributive lattice. Quantum logic, being the orthocomplemented modular lattice geometry of closed subspaces of Hilbert space $\mathcal{H}$ (or of orthogonal projections on the subspaces), embeds an infinite number of complete orthonormal frames in $\mathcal{H}$. These frames, as sets of orthonormal vectors, can be interpreted as Boolean lattices which are the classical logics of possible events (readings) occurring during the measurements of fixed physical observables. So the quantum experiments, discovered mainly in the XX-th century, disclosed the rotational degrees of freedom of the classical logic (unitary rotations between eigenframes of noncommuting observables) in a new world of $\mathcal{H}$ by means of the group of unitary transformations.

For simplicity and for lack of time we avoid here definitions of the important concepts of further progress of quantum logic, such as compatibility and orthomodularity, see.,e.g., (Ludwig 1983 Appendix I, p. 343–352). We also avoid the discussion of such basic discoveries in the statistical structure of quantum mechanics as Gleason's theorem 1957, the quantum concept of entropy or information, see (Ingarden and Kossakowski 1968), (Ohya and Petz 1993) and (Ingarden et al. 1997), the concepts of quantum dynamical semigroup and quantum open system (Kossakowski 1972), (Davies 1976), the important concept of effect (Günther Ludwig 1970, see (Kraus 1983)), and the related generalization of the concept of observable from that of a self–adjoint operator or projection operator-valued measure (PV measure or spectral measure) to that of positive operator-valued measure (POV measure or semispectral measure) 1970, see (Grabowski 1990) and (Busch et al. 1997), aa well as the introduction, by means of the latter, of the concepts of indirect unsharp (fuzzy) measurement (Holevo 1982) and also of unsharp objectification (Busch et al. 1991), p. 127. We avoid also the concepts of continuously observed systems, posterior states, quantum filtering and nondemolition measurements (Staszewski 1993), (Belavkin 1994), the EPR effect (Busch et al. 1997), Bell's inequalities, entanglement and teleportation, see (Horodecki 1996). We assume that all these concepts are now more or less well-known among most of mathematical physicists and we shall use some of them later on. For the present state of quantum logic we can refer, for example, to the review papers and books by C. Piron (Piron 1972), (Piron 1976), J. M. Jauch (Jauch 1968), V. S. Varadarajan (Varadarajan 1985), E. Beltrametti and G. Cassinelli (Beltrametti and Cassinelli 1981), P. Mittelstaedt (Mittelstaedt 1978).

Here we briefly mention only one mathematical and one physical question. Mathematically, there were objections that the Birkhoff–von Neumann logic is not a logic in the proper sense since implication is merely a relation (of order) there, and not an operation. This objection was rejected by the Torunian logician Jerzy Kotas, (Kotas 1967), (Kotas 1971), who showned that there are 6 operations in quantum logic correspoding to the classical connectives, especially to implication, of which 5 are identical and all of the latter are equivalent to the implication relation of Birkhoff–von Neumann. Kotas also presented two equivalent implication–negation axiom systems which are equivalent to the modular logic of Birkhoff–von Neumann.

The physical question is perhaps the most important problem and the hope of many quantum logicians, namely, the representation problem, i.e., whether the modular lattice with possibly some additional conditions can always be represented in a Hilbert space. Many believed that they proved this. For example, B. C. van Fraassen wrote optimistically in 1991 (van Frassen 1991, p. 199), as follows: "The postulates of quantum logic narrowed down the class of models step by step, and eventually the representation theorems showed that the state space was constrained to be a Hilbert space." But in the same year other experts, Busch, Lahti and Mittelstaedt, in the book mentioned above wrote pessimistically about the same problem (Busch et al. 1991, p. 4):

> Each of the so-called axiomatic approaches has deepened our understanding of the mathematical and conceptual structures of quantum mechanics. However, none of them led to a thorough justification of the ordinary *Hilbert space quantum mechanics*. In particular, the quantum logical lattice approach is not sufficient for a reconstruction of this theory.

In this point they quoted the book by G. Kalmbach *Measures and Hilbert Lattices* of 1986 (Kalmbach 1986) meaning perhaps that the Hilbert space property of the lattice should be directly assumed. The problem should be finally cleared up by mathematicians and logicians. As a physicist, I think that here we have a similar situation as with the classical mechanics. The principles of classical mechanics are not contained in the Boolean logic, we cannot expect another situation in the quantum case. So some additional physical postulates are unavoidable. Also Hilbert space theory is based on abstract axioms, not on a concrete representation as a model. For physicists it seems that any complete mathematical or physical theory should be axiomatic, otherwise we do not know what is basic and what can be derived from simpler assumptions. But sometimes the pedantic axiomatization, especially in physics, can be also considered a nuisance or the question only of mathematical elegance which is not always urgent and convenient, or necessary. We are not shoemakers, as Kant said. But there is also an old English saying of Thomas Dekker (ca.1572–1632): "Brave shoemakers, all gentlemen of the gentle craft."

However, from the general logical and linguistic point of view, axiomatization, although very useful, is neither necessary nor sufficient for a complete theory. Linguistically, a syntax requires a semantics. Logically, an axiom system needs a model to check the consistency and independence of the axioms by means of constructing an interpretation. This interpetation can be two-valued for a many-valued theory, or conversely. (E.g., H. Rasiowa proved the independence of the axioms of the Tarski-Bernays 2-valued axiom system by means of a 4-valued logics (Rasiowa 1955), for a general discussion of these problems see (Zinowiew 1963, Chap. 3).)

There are yet two brief comments which we should like to add here. First, that there is actually no sharp contradiction between the idea of many-valued logic and nondistributive logic. Namely, Jammer said that in the modular logic there are only two logical values (bivalence) since the projection operator on a closed subspace a of $\mathcal{H}$ has only two eigenvalues, 0 and 1. But, as we mentioned, there is a possibility of infinitely many unitary rotations of this subspace around zero vector O in the Hilbert space $\mathcal{H}$ giving other pairs of eigenvalues, 0 and 1. Of course, according to the gauge invariance with respect to a constant (initial or *absolute*) phase of the state representation, the rotated subspaces can be also taken as a representation of the given proposition. But if the representation (i.e., the initial phase) is once fixed as some subspace a, interpreted as an eigenstate found in a measurement of some observable, the *relative* phases between a and its rotated subspaces have already a physical meaning (e.g., as a Berry phase). Then, all the rotated subspaces, except all the ones compatible with a, i.e., all the subspaces "skew" with a, are not true or false, but "undetermined" propositions when a is fixed as true in the given measurement. They can be true or false only when another specially adjusted measurement, incompatible with the previous one, is done (when the corresponding observables do not commute). Saying otherwise, we have in $\mathcal{H}$ as if infinitely many incompatible bivalent logics, forming together one infinitely-many-valued logic corresponding to the Birkhoff–von Neumann modular logic. On the other hand, also the phenomenon of a 'quantum jump' can be considered as corresponding to the logical value 'change' suggested by Słupecki in his interpretation of the third Łukasiewicz's logical value. But since there are, in general, infinitely many different possible quantum jumps of a quantum system, also here there are, actually, infinitely many logical values. Thus, in principle, both of the mentioned logical approaches to quantum mechanics are not contradictory and both are physically correct. But the lattice method is much more detailed and specific and introduces a new algebraic algorithm which can be further developed as Kotas and others have shown. The problem, however, arises if this method is sufficient for full understanding of quantum mechanics. We shall discuss this question later on.

The second comment is that the condition of many-valuedness is not yet sufficient for characterizing the quantum character of logic. For example, the classical (non-quantum) probability can be defined in an infinitely-many-valued, but non-quantum logic, see (Zawirski 1935), while the essentially different, although analogous, quantum probability can be formalized in another non-equivalent to the former, infinitely-many-valued, but quantum logic. The logical differences between these cases are just those between distributivity and modularity (or orthomodularity).

2 Classical (Non–Quantum) Modality

Modality is just this new idea which we should like to add to the concept of quantum logic in the interpretation of quantum mechanics. The theory of modality can be compared with many parts of grammar of a natural language. It is as if morphology, or better pragmatics of logic, as we shall explain later on. In other words, we should like to bring the concept of logic nearer to that of natural language. Modality "conjugates" logical concepts, so to say, makes them more elastic and more sophisticated, as if more cultural. Thus we propose to go as if from the elementary school of the so-called extensional logic to the higher school of the so-called modal or intensional logic. Among logicians this approach is not new. In fact, it started already in Antiquity, in the logical works of Aristotle, but developed rather slowly till the XX-th century when its progress became very intensive and rapid, especially in the last 5–6 decades, also in Poland. (Polish logicians Jan Łukasiewicz, Alfred Tarski, Mordchaj Wajsberg, Jerzy Słupecki, Roman Suszko, Jerzy Łoś, Stanisław Jaśkowski, Jerzy Kotas, Jerzy Perzanowski and others, and to some extent also mathematicians Helena Rasiowa and Roman Sikorski, see (Rasiowa and Sikorski 1970), contributed directly or indirectly to modal logic.)

To begin with, we define in a rather informal and brief way the classical (non-quantum or distributive) modal logic. *Modality* (from the Latin word *modus* meaning a 'measure, quantity, rhytm, limit, restricton, end, method, way') in logic means a 'logical valuation' of something, an expression of the relation or the attitude of the speaker to something he is speaking about, an opinion or an estimation or assessment of something from some point of view by somebody or by an observer or an observing device. This is a generalization of the assessment of truth or falsity of a sentence. For example, this point of view may be that of truth and probability, especially: of possibility or necessity. But not only, it may be also the point of view of obligation, of knowledge and believing, of potentiality and causality, of time and space, of existence, etc. Since G. H. von Wright (1951) and A. N. Prior (1967) one distinguishes more or less formalized branches of modal logic:

- alethic logic — about truth and probability,
- special alethic logic, or modal logic in the narrower sense — only about possibility and necessity (historically, it was the first form of modal logic),
- deontic logic — about obligation, permission, prohibition, in future maybe also: sin, crime, kindness, good and bad, the questions usually considered by ethics or axiology,

- causal logic — about potentiality and causality,
- epistemic logic — about knowledge, falsification, measurability,
- spatio-temporal logic — about space and time,
- existential logic – about existence, ontology, *modi existentiae* as in Middle Age philosophy.

In natural languages, a kind of modal logics is formed by grammatical categories of personal pronouns, of family relation terms, of kindness expressions, modal verbs in English called *modals* (*can, could, may, might, will, would, shall, should, must, ought, need*), modal adverbs as *certainly, surely, perhaps, maybe, possibly, probably* etc., modes and temporal conjugation of verbs, personal conjugation of verbs in Polish, etc. Estimation of space and time as a condition of truth can be generalized to any 'indexing' or 'referring to a person or an observer', 'point of reference', 'occasionality', 'relativity', or 'context of use'.

The first theory of modality in the narrower sense was the syntactical one developed by Clarence Irving Lewis (1883–1964) who proposed in 1918 and later five axiom systems, more and more general, for the concepts (functors) of possibility $\Diamond$ and necessity $\Box$: S1, S2, S3, S4, S5, (Lewis and Lanford 1932, 1959), (Fays 1965), (Perzanowski 1989). These axiomatic studies were deepened by important algebraic investigations (with applications to topology) by Alfred Tarski (Tarski 1938) and J.C.C. McKinsey (McKinsey 1941). The second theory of modality was the semantical theory of Saul Kripke who defined in 1959 the so-called *Kripke structure* $\mathcal{K} = (G, K, Q, R)$, where K is the set of possible worlds called universe, $G \in K$ is the real world, $Q \subseteq K$ is the set of anomalous worlds, and $R \subseteq K^2$ is a binary relation in K which enables to decide which of the axiom systems of Lewis is valid, cf. (Fays 1965), (Marciszewski 1987 p. 328–330). The most recent and general is the pragmatic theory of modality developed by Richard Montague (1930–1971) in 1968 (Montague 1974) and by his coworkwers Dana Scott, Daniel Kaplan, M. J. Cresswell and pupil Daniel Gallin (Gallin 1975), see also the books in Polish: (Tokarz 1993), (Marciszewski 1987). The point of departure for Montague was the classification of semiotics (linguistics or informatics in the general sense) by Charles W. Morris (1938). Montague has written about this in (Montague 1974, p. 95), as follows:

> The study of language (or *semiosis* or *semiotic*) was partitioned in Morris [book 1938] into three branches — syntax, semantics, and pragmatics — that may be characterized roughly as follows. Syntax is concerned solely with relations between linguistic expressions; semantics with relations between expressions and the objects to which they refer; and pragmatics with relations among expressions, the objects to which they refer, and the users of contexts of use of the expressions.
>
> Syntax had already been extensively developed at the time at which Morris wrote, largely by Tarski, Gödel, and members of the Hilbert group. [..] Most contemporary work in syntax falls into one

> of the subfields — proof theory or the as yet rather tentative field of mathematical linguistics.
>
> The foudations of semantics had also been completely laid (in Tarski 1933) by the time of Morris remarks; but its most extensive development has occurred since then under the name 'model theory'. [...]
>
> Pragmatics, however, was still futuristic at the time of Morris monograph. It was suggested in Bar–Hillel (1954) that pragmatics concern itself with what C.S. Peirce had in the last century called *indexical expressions*, that is, words and sentences of which the reference cannot be determined without knowledge of the context of use; examples are 'I' and 'here', as well as sentences involving tenses. Other terms for these expresssions are 'egocentric particulars' (Russell), 'token-reflexive expressions' (Reichenbach), 'indicator words' (Goodman), 'non-eternal sentences' (Quine).

Since Gottlob Frege (1848–1925) one distinguishes in logic for any linguistic expression (name, predicate or sentence) ζ its *extension* or denotation or reference Ext$[\zeta]$, and its *intension* or connotation or sense. Please notice the essential difference of meaning between the word "intension" connected with "intense" meaning 'high quality, high degree', and the word "intention" connected with "intent" meaning 'purpose, goal'. Following Rudolf Carnap (1891–1970), one defines the intension of ζ, Int$[\zeta]$, by the equation

$$\mathrm{Int}[\zeta](i) = \mathrm{Ext}[\zeta, i], \tag{5}$$

where $i \in I$ is an index or context of use, and I is the set of indices. Expressions for which the dependence on i is essential, i.e., the set I has more than one element, are called *indexicals* or *modal expressions*. For a name (term) its extension is an element of a set or a class denoted (or referred to) by this name (e.g. dog, number), while according to Frege extension of a sentence is its truth value, 0 or 1. (This is the usual simplification of the 2-valued extensional propositional calculus, drastically insufficient in the case of implication. In 1967 Roman Suszko, inspired by the philosophy of L. Wittgenstein (Wittgenstein 1961) interpreted by B. Wolniewicz (Wolniewicz 1968) as an ontology of situations, proposed a "non-Fregean logic" in which denotation of a sentence is a 'situation', in German 'Sachlage' (Suszko 1968).) If one considers logical functors, i.e., sentence-valued functions of sentences, one distinguishes between *extensional functors* depending only on the truth values of their arguments, 0 and 1, i.e., only on their extensions, and *intensional functors* or *modal functors* depending also on their intensions (contents, sense). Extensional functors are, for example, the usual logical constants or connectives of classical bivalued propositional calculus, as $\neg$ 'non' (negation), $\cup$ 'or' (join, alternative), $\cap$ 'and' (meet, conjunction). Examples of modal functors are: $\Diamond$ (denoted also by M or P) 'it is possible that', $\Box$ (or L or N) 'it is necessary that', 'I believe that', 'I know that' etc. Functors $\Diamond$ and $\Box$ are mutually connected by the relations

$$\Box p \equiv \neg\Diamond\neg p, \quad \Diamond p \equiv \neg\Box\neg p. \tag{6}$$

Extensional functors can be treated by the usual classical Boolean logic as *extensional logic*, but modal functors are not contained at all in Boolean logic and require many-valued nonclassical logic, such as *intensional logic* or *modal logic*. This cannot be easily seen by the usual axiomatic way of developing propositional calculus, only by means of the so-called matricial method (of logical-value matrices generalizing the truth-value matrices) of defining propositional functors and checking logical theorems about them. E.g., in the case of the possibility and necessity functors defined by S4 or S5 axiom systems, cf. also (Fays 1965), (Rasiowa and Sikorski 1970 Chap. XI), an infinitely-many-valued logic is needed. We see that, by introduction of many logical values, the concept of essentially extensional functors can be generalized, and the modal functors can be considered as extensional in this generalized weaker sense (they may be called, for example, *semiextensional*). For example, Łukasiewicz has defined many 3- and 4-valued modalities. With higher number of logical values the problem becomes much more complicated, but many theoretical insights are obtained by Słupecki, Łoś and others.

If a modality concerns not a sentence, but an object, a person or an action, i.e., if it belongs to objective language, not to metalanguage, then one calls it, according the Middle Age Latin terminology, *modalitas de re*. For example, "This function is a possible solution." or "He is a probable candidate." or "Sherlock Holmes could live in London.". If a modality is about a sentence, it belongs to the metalanguage, as "It is possible that this function is a solution." or "It is probable that he is a candidate." or "It is possible that Sherlock Holmes lived in London.", then one traditionally calls it *modalitas de dicto*.

3 Quantum Modality

What we said about modality up to now was a part of non-quantum, although many-valued logic. General quantum modal logic is not yet elaborated in detail, except the concepts of special modalities, possibility and necessity. The latter problem has been solved by the mentioned Dutch-American philosopher and physicist, Bas V. van Fraassen (now Princeton University, Princeton, N. J.). The first idea has been presented by him in 1981 (van Frassen 1981) and then developed in 1990 (van Frassen 1991) and in his book 1991 (van Frassen 1991a), cf. also the brief presentation of his special modal interpretation in the quoted book by Busch–Lahti–Mittelstaedt 1991 (Busch et al. 1991 p. 118–122). The latter authors described van Fraassen's idea in the following words:

> A further interpretation of quantum mechanics which goes beyond the minimal interpetation but which avoids the objectification problem is a modal interpretation of quantum mechanics developed by van Fraassen (1981, 1990). This interpretation considers quantum mechanics as "*a pure theory of possible*, with testable, empirical implications for what actually happens".

Here by the "objectification problem" one means the question of wheather each quantum measurement should lead to a definite sharp result, see (Busch et al. 1991) p. 33. As we know a measurement of a generalized observable (POV measure) leads, in general, to an unsharp result (for example, a simultaneous measurement of momentum and position of a particle in the Heisenberg state with the minimal uncertainty of position and momentum). Thus the objectification problem is denied or avoided. What regards the term "minimal interpretation", it is defined as follows (Busch et al. 1991, p. 11):

> The number $E_T(X) = tr[TE(X)]$ is the probability that a measurement of the observable [as PV or POV measure] E performed on the system $\mathcal{S}$ in the state T leads to a result in the set X [e.g., $X \in \mathcal{B}(\mathcal{R})$ in the real Borel measure space $(\mathcal{R}, \mathcal{B}(\mathcal{R}))$].

To show the fundamental additional assumption of van Fraassen (for simplicity we avoid his further assumptions) we quote its brief presentation from the same book (Busch et al. 1991, p. 119):

> Consider a physical system $\mathcal{S}$, and let E and T be any of its observables and states. The modal interpretation distinguishes between two types of propositions: *the value-attributing propositions* — an observable E has value X, to be denoted (E, X) — and *the state-attributung propositions* — a measurement of E leads to a result in X, to be denoted $[E, X]$. In accordance with the minimal interpretation a state T *makes* the state-attributing proposition $[E, X]$ *true* if $E_T(X) = 1$. The heart of the interpretation is to characterize the truth of a value-attributing proposition, that is, to answer the question which value-attribitions (E, X) are true in a given state T. The truth of (E, X) will not be identified with the truth of $[E, X]$. To explain this, the following definition is needed: a state T' is *possible relative to* T [van Fraassen denotes this relation by TRT'] if and only if $tr[T'P] = 1$ whenever $tr[TP] = 1$ for any projection operator P. A vector state $P[\varphi]$ is thus possible relative to T exactly when φ is in the range of T. In particular, if $P[\varphi]$ occurs in a decomposition of T, then $P[\varphi]$ is possible relative to T. [...] The modal interetation then starts with the following postulate.
>
> (P) *Given that system $\mathcal{S}$ is in a state T, there is a certain pure state $P[\varphi]$ which is possible relative to T, and such that for all observables E pertaining to $\mathcal{S}$:*

a) a state–attribution $[E, X]$ *is true if and only if* T *makes it true;*
b) a value–attribution (E, X) *is true if and only if* $P[\varphi]$ *makes* $[E, X]$ *true.*

Above we had only the definition of relative prossibility R in van Fraassen's interpretation. Now we give his general definitions of functors of *necessity* $\Box$ and *possibility* $\Diamond$ by means of relation R. These functors have been called by van Fraassen *modal operators* and formulated in his book in definition D12, p. 314, as follows (we remind that propositions or events, or image spaces — ranges — of mixed state operators, are represented as closed subspaces of $\mathcal{H}$; when the state operator T has image space x and the state operator T' image space y, TRT' or xRy corresponds to $y \subseteq x$):

If q is any proposition, then

$$\Box q = \{w : \text{ for all } w', \text{ if } wRw' \text{ then } w' \in q\}, \tag{7}$$

$$\Diamond q = \{w : \text{ for some } w', wRw' \text{ and } w' \in q\}. \tag{8}$$

For lack of time we resign from further development and discussion of van Fraassen's special modal interpretation. We think that it is an important contribution to understanding of quantum mechanics. It rose some interest in literature (papers by Lahti, Cassinelli and others), but not yet proportional to its importance. The reason is, perhaps, that the considered modality is only special and that the general concept of modality is not yet sufficiently well-known among physicists, and in part also among philosophers. What we would like to propose is to generalize this interpretation and to show its connections with classical physics which actually unconsciously used the modal concepts since Antiquity. The reason is clear: modality is an integral part of human language and without it we cannot understand the world and ourselves in the world. But the Western philosophy and culture, in contradistinction to the Far Eastern one, as Indian, Chinese and Japanese, is concentrated more on sharpness, categorical statements, absoluteness, and does not have sufficient feeling for modalities: fuzziness, shades and relativity.

4 Modal Interpretation of Classical and Quantum Physics

The problem of modality in physics, classical and quantum, has been briefly discussed by the present author in the Introduction to his recent book with Kossakowski and Ohya 1997 (Ingarden et al. 1997), and in his lecture at the 6-th Polish Congress of Philosophy in Toruń 1995 (Ingarden 1997a). These interests of the author have been prompted by his contacts with the Japanese language and with the Far-Eastern Philosophy during his multiple stays in Japan as visiting professor. Indeed, the Japanese language is perhaps one of the richest in modal expressions and grammatical modal forms. It is a language of many styles of speaking and writing and of many shades of meaning and steps of kindness. In Buddhist and Jainist philosophy there is a distinction between two philosophical views of reality: the *substance view* (in Sanskrit *dravyarthika naya*) represented, for example, by the Hinduistic Vedanta and in Europe by the Aristotelian philosophy, and the *modal view* (*paryayarthika naya*) represented by the Buddhistic philosophy of the "middle path" (*madhyama pratipad*), the so-called *Madhyamika* system developed especially by the Indian Buddhist philosopher Nagarjuna (II–III century), see (Murti 1987). This point of view is connected with the Buddhist understanding of the emptiness, vacuum, the Nirvana, about which one of the Buddhist modern Japanese philosophers, Nagao, said "Emptiness is not, however, simply nothingness. It is also immediately and necessarily the being of dependent co-arising", (Nagao 1989, p. 4). The "dependent co-arising" (in Sanskrit *pratitya-samutpada*) is understood as the so-called Samsara or the phenomenal world of experience. It can neither be said about this world that it exists nor that it does not exist, it exists only potentially, virtually, as probability or propensity. All these old philosophical ideas seem to be very near to the ideas of modern quantum physics. The problem is only to express them mathematically.

Just with the idea of the "mathematical expression" the question is connected of whether a modal concept, for example, of something unsharp as a quantum non-classical "unsharp object", can be described by a model in a bivalued Boolean logic? Of course, yes, an example is a representation in a Hilbert space of "probability amplitudes". This was discovered by Born and Heisenberg in the 20-ths. But in place of the two logical values of classical logic there appear to be infinitely many new values of real positive and normalized probabilities or of complex probability amplitudes. When we speak about complex or, possibly, quaternion "amplitudes" of probability, I propose to call this approach *submodality*, since we speek about as if non-real "roots" x of probability p, in the sense that $p = \bar{x}x$, where $\bar{x}$ is the complex or quaternion number conjugated to x. But when we speak about truth of quantum mechanics or quantum field theory, i.e., in the metalanguage (as in the mentioned Margenau criticism), not about reality, but about sentences about reality, in the logic of the 2nd or higher order, then we can use the term *supermodality*. This new modality can be the zero-modality, i.e., the limit categorical language of the bivalued Boolean logic, or it may be again a proper many-valued modality. This procedure can be, in principle, repeated. As is well-known, distinguishing among many logical types (orders) of sentences is necessary for avoiding antynomies, see, e.g., (Mostowski 1948, Chap. VIII). I think that the vectors of a complete orthonormal base of a complex separable Hilbert space $\mathcal{H}$ of quantum mechanics (these vectors correspond to the atoms of a quantum lattice) can be considered directly as generalized logical values of this submodal quantum logic. They are complex, but they are contained in the unit Hilbert ball, since pure state vectors are normalized to 1. Only for linear operations we can sometimes go outside of the unit ball, but in the end we have to come back to it. Among the linear operators of a Hilbert space $\mathcal{H}$ only the effects and POV measures constructed by them as generalized observables, and as special cases probability (density) operators as mixed and pure states, together with van Fraassen special modal operators, have a physical meaning and can be called *general modal operators*. Reminding that effects E as operators are defined by the conditions

$$O \leq E \leq I, \tag{9}$$

we see here the quantum correspodence and a generalization of the classical probability condition $0 \leq p \leq 1$ with a continuous infinity of values in the real interval $[0,1]$. But the quantum probability and modality is much more general and powerful than the classical probability and modality. In both cases probability is the most important and general of all the modalities. The mentioned quantum modalities are only of one vector argument, but in principle also many-argument operators can be considered. It is a far going generalization, but I think that modern physics compels us to go there. Of course, this problem requires further studies from the logical point of view in the sense of the semi-extensional logic.

But not only quantum physics requires modality. Actually, since Antiquity many modal concepts have been used freely in physics. Namely, causality is a kind of necessity, potentiality and probability are kinds or generalizations of possibility. Also energy as containing potential energy is a modal concept, of the possibility kind, since it is not work but possibility of work, see also (Ingarden 1997a) for further examples and discussion. But perhaps the most important for physics is the concept of relativity, for example, the relativity with respect to a space-time frame, and, in general, the role of the observer in relativity theory and quantum mechanics, as examples and generalizations of the ideas of index and indexical in the sense of Peirce–Bar-Hillel–Carnap–Montague. The essential point is that all physical measurements have a reference frame, a measuring device and a surrounding as a "context of use". In principle, they are always done in an open system interacting more or less strongly with the measured system, even if it is energetically or informationally (in entropy, e.g., thermally) isolated, see (Ingarden et al. 1997). Also in classical physics we have the problem of unsharp results of measurement, but only in quantum physics it is acute and cannot be reduced to the question of lack of knowledge, an error or approximation, see (Busch and Lahti 1990). The concept of fuzzy set has been introduced to mathematics by a purely classical motivation, namely, the fact of existence of unsharp concepts in natural languages, independently by K. Menger in 1951 (Menger 1951) and L. A. Zadeh in 1965 (Zadeh 1965). Only up to 1984 there appeared above 2400 papers about fuzzy set theory, see (Kaufmann 1972), (Drewniak 1984). This theory was started as if anew by S. Bugajski in application to probability and quantum physics, see, e.g., (Bugajski 1996). It may be remarked that the quantum theory of pressure broadening of spectral lines of Aleksander Jabłoński in the 30's and 40's, (Jabłoński 1937) (Jabłoński 1945), used actually the idea of statistics of spectral lines, i.e., of statistical hamiltonians as the 2nd order statistical theory, statistics of statistics, supermodality, although this principal point of view was not especially elaborated by him. It seems that now some elaborations of this problem already exist in the literature, but, perhaps, some further studies of this question would be desirable.

We remark, however, that by far not always quantum physics leads to fuzziness. For example, Niels Bohr corrected his previous false idea that the energy and momentum conservation laws are not sharply fulfilled in quantum mechanics, only approximately, statistically. Actually, they are conserved sharply (in an isolated system), with necessity, i.e., as a special modality. (Also entropy-information is sharply conserved in an isolated system.) It is very likely that, in the process of massive pair formation in vacuum near black holes, these conservation laws are reasons of the repulsion force of the expanding universe, to equilibrate the gravitation attractive force. Maybe this quantum effect can explain not only the Hawking radiation, but also the pressure caused by this radiation and the value of the cosmic constant Λ of Einstein which has been recently reintroduced in relativistic cosmology?

I finish with the hope that, about the end of this century, the modal point of view may be usuful in relaxing the yet existing difficulties in the interpretation and understanding of quantum mechanics.

Acknowledgements

The author would like to thank cordially Professor Rudolf Haag for an interesting discussion on the physical aspects of the present paper after its presentation at the 12-th Born Symposium in Wrocław on September 23, 1998, and to Professor Jerzy Perzanowski for a detailed discussion held later in Toruń and many useful remarks on the logical and philosophical aspects of the paper.

References

Belavkin, V.P., (1994): Nondemolition principle of quantum measurement theory, Foundations of Physics **24**, 685–714

Beltrametti, E., and G. Cassinelli, (1981): *The Logic of Quantum Mechanics*, Addison–Wesley, Reading, Mass.

Bugajski, S., (1996): Fundamentals of fuzzy probability theory, Intern. J. of Theoretical Physics **35**, 2229–2244

Busch, P., Lahti, P.J., (1990): Some remarks on unsharp quantum measurements, quantum non-demolition, and all that, Annalen der Physik **47**, 369–382

Busch, P., Lahti, P.J., Mittelstaedt, P., (1991): *The Quantum Theory of Measurement*, Springer, Berlin

Busch, P., Grabowski, M., Lahti, P.J., (1997): *Operational Quantum Physics*, 2nd ed., Springer, Berlin

Davies, E.B., (1976): *Quantum Theory of Open Systems*, Academic Press, London

Davies, N., (1998): *Europa. Rozprawa historyka z historią* (In Polish: Europe. A History), transl. from English by E. Tabakowska, Znak, Kraków

Drewniak, J., (1984): *Podstawy teorii zbiorów rozmytych*, (In Polish: Foundations of the Theory of Fuzzy Sets), UŚ, Katowice

Fays, R., (1965): *Modal Logic*, Gauthier-Villars, Paris

Gallin, D., (1975): *Intensional and Higher Order Modal Logic*, North-Holland, Amsterdam

Grabowski, M., (1990): *Miary półspektralne w nierelawistycznej mechanice kwantowej* (in Polish: Semispectral Measures in Nonrelativistic Quantum Mechanics), UMK, Toruń

Holevo, A.S., (1982): *Probabilistic and Statistical Aspects of Quantum Theory*, North–Holland, Amsterdam

Horodecki, R., (1996): *Correlation and Information–theoretic Aspects of Quantum Nonseperability of Mixed States*, UG, Gdańsk

Ingarden, R.S., Kossakowski, A., (1968): An axiomatic definition of information in quantum mechanics, Bull. Acad. Polon. Sci. Sér. math. astron. phys. **16**, 61–65

Ingarden, R.S., Kossakowski, A., M. Ohya, (1997): *Information Dynamics and Open Systems. Classical and Quantum Approach*, Kluwer, Dordrecht

Ingarden, R.S., (1997): Modalność w fizyce i językoznawstwie (In Polish: Modality in physics and linguistics), in: J. Perzanowski and A. Pietruszczak *Byt, Logos, Matematyka. FLFL 1995*, UMK, Toruń, 63–70

Jabłoński, A., (1937): Über die wellenmechanische Behandlung der Linienverbreitung, Acta Physica Polonica **6**, 371–391

Jabłoński, A., (1945): General theory of pressure broadening of spectral lines, Physical Review **68**, 78–93

Jammer, M., (1977): *The Philosophy of Quantum Mechanics. The Interpretations of Quantum Mechanics in Historical Perspective*, Wiley-Interscience, New York

Jauch, J.M., (1968): *Foundations of Quantum Mechanics*, Addison–Wesley, Reading MA

Kalmbach, G., (1986): *Measures and Hilbert Lattices*, World Scientific, Singapore

Kaufmann, A., (1972): *Theory of Fuzzy Sets*, Masson, Paris

Kossakowski, A., (1972): On quantum statistical mechanics of non-Hamiltonian systems, Reports on Mathematical Physics **3**, 247–274

Kotas, J., (1967): An axiom system for the modular logic, Studia Logica **21**, 17–38

Kotas, J., (1971): The modular logic as a calculus of logical schemata, Studia Logica **27**, 73–79

Kraus, K., (1983): *States, Effects and Operations*, Springer, Berlin

Lewis, C.I., Lanford, C.H., (1932), (1959): *Symbolic Logic*, Dover, New York, new ed. with Appendix III 1959

Ludwig, G., (1983): *Foundations of Quantum Mechanics I*, Springer, New York

Marciszewski, W. (ed.), (1987): *Logika formalna. Zarys encyklopedyczny z zastosowaniem do informatyki i lingwistyki* (in Polish: Formal Logic. Encyclopedic Outline with Application to Informatics and Linguistics), PWN, Warsaw

McKinsey, J.C.C., (1941): A solution of the decision problem for the Lewis system S2 and S4 with an applicaion to topology, *J. of Symbolic Logic* **6**, 117–134

Menger, K., (1951): Ensembles flous et fonctions aléatoires, C. R. Acad. Sci. Paris **232**, 2001–2003

Mittelstaedt, P., (1978): *Quantum Logic*, Reidel, Dordrecht

Montague, R., (1974): *Formal Philosophy*, Selected papers ed. by R. H. Thomason, Yale Univ. Press, New Haven, 3rd printing 1979

Mostowski, A., (1948): *Logika matematyczna* (In Polish: Mathematical Logic), UW, Warszawa

Murti, T.R.V., (1987): *The Central Philosophy of Buddhism. A Study of the Madhyamika System*, Unwin, London

Nagao, G., (1989): *The Fundamental Standpoint of Madhyamika Philosophy*, transl. from Japanese by J. P. Keenan, State Univ. of N. Y., Albany, N. Y.

Ohya, M., Petz, D., (1993): *Quantum Entropy and its Use*, Springer, Berlin

Perzanowski, J., (1989): Logiki modalne a filozofia (In Polish: Modal logics and philosophy), in: J. Perzanowski (ed.), *Jak filozofować? Studia z metodologii filozofii* (In Polish: How to Philosophize? Studies from Methodology of Philosophy), PWN, Warszawa

Piron, C., (1972): Survey of general quantum physics, Foundations of Physics **2** 287–314, reprinted in: C. Hooker (ed.), *The Logico-Algebraic Approach to Quantum Mechanics*, Reidel, Dordrecht 1975, 513–544

Piron, C., (1976): *Foundations of Quantum Physics*, Benjamin, Reading, Mass.

Rasiowa, H., (1955): O pewnym fragmencie implikacyjnego rachunku zdań (in Polish with Engl. and Russ. summaries: On a fragment of the implicative propositional calculus), *Studia Logica* **3**, 208–226

Rasiowa, H., Sikorski, R., (1970): *The Mathematics of Metamathematics*, 3rd ed., PWN, Warszawa

Staszewski, P., (1993): *Quantum Mechanics of Continuously Observed Systems*, UMK, Toruń

Suszko, R., (1968): Ontology in Tractatus of L. Wittgenstein, *Notre Dame J. of Formal Logic* **9**, 7–33, in Polish in: R. Suszko, *Wybór Pism* (Selected Papers), Pol. Tow. Semiotyczne, Warszawa 196–224

Tarski, A., (1938): Der Aussagenkalkül und die Topologie, *Fundamenta Math.* **25**, 103–134

Tokarz, M., (1993): *Elementy pragmatyki logicznej* (in Polish: Elements of Logical Pragmatics), PWN, Warszawa

van Frassen, B.C., (1981): A modal interpretation of quantum mechanics, in: E. Beltrametti and B. C. van Frassen (eds.), *Current Issues in Quantum Logic*, Plenum, New York, 229–258

van Frassen, B.C., (1991): The modal interpretation of quantum mechanics, in: P. Lahti and P. Mittelstaedt (eds.), *Symposium on the Foundations of Modern Physics Joensuu 1990*, World Scientific, Singapore, 440–460

van Frassen, B.C., (1991): *Quantum Mechanics: An Empiricist View*, Clarendon, Oxford

Varadarajan, V.S., (1985): *Geometry of Quantum Theory*, 2nd ed., Springer, New York

Wittgenstein, L., (1961): *Tractatus Logico-Philosophicus*, German text and Engl. transl. by D. F. Pears & B. F. McGuinness, Routledge & Kogan Paul, London

Wolniewicz, B., (1968): *Rzeczy i Fakty. Wstęp do pierwszej filozofii Wittgensteina*, PWN, Warszawa

Zadeh, L.A., (1965): Fuzzy sets, Information and Control **8** 338–353

Zawirski, Z., (1935): Über das Verhältnis der mehrwertigen Logik zur Wahrscheinlichkeitsrechnung, Studia Philosophica **1** 407–442

Zinowiew, A., (1963): *Filozoficzne Problemy Logiki Wielowartościowej* (in Polish: Philosophical Problems of Many-valued Logic), transl. from Russian by J. Jaroń (Russ. ed. by Acad. Sci. USSR 1960), PWN, Warszawa

On Localisation in Relativistic Quantum Mechanics

Th.W. Ruijgrok

Instituut voor Theoretische Fysica
Princetonplein 5, P.O. Box 80.006
3508 TA Utrecht, The Netherlands

Abstract. In nonrelativistic quantum mechanics causality is violated in an obvious way. The hope that this acausality would disappear in relativistic theories, in which the speed of propagation is finite, has turned out to be an idle hope. A localised state spreads over all space under a time translation or a boost.

In this paper it is suggested that this strange behaviour is actually a semantic problem. The eigenstates of the Newton-Wigner position operator will be considered as single particle states, which are localised with an accuracy equal to their Compton wavelength. Correspondingly the nonlocality of a two-particle potential will not extend beyond the Compton wavelength of the particles and can therefore still be called local.

These ideas will be elaborated in the framework of a previously formulated relativistic quantum theory. With this theory it will be shown that the sharp edge of a hard sphere interaction between two particles can still be determined with any accuracy by measuring the cross section in a high energy experiment.

1 Introduction

In nonrelativistic quantum mechanics the wave function of a particle, which initially is confined to a finite region of space, will spread immediately over all space. This is understood by the fact that, when written as a superposition of plane waves, the wave function is seen to contain components with arbitrarily large momentum.

For a long time it was generally believed that this acausal behaviour would disappear once the theory would have been put in a proper relativistic invariant form. Since no signal travels faster than light this behaviour would be prohibited. This belief, however, turned out not to be justified.

In 1949 Newton and Wigner (Newton and Wigner 1949) constructed localised single particle states, which deserved this name, because they satisfy a number of plausible conditions: they transform properly under rotations and under space translations and they are eigenstates of an hermitian operator, which in the limit where the speed of light becomes infinite, reduces to the correct nonrelativistic position operator.

These localised states, however, turned out to have some very peculiar properties. When being observed from a moving coordinate system or when subjected to a time translation, they were no longer localised, and the probability to observe the particle anywhere in space, was not zero. Later it was shown by Hegerfeldt (Hegerfeldt 1974, 1998) that a much weaker form of localisation of massive relativistic particles with positive energy, always gives rise to acausal behaviour. In spite of the fact that Ruijsenaars (Ruijsenaars 1981) demonstrated that the probability to detect a superluminal electron is less than 10^{-1000}, the discussion about the violation of causality goes on.

The present paper is a contribution to this discussion. It will be shown that all previous considerations, which applied to single particles, keep their validity also for systems of interacting particles. This will be done, not by extending the general proofs, but by first constructing a relativistic invariant particle theory, and then applying this to the calculation of some nonlocal and acausal effects. The conclusion will be the same as that of Ruijsenaars: the effects are too small to be observed.

This will be discussed in Section 3. First, however, a brief description will be given of the theory mentioned above. A full exposé has recently been published in (Ruijgrok 1998).

2 Relativistic Quantum Mechanics

2.1 Classical Theory

Since in 1949 Dirac (Dirac 1949) showed that the incorporation of relativity into classical Hamiltonian mechanics presents a formidable problem, the development of relativistic dynamical theories, both classical and quantum mechanical, was strongly influenced by his considerations. Kerner's collection of reprints (Kerner (ed.) 1972) gives a good idea of the progress that was made on this subject up to 1970. Later developments are discussed in (Llosa (Ed.) 1982), especially by Hill (Hill 1982, p. 104).

A way of presenting the difficulties, which arise when trying to make the transition from nonrelativistic– to relativistic classical mechanics, is the following.

Consider the problem of how to derive Newton's equations for two or more interacting particles, from the requirement that these equations be invariant under space and time translations, as well as under rotations and boosts of the coordinate system. This problem can be phrased as: "How to construct all possible realisations of the Galilei group?"

The solution is given by Cisło Łopuszański and Stichel (Cisło et al. 1998). As a special case it was shown by Sudarshan and Mukunda (Sudarshan and Mukunda 1974) that it is possible to introduce the interaction between particles in such a way that only the generator for infinitesimal time translations is affected.

This means that the Hamiltonian gets an extra term, which is the translational– and rotational invariant potential energy, while all other generators remain the same. In particular the generator for translations is unaltered and is equal to the sum of the three-momenta of the individual particles.

Dirac (Dirac 1949), in his article "Forms of Relativistic Dynamics", considered the same problem for the Poincaré group. For the "point" form he concludes that again the generators for rotations and boosts are unchanged and that the zero component of the four-momentum should get an extra term, after switching on the interaction. Now, however, also the total three-momentum acquires an additional term. "These [extra terms] cause the real difficulty in the problem of constructing a theory of a relativistic dynamical system...", and half a century after Dirac wrote these words, there is still no practical theory, which could be used for the calculation of high energy collisions or bound states of classical relativistic particles.

The delicacy of the problem becomes clear from the existence of a no-go theorem(Currie 1963),(Currie et al. 1963),(Leutwyler 1965). This theorem states that there can be no interaction at all when too stringent additional assumptions are made, e.g., about the canonical character of the Lorentz transformations and of the position coordinates.

The remarks made so far seem to suggest that it may be impossible to construct an elegant relativistic theory with a direct particle interaction, i.e., without using an intervening field with an infinite number of degrees of freedom. This, however, is not the case, as was shown for instance in the papers by Currie (Curie 1966) and by Van Dam and Wigner (Van Dam and Wigner 1965), (Van Dam and Wigner 1966). The latter construct a Poincaré-invariant theory, in which the conserved total four momentum is split into a kinetic part and a part in transit, which vanishes long before and long after the collision.Therefore, during the collision, not only the kinetic energy $P^0 = p_1^0 + p_2^0$ is not conserved, but also the three-momentum $\vec{\mathbf{P}} = \vec{\mathbf{p}}_1 + \vec{\mathbf{p}}_2$ will vary in time. After the collision has been completed the four-vector P^μ has again the same value as before.

This time dependence of all four components of the sum of the particle momenta is not particular for the theories mentioned above,but is valid in a much wider framework, as was already discussed by Møller (Møller 1952).

This may be an acceptable situation for a classical theory, but for the quantised version it is not. If the quantum theory is to have a form like the Lippmann-Schwinger equation, a sum or integral over intermediate states is going to appear, which, usually, is restricted by the kinetic three-momentum $\vec{\mathbf{P}}$ having the same value for the intermediate states as for the initial and final state. Since it was shown by Van Dam and Wigner (Van Dam and Wigner 1965), (Van Dam and Wigner 1966) that for a relativistic theory this cannot be realised, another three-vector must be found, which *is* conserved.

The obvious candidate is the conserved three-vector of the total momentum, which includes the momentum in transit during the collision. This, however, is an extremely complicated expression, containing the particle momenta along both world lines.

For that reason we will now define a much simpler three-vector, viz. the velocity of the total system as given by the kinetic momenta only:

$$\vec{\mathbf{v}} = \frac{\vec{\mathbf{p}}_1 + \vec{\mathbf{p}}_2 + \cdots + \vec{\mathbf{p}}_N}{p_1^0 + p_2^0 + \cdots + p_N^0} = \frac{\vec{\mathbf{P}}}{P^0} \tag{1}$$

with the free particle relation

$$p_i^0 = +\sqrt{|\vec{\mathbf{p}}_i|^2 + m_i^2} \qquad i = 1, \ldots, N \tag{2}$$

between the momentum components and the restmass m_i. It will now be assumed that this vector is time independent, although $\vec{\mathbf{P}}(t)$ and $P^0(t)$ do vary during the collision. In a classical two-particle system it is almost obvious, at least it was to Møller (Møller 1952), that this total particle velocity is constant. It must be stressed, however, that this is not a mathematical identity, but rather a physical hypothesis.

2.2 The Equation

After these preliminaries the new quasipotential theory for relativistic scattering amplitudes is now defined by the following generalisation of the Lippmann-Schwinger equation (Lippmann and Schwinger 1950)

$$M_{\alpha\beta}(s) = V_{\alpha\beta} - \int_\gamma V_{\alpha\gamma} L_\gamma(\vec{\mathbf{v}}, s) M_{\gamma\beta}(s) \qquad \text{for} \qquad \vec{\mathbf{v}}_\alpha = \vec{\mathbf{v}}_\beta \equiv \vec{\mathbf{v}}\,. \tag{3}$$

The integration element for the intermediate state $\gamma = (p_1, \cdots, p_n)$ is

$$\int_\gamma \cdots = \int dp_1 \ldots dp_n \Pi_{j=1}^n \delta(p_j^2 - m_j^2)\theta(p_j^0) \cdots \tag{4}$$

and the velocities $\vec{\mathbf{v}}_\alpha$ and $\vec{\mathbf{v}}_\beta$ are defined by (1).
The Lorentz invariant propagator on the upper rim of the unitarity cut is taken as

$$L_\gamma(\vec{\mathbf{v}}, s_0 + i0) = \int_0^\infty \frac{ds'}{s' - s_0 - i0} \delta_4(P_\gamma - \frac{s'}{s_0} P_0)\,, \tag{5}$$

in which the four-momenta P_γ and P_0 are equal to

$$P_\gamma = \sqrt{\frac{s_\gamma}{1 - |\vec{\mathbf{v}}_\gamma|^2}}\,(1, \vec{\mathbf{v}}_\gamma) \quad \text{and} \quad P_0 = \sqrt{\frac{s_0}{1 - |\vec{\mathbf{v}}|^2}}\,(1, \vec{\mathbf{v}})\,, \tag{6}$$

so that $s_\gamma = P_\gamma^2$ and $s_0 = P_0^2$.
The unitarity of the S-matrix is guaranteed by the hermiticity of $V_{\alpha\beta}$ and by the equation

$$\lim_{s \to s_0 + i0} \text{Im}\ L_\gamma(\vec{v}, s) = \pi \delta_4(P_\gamma - P_0)\,. \tag{7}$$

The form (5) of the propagator furthermore ensures the equality of the total velocities in the initial and intermediate state $\vec{v} = \vec{v}_\gamma$. This can be seen explicitly by performing the integration in (5), which leads to

$$L_\gamma(\vec{v}, s) = L_\gamma^0(s)\, L^3(\vec{v}_\gamma, \vec{v}) \tag{8}$$

with

$$L_\gamma^0(s) = \frac{1}{s_\gamma^{3/2}(\sqrt{s_\gamma} - \sqrt{s})} \qquad \text{and} \qquad L^3(\vec{v}_\gamma, \vec{v}) = (1 - |\vec{v}|^2)^2 \delta_3(\vec{v}_\gamma - \vec{v})\,. \tag{9}$$

Both factors in (8) are Lorentz invariant.

The total cross section for the scattering of two particles in the state β is calculated in the standard way

$$\sigma_{tot}(\beta) = \frac{(2\pi)^4}{2\sqrt{\lambda(s, m_1^2, m_2^2)}} \int_\alpha |M_{\alpha\beta}(s + i0)|^2 \delta_4(P_\alpha - P_\beta) \tag{10}$$

with

$$s = s_\beta = P_\beta^2 \qquad \text{and} \qquad \lambda(x, y, z) = x^2 + y^2 + z^2 - 2xy - 2yz - 2zx\,. \tag{11}$$

In the same way the decay rate of an unstable particle–by definition in its own rest system–is given by

$$\frac{1}{\tau} = \frac{(2\pi)^4}{2m_u} \int_\alpha |M_{\alpha\beta}(+)|^2 \delta_4(P_\alpha - P_\beta)\,,$$

where m_u is its mass and P_β its four-momentum.

For later purpose it is useful to define the Hilbert space spanned by the products of single particle states. Instead, however, of labelling these states by the three-momentum and the mass of the individual particles, we will use the three-velocity and the mass. Suppressing the masses, a many particle state will therefore be written as $|\alpha\rangle = |\vec{v}_1, \cdots, \vec{v}_{n_\alpha}\rangle$. For later convenience we choose the relativistic normalisation as

$$\langle \alpha' | \alpha \rangle = \begin{cases} \prod_{i=1}^{n_\alpha} L^3(\vec{v}'_i, \vec{v}_i) & \text{if } n_{\alpha'} = n_\alpha \\ 0 & \text{if } n_{\alpha'} \neq n_\alpha\,. \end{cases} \tag{12}$$

Using the transformation formula

$$\gamma^4(v) d\vec{v} = \frac{d\vec{p}}{m^2 p^0} \qquad \text{with} \qquad \gamma(v) = \frac{1}{\sqrt{1 - |\vec{v}|^2}} \tag{13}$$

it can be shown that for single particle states the connection with the standard normalisation in momentum space, is given by

$$\langle \vec{\mathbf{p}}' | \vec{\mathbf{p}} \rangle = \frac{16\pi^3 p^{02}}{m^4} \langle \vec{\mathbf{v}}' | \vec{\mathbf{v}} \rangle .$$

It is convenient to change the integration of (4), into an integration over the velocities. Disregarding the internal degrees of freedom and fixing the number of particles, we get

$$\int_\alpha \cdots = \mu_\alpha^2 \overset{*}{\int_\alpha} \cdots \qquad \text{with} \tag{14}$$

$$\mu_\alpha = \prod_{i \epsilon \alpha} \frac{m_i}{\sqrt{2}} \quad \text{and} \quad \overset{*}{\int_\alpha} \cdots = \int \gamma^4(v_1) d\vec{\mathbf{v}}_1 \cdots \gamma^4(v_{n_\alpha}) d\vec{\mathbf{v}}_{n_\alpha} \cdots .$$

A simple consequence is that the unit operator can be written as

$$\overset{*}{\int_\gamma} |\gamma\rangle\langle\gamma| = 1 , \tag{15}$$

where a summation over the number of particles and over the internal degrees of freedom is also implied.

2.3 The Potential

We now want to show how to construct the potential $V_{\alpha\beta}$, occurring in the main equation (3). Since it must be Lorentz invariant, it should be a function of the scalars that can be formed with the four-momenta of the particles, which make up the states α and β. As most important example we first consider the elastic scattering of two spinless particles with masses m_1 and m_2. The initial and final states are therefore labelled by $\beta = (\vec{\mathbf{p}}_1, m_1; \vec{\mathbf{p}}_2, m_2)$ and $\alpha = (\vec{\mathbf{p}}'_1, m_1; \vec{\mathbf{p}}'_2, m_2)$. For the time being we shall not use the velocity basis, defined in (14).
Suppose, e.g., that in the nonrelativistic limit we want the interaction to reduce to an attractive Yukawa potential, as derived from the one-boson exchange diagram (Fig. 1).

In the nonrelativistic theory and in the coordinate representation, this attractive Yukawa potential is given by

$$V^{NR}(r) = -\alpha\hbar c \frac{e^{-\mu r}}{r} , \tag{16}$$

where $r = |\vec{\mathbf{r}}|$ is the distance between the particles and α is a positive and dimensionless coupling constant. In what follows we will take units such that $\hbar = 1$ and $c = 1$. In order to calculate the potential in the momentum representation, the momenta, which are canonically conjugate to the centre of mass coordinates and the relative coordinates must first be defined

$$\vec{\mathbf{P}}_\beta = \vec{\mathbf{p}}_1 + \vec{\mathbf{p}}_2\,, \qquad \vec{\mathbf{p}}_\beta = \frac{m_2\vec{\mathbf{p}}_1 - m_1\vec{\mathbf{p}}_2}{m_1+m_2}\,, \tag{17}$$

$$\vec{\mathbf{P}}_\alpha = \vec{\mathbf{p}}'_1 + \vec{\mathbf{p}}'_2\,, \qquad \vec{\mathbf{p}}_\alpha = \frac{m_2\vec{\mathbf{p}}'_1 - m_1\vec{\mathbf{p}}'_2}{m_1+m_2}\,, \tag{18}$$

with the inversion

$$\vec{\mathbf{p}}_1 = \frac{m_1}{m_1+m_2}\vec{\mathbf{P}}_\beta + \vec{\mathbf{p}}_\beta\,, \qquad \vec{\mathbf{p}}_2 = \frac{m_2}{m_1+m_2}\vec{\mathbf{P}}_\beta - \vec{\mathbf{p}}_\beta\,,$$

$$\vec{\mathbf{p}}'_1 = \frac{m_1}{m_1+m_2}\vec{\mathbf{P}}_\alpha + \vec{\mathbf{p}}_\alpha\,, \qquad \vec{\mathbf{p}}'_2 = \frac{m_2}{m_1+m_2}\vec{\mathbf{P}}_\alpha - \vec{\mathbf{p}}_\alpha\,. \tag{19}$$

In the nonrelativistic case the total momentum is conserved, $\vec{\mathbf{P}}_\alpha = \vec{\mathbf{P}}_\beta$, so that also

$$\vec{\mathbf{p}}'_1 - \vec{\mathbf{p}}_1 = \vec{\mathbf{p}}_2 - \vec{\mathbf{p}}'_2 = \vec{\mathbf{p}}_\alpha - \vec{\mathbf{p}}_\beta \equiv \vec{\mathbf{q}}\,. \tag{20}$$

The nonrelativistic potential in momentum space can be written as a function of $\vec{\mathbf{q}}$:

$$V^{NR}_{\alpha\beta} = \frac{1}{(2\pi)^3}\int e^{-i\vec{\mathbf{q}}\cdot\vec{\mathbf{r}}}\,V^{NR}(r)\,d\vec{\mathbf{r}} = -\frac{\alpha}{2\pi^2(|\vec{\mathbf{q}}|^2+\mu^2)}\,, \tag{21}$$

so that the full potential must approach

$$V_{\alpha\beta} \approx -\frac{2\alpha m_1 m_2}{\pi^2(|\vec{\mathbf{q}}|^2+\mu^2)} \tag{22}$$

in the nonrelativistic limit. An extra factor $4m_1m_2$ had to be included, because of the different normalisation of the states. The standard way to satisfy condition (22) is, first to introduce the Mandelstam variables

$$s = (p_1+p_2)^2 = (p'_1+p'_2)^2\,, \qquad t = (p'_1-p_1)^2 = (p'_2-p_2)^2\,, \tag{23}$$
$$u = (p'_1-p_2)^2 = (p'_2-p_1)^2\,, \qquad s+t+u = 2(m_1^2+m_2^2)\,,$$

and then to observe that in the nonrelativistic limit

$$t \overset{\mathrm{NR}}{\Longrightarrow} -|\vec{\mathbf{q}}|^2 + \left[\frac{|\vec{\mathbf{p}}'_1|^2 - |\vec{\mathbf{p}}_1|^2}{2m_1}\right]^2 + \cdots\,. \tag{24}$$

The NR-limit is defined by $\frac{|\vec{\mathbf{p}}_1|^2}{2m_1} \ll m_1$, or $|\vec{\mathbf{v}}_1| \ll 1$.
Combination of (22) with (24) strongly suggests to take for the relativistic attractive Yukawa potential the following expression

$$V_{\alpha\beta} = \frac{2\alpha m_1 m_2}{\pi^2(t-\mu^2)}\,. \tag{25}$$

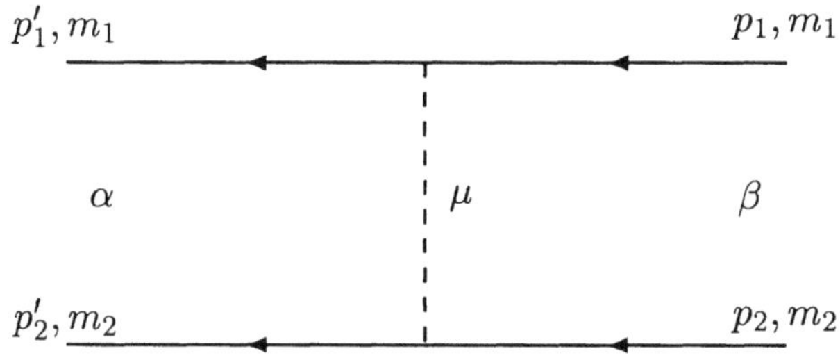

Fig. 1. One-boson exchange diagram for Yukawa interaction.
The mass of the exchanged particle is μ.

This is indeed the same as the Born approximation in a simple field theory, given by the one-boson exchange diagram (Fig. 1).

In the derivation as given above, the conservation of energy and momentum $p_1 + p_2 = p'_1 + p'_2$ played a crucial rôle. In our theory this is replaced by velocity conservation, expressed by

$$\frac{\vec{\mathbf{p}}_1 + \vec{\mathbf{p}}_2}{p_1^0 + p_2^0} = \frac{\vec{\mathbf{p}}'_1 + \vec{\mathbf{p}}'_2}{p'^0_1 + p'^0_2} \tag{26}$$

or, what amounts to the same, by

$$\frac{p_1 + p_2}{\sqrt{s}} = \frac{p'_1 + p'_2}{\sqrt{s'}} \quad \text{with} \quad s = (p_1 + p_2)^2 \quad \text{and} \quad s' = (p'_1 + p'_2)^2 \,. \tag{27}$$

The four-vectors of the relative momenta are again defined as in (17),

$$p_\beta = \frac{m_2 p_1 - m_1 p_2}{m_1 + m_2} \quad \text{and} \quad p_\alpha = \frac{m_2 p'_1 - m_1 p'_2}{m_1 + m_2} \,. \tag{28}$$

The four-vectors q_1 and q_2 for the energy-momentum transfer, and defined as before by

$$q_1 = p'_1 - p_1 \quad \text{and} \quad q_2 = p_2 - p'_2 \,,$$

are now different from each other, except if $s' = s$, when (27) again expresses energy- and momentum conservation. Therefore, it is not clear which of the two possible forms for t in (23) should be used in a Lorentz invariant expression for the potential $V_{\alpha\beta}$. However, the Mandelstam variables can also be written in the form

$$\begin{aligned}
\bar{s} &= (p'_1 + p'_2) \cdot (p_1 + p_2) \,, \\
\bar{t} &= (p'_1 - p_1) \cdot (p_2 - p'_2) = q_1 \cdot q_2 \,, \\
\bar{u} &= (p'_1 - p_2) \cdot (p_1 - p'_2) \,.
\end{aligned} \tag{29}$$

They have been given different names, because now they are also defined when $q_1 \neq q_2$. They satisfy the more general relations

$$\bar{s} = \sqrt{s's} \quad \text{and} \quad \bar{s} + \bar{t} + \bar{u} = 2(m_1^2 + m_2^2) - (\sqrt{s'} - \sqrt{s}\,)^2 \,. \tag{30}$$

Notice that for $s' = s$, i.e., on the mass shell, $\bar{t} = t$, because then energy and momentum are again conserved (27).
The definition of the relativistic Yukawa potential is now taken as an extension of (25):

$$V_{\alpha\beta} = \frac{2\alpha m_1 m_2}{\pi^2(\bar{t} - \mu^2)} \,. \tag{31}$$

It is equal to (25) for the standard theory in which $s' = s$, while it is a definition if the states α and β have different s-values, which is allowed in the present theory.

All ingredients of the basic equation (3) are now well defined and scattering amplitudes can in principle be calculated, as well as the masses of bound states.

The discrete spectrum of invariant masses M_n is defined by the poles $s_n = M_n^2$ of the Green function.

If the wave functions ψ_γ^n of the eigenstates $|n, \vec{\mathbf{v}}\rangle$ are defined by

$$\langle \gamma | n, \vec{\mathbf{v}} \rangle = \psi_\gamma^n L^3(\vec{\mathbf{v}}_\gamma, \vec{\mathbf{v}}) \,, \tag{32}$$

the eigenstates can be expanded in free-particle states

$$|n, \vec{\mathbf{v}}\rangle = \int_\gamma^{*} \psi_\gamma^n L^3(\vec{\mathbf{v}}_\gamma, \vec{\mathbf{v}}) |\gamma\rangle \,. \tag{33}$$

The eigenvalue equations for the wave functions become

$$\left(\sqrt{P_\beta^2} - M_n\right)\psi_\beta^n + \int_\gamma^{*} \frac{\mu_\beta \mu_\gamma}{(s_\beta s_\gamma)^{3/4}} V_{\beta\gamma} L^3(\vec{\mathbf{v}}_\gamma, \vec{\mathbf{v}}) \psi_\gamma^n = 0 \quad \text{for} \quad \vec{\mathbf{v}}_\beta = \vec{\mathbf{v}} \tag{34}$$

in which μ_β is given by (14). The hermiticity of $V_{\alpha\beta}$ guarantees the orthogonality of the eigenfunctions. If we require the states $|n, \vec{\mathbf{v}}\rangle$ to be normalised in the same way as the single free–particle states (see(12)), then we should have

$$\langle n', \vec{\mathbf{v}}' | n, \vec{\mathbf{v}} \rangle = \delta_{n',n} L^3(\vec{\mathbf{v}}', \vec{\mathbf{v}}) \,. \tag{35}$$

This implies for the wave functions the following normalisation

$$\int_\gamma^{*} \psi_\gamma^{n'*} \psi_\gamma^n L^3(\vec{\mathbf{v}}_\gamma, \vec{\mathbf{v}}) = \delta_{n',n} \,. \tag{36}$$

The equation (34) is the basic manifestly Lorentz invariant equation, from which the mass spectrum can be calculated.

The theory outlined in this section, has been successfully applied to a number of relativistic systems (de Groot and Ruijgrok 1975), (de Groot and Ruijgrok 1986), (Hersbach and Ruijgrok 1991)–(Hersbach 1994), (van Meijgaard and Ruijgrok 1985), (Ruijgrok 1976)–(Ruijgrok 1996).

2.4 Poincaré Invariance

Because of the equations (13) and (14) the integration elements in the basic equations (3) and (34) are invariant under homogeneous Lorentz transformations. As a consequence the scattering amplitude $M_{\alpha\beta}$ will transform in the same way as the potential $V_{\alpha\beta}$. For spinless particles the potential should therefore be a Lorentz scalar, constructed from the momentum four-vectors of the particles in the states α and β. In general the invariance under rotations and Lorentz boosts can be guaranteed by requiring that the potential satisfy

$$[\widehat{J}_{\mu\nu}, V] = 0\,, \tag{37}$$

where V is defined by its matrix elements

$$\langle\alpha|V|\beta\rangle = V_{\alpha\beta}L^3(\vec{v}_\alpha, \vec{v}_\beta) \tag{38}$$

and $\widehat{J}_{\mu\nu}$ are the generators of infinitesimal transformations, of which it is known how they operate on the free-particle states $|\alpha\rangle$, and for which

$$[\widehat{J}_{\mu\nu}, \widehat{J}_{\lambda\sigma}] = i(g_{\mu\lambda}\widehat{J}_{\sigma\nu} - g_{\mu\sigma}\widehat{J}_{\lambda\nu} + g_{\nu\lambda}\widehat{J}_{\mu\sigma} - g_{\nu\sigma}\widehat{J}_{\mu\lambda})\,. \tag{39}$$

For particles with spin the potential may contain factors of the form $D = \overline{u}_{\lambda'}p_\mu\gamma^\mu u_\lambda$, where u_λ is a Dirac spinor satisfying the free-particle Dirac equation. This latter property is essential for proving (37) also for this case. In most other theories this creates difficulties, because in intermediate states particles go off mass-shell, while in the present case the particles are always on shell.

It should be emphasised that the operators $\widehat{J}_{\mu\nu}$ are not modified if there is interaction between the particles.

When the Poincaré group is considered, the generators $\widehat{P}_\mu$ for infinitesimal translations in time and space cannot be the same as for noninteracting particles, because the total kinetic energy and the total kinetic momentum are not conserved in the interaction. The latter was replaced by conservation of total velocity, described by the factor $L^3(\vec{v}_\gamma, \vec{v})$ in (3) and (34).

The correct form of $\widehat{P}_\mu$ can be shown to be the following (in terms of its matrix elements between free-particle states)

$$\langle\alpha|\widehat{P}_\mu|\beta\rangle = (s_\alpha s_\beta)^{1/4}[\langle\alpha|\beta\rangle + \frac{\mu_\alpha\mu_\beta}{s_\alpha s_\beta}\langle\alpha|V|\beta\rangle]u_\mu(\vec{v}) \quad \text{with} \quad \vec{v}_\alpha = \vec{v}_\beta = \vec{v}\ , \tag{40}$$

where μ_α was given in (14) and $u_\mu(\vec{v})$ are the components of the four-vector

$$u(\vec{v}) = \gamma(v)(1, \vec{v}) \,. \tag{41}$$

The proof, the details of which are given in (Ruijgrok 1998), goes along the following lines:

1. Define the stationary scattering states

$$|\beta\rangle^+ = |\beta\rangle - \int\limits_\gamma^* \mu_\gamma \mu_\beta (\frac{s_\gamma}{s_\beta})^{3/4} M_{\gamma\beta}(+) L^0_\gamma (s_\beta + i0) L^3(\vec{v}_\gamma, \vec{v})|\gamma\rangle \tag{42}$$

   with $\vec{v}_\beta = \vec{v}$.
2. Prove that, together with the bound states (33), they form a complete orthonormal set. This is guaranteed by the unitarity of the S-matrix, which follows from the hermiticity of $V_{\alpha\beta}$.
3. Prove that these stationary scattering states and bound states are eigenstates of $\widehat{P}_\mu$ defined by (40), i.e.,

$$\widehat{P}_\mu |\beta\rangle^+ = \sqrt{s_\beta} u_\mu |\beta\rangle^+ \quad \text{and} \quad \widehat{P}_\mu |n, \vec{v}\rangle = M_n u_\mu |n, \vec{v}\rangle \,. \tag{43}$$

   Here it is essential that $M_{\alpha\beta}$ is the solution of (3) and ψ^n_γ satisfy (34).
4. From the completeness and from (43) it follows immediately that

$$[\widehat{P}_\mu, \widehat{P}_\nu] = 0 \,. \tag{44}$$

The remaining condition for the full Poincaré invariance is

$$[\widehat{P}_\mu, \widehat{J}_{\nu\lambda}] = i(g_{\mu\nu}\widehat{P}_\lambda - g_{\mu\lambda}\widehat{P}_\nu) \,. \tag{45}$$

This can be proved by writing $\widehat{P}_\mu$, defined by (40), as the sum of the four-momentum of the free particles and the four-momentum residing in the potential

$$\widehat{P}_\mu = \widetilde{P}_\mu + CV\widetilde{P}_\mu \tag{46}$$

in which C is an operator which commutes with $\widehat{J}_{\nu\lambda}$. Using the fact that (45) is satisfied if $\widehat{P}_\mu$ is replaced by $\widetilde{P}_\mu$, substitution of (46) into (45) leads to the condition (37), which is the condition for the potential to be Lorentz invariant. This ends the proof of the Poincaré invariance of the theory.

It is of the so called point-form as defined by Dirac (Dirac 1949), because only $\widehat{P}_\mu$ is changed by the interaction, while the $\widehat{J}_{\mu\nu}$ are left untouched.

3 Locality and Causality

3.1 The Position Operator

In order to compare the results of an experiment with a theoretical prediction, it will in most cases be sufficient to calculate a differential cross section, a mean life-time or the spectrum of bound states. These quantities can in principle be calculated by the type of theory, described in this paper.

If it comes, however, to measuring the position of a particle, it will be necessary to give a definition of the concept of a particle being confined to a finite region in space. This definition must be given in terms of already existing notions. The eigenstates $|\vec{r}\rangle$ of the position operator $\vec{Q}$ of Newton and Wigner (Newton and Wigner 1949) provide such a definition, as will now be shown.

In the velocity reprensentation the states $|\vec{r}\rangle$ are defined by

$$\langle\vec{v}|\vec{r}\rangle = \frac{m}{(2\pi)^{3/2}}\sqrt{p^0}\, e^{-i\vec{p}\cdot\vec{r}}\,, \tag{47}$$

where $p = (p^0, \vec{p}) = m\gamma(v)(1, \vec{v})$. Using (13)-(15), one easily proves the orthonormality

$$\langle\vec{r}'|\vec{r}\rangle = \int \langle\vec{r}'|\vec{v}\rangle\langle\vec{v}|\vec{r}\rangle\gamma^4(v)\, d\vec{v} = \delta(\vec{r}' - \vec{r}) \tag{48}$$

and the completeness

$$\int |\vec{r}\rangle\, d\vec{r}\,\langle\vec{r}| = 1. \tag{49}$$

Taking for the position operator in the momentum representation

$$\vec{Q} = i\left(\frac{\partial}{\partial\vec{p}} - \frac{\vec{p}}{2p^{02}}\right), \tag{50}$$

one finds on substitution of (47)

$$\langle\vec{v}|\vec{Q}|\vec{r}\rangle = i\left(\frac{\partial}{\partial\vec{p}} - \frac{\vec{p}}{2p^{02}}\right)\langle\vec{v}|\vec{r}\rangle = \vec{r}\langle\vec{v}|\vec{r}\rangle\,, \tag{51}$$

so that $|\vec{r}\rangle$ is indeed an eigenvector

$$\vec{Q}|\vec{r}\rangle = \vec{r}|\vec{r}\rangle\,. \tag{52}$$

If $|\vec{r}, \vec{a}\rangle$ is the state obtained by applying a translation $\vec{a}$ to a state $|\vec{r}\rangle$, then

$$\begin{aligned}\langle\vec{r}'|\vec{r}, \vec{a}\rangle &= \langle\vec{r}'|e^{-i\hat{\vec{P}}\cdot\vec{a}}|\vec{r}\rangle \\ &= \int \langle\vec{r}'|e^{-i\hat{\vec{P}}\cdot\vec{a}}|\vec{v}\rangle\langle\vec{v}|\vec{r}\rangle\gamma^4(v)\, d\vec{v} \\ &= \delta(\vec{r}' - \vec{r} - \vec{a}) = \langle\vec{r}'|\vec{r} + \vec{a}\rangle\,,\end{aligned} \tag{53}$$

hence

$$|\vec{r}, \vec{a}\rangle = |\vec{r} + \vec{a}\rangle . \tag{54}$$

For Newton and Wigner (Newton and Wigner 1949) this was one of the required properties of $\vec{Q}$.

They define "the wave function in coordinate space" by the Fourier transform (their equation (2)). The wave function of the state $|\vec{v}\rangle$ therefore is

$$\langle \vec{x} \,|\, \vec{v}\rangle = \frac{1}{(2\pi)^{3/2}}\, e^{i\vec{p}\cdot\vec{x}} \tag{55}$$

and the localised state $|\vec{r}\rangle$ has as wave function

$$\begin{aligned}\psi^{NW}_{\vec{r}}(\vec{x}) &= \int \langle \vec{x}\,|\,\vec{v}\rangle\langle \vec{v}\,|\,\vec{r}\rangle \gamma^4(v)\, d\vec{v} \\ &= \frac{1}{m(2\pi)^{3/2}} \int \frac{d\vec{p}}{\sqrt{p^0}}\, e^{i\vec{p}\cdot(\vec{x}-\vec{r})} .\end{aligned} \tag{56}$$

This integral can be evaluated and for $\vec{r} = 0$ is found to be equal to ((Newton and Wigner 1949) (9a))

$$\psi^{NW}_{\vec{r}=0}(\vec{x}) = \frac{\Gamma(3/4) m^{3/2}}{(2\pi^2)^{5/4}} \frac{K_{5/4}(mx)}{(mx)^{5/4}} . \tag{57}$$

The behaviour for $x \to 0$ and for $mx >> 1$ is given by

$$\psi^{NW}_{\vec{r}=0}(\vec{x}) \sim (mx)^{-5/2} \quad \text{and by} \quad \psi^{NW}_{\vec{r}=0}(\vec{x}) \sim \frac{e^{-mx}}{(mx)^{7/4}} . \tag{58}$$

Although this NW-wave function has an extension of the size of the Compton wavelength, the states $|\vec{r}\rangle$ will be referred to as localised states. The states $|\vec{x}\rangle$, defined by (55), are not orthogonal, and therefore the Newton-Wigner wave functions $\psi^{NW}(\vec{x}) = \langle \vec{x}|\psi\rangle$ cannot be interpreted as probability amplitudes. Moreover $|\vec{x}\rangle$ does not transform correctly, i.e., like (54), under a spatial translation.

It therefore seems better to define the coordinate representation of a state $|\psi\rangle$ by using the localised states $|\vec{r}\rangle$:

$$\psi(\vec{r}) = \langle \vec{r}\,|\psi\rangle . \tag{59}$$

If $|\psi\rangle$ is normalised to unity, this allows $\psi(\vec{r})$ to be seen as a probability amplitude:

$$1 = \langle\psi|\psi\rangle = \int \langle\psi|\vec{r}\rangle\, d\vec{r}\, \langle\vec{r}\,|\psi\rangle = \int |\psi(\vec{r})|^2\, d\vec{r} . \tag{60}$$

This total probability is the same for any Poincaré transformed observer, because the group representation is unitary. The wave function $\psi(\vec{r})$ (59)

may, however, change in an unexpected way. Under a space translation nothing strange happens:

$$\psi^{\vec{a}}(\vec{r}) \equiv \langle \vec{r}|e^{-i\hat{\vec{P}}\cdot\vec{a}}|\psi\rangle = \langle \vec{r}+\vec{a}|\psi\rangle = \psi(\vec{r}+\vec{a})\,. \tag{61}$$

The time shifted wave function of a localised state, however, behaves in an abnormal way. It is given by

$$\langle \vec{r}\,'|\vec{r},t\rangle \equiv \langle \vec{r}\,'|e^{i\hat{P}^0 t}|\vec{r}\rangle = \frac{1}{(2\pi)^3}\int e^{i[p^0 t-\vec{p}\cdot(\vec{r}-\vec{r}\,')]}\,d\vec{p}\,. \tag{62}$$

This integral is found (Prudnikov et al. 1986) to be equal to

$$\langle \vec{r}\,'|\vec{r},t\rangle = \frac{-im^4 t}{2\pi^2 z^2}K_2(z) \quad \text{for} \quad |\vec{r}\,'-\vec{r}| > t > 0 \tag{63}$$

and

$$\langle \vec{r}\,'|\vec{r},t\rangle = \frac{m^4 t}{4\pi z^2}[J_2(z)+iY_2(z)] \quad \text{for} \quad t > |\vec{r}\,'-\vec{r}| > 0 \tag{64}$$

with $z = m\sqrt{||\vec{r}\,'-\vec{r}|^2 - t^2|}$ and K_2, J_2 and Y_2 being the usual Bessel functions.

Because in (63) the wave function does not vanish if $|\vec{r}\,'-\vec{r}| > t$, this is acausal behaviour. It is, however, negligibly small, already at a distance of a few Compton wavelengths into the acausal region. Taking $|\vec{r}\,'-\vec{r}| = t + \frac{f}{m}$ and writing $T = mt$, it is found that for large times the wave function is proportional to

$$\langle \vec{r}\,'|\vec{r},t\rangle \sim \frac{e^{-\sqrt{2fT}}}{T^{5/4}}\,. \tag{65}$$

Since T is the number of Compton wavelengths travelled by light in the time t, it is clear that the width of this acausal front will shrink to zero with increasing time. Even within microscopic times it will become less than the Compton wavelength, and is therefore not measurable.

Another strange phenomenon–already mentioned by Newton and Wigner (Newton and Wigner 1949)–is the loss of locality when the coordinate system is boosted to a moving frame. The magnitude of this effect could be calculated in the same way as for a time translation, but this has not yet been done.

The conclusion, so far, is that when measuring the position of a particle, no observable acausal effects exist, in spite of the fact that the operator $\vec{Q}$ "has no simple covariant meaning under relativistic transformations" (Newton and Wigner 1949).

3.2 Nonlocal Interactions

In this section the question will be considered as to whether the use of the localised states $|\vec{r}\rangle$, as basis for the coordinate representation, will lead to nonlocal interactions.

Recall the way in which the potential $V_{\alpha\beta}$, as it occurs in the equations (3) and (34), is constructed from the nonrelativistic potential $V^{NR}(r)$. (Restriction to two particles with an interaction, which only depends on the distance between the particles.)

First calculate the Fourier transform of $V^{NR}(r)$ as function of the square of the momentum transfer $|\vec{q}|^2$ and then replace $-|\vec{q}|^2$ by the extended Mandelstam variable $\bar{t}$, eq(29). Or, alternatively, calculate the Feynman diagrams one wishes to take into account, eg. only the one-boson exchange diagram of Fig. 1, and replace the dependence on the Mandelstam variable t by $\bar{t}$. In this way one arrives at the potential $V_{\alpha\beta}$, which in operator form can be written as in (38).

In the coordinate representation, defined by the states $|\vec{r}\rangle$, the potential becomes, using eqs. (14) and (15):

$$\langle \vec{r}_1', \vec{r}_2'|V|\vec{r}_1, \vec{r}_2\rangle = \int \langle \vec{r}_1', \vec{r}_2'|\alpha\rangle\langle\alpha|V|\beta\rangle\langle\beta|\vec{r}_1, \vec{r}_2\rangle \gamma^4(v_1')d\vec{v}_1' \cdots \gamma^4(v_2)d\vec{v}_2 \tag{66}$$

with

$$|\alpha\rangle = |\vec{v}_1', \vec{v}_2'\rangle \quad \text{and} \quad |\beta\rangle = |\vec{v}_1, \vec{v}_2\rangle \tag{67}$$

and $\langle\alpha|V|\beta\rangle$ given by (38). Upon substituting the two-particle version of (47) into (66), and using (13), the potential becomes

$$\begin{aligned}\langle \vec{r}_1', \vec{r}_2'|V|\vec{r}_1, \vec{r}_2\rangle & \\ \sim \int & \frac{d\vec{p}_1' d\vec{p}_2' d\vec{p}_1 d\vec{p}_2}{\sqrt{p_1'^0 p_2'^0 p_1^0 p_2^0}} e^{i(\vec{p}_1'\cdot\vec{r}_1' + \vec{p}_2'\cdot\vec{r}_2' - \vec{p}_1\cdot\vec{r}_1 - \vec{p}_2\cdot\vec{r}_2)} \\ & \times V_{\alpha\beta}(\bar{t}) L^3(\vec{v}_\alpha, \vec{v}_\beta)\,.\end{aligned} \tag{68}$$

This integral can be reduced considerably by taking the static limit, in which $m_2 \to \infty$. In this case it can be shown (Ruijgrok 1998) that

$$\langle \vec{r}_1', \vec{r}_2'|V|\vec{r}_1, \vec{r}_2\rangle \sim \delta(\vec{r}_2' - \vec{r}_2)\frac{K_1(m_1 D)}{m_1 D} V^{NR}(z) \tag{69}$$

in which

$$\vec{D} = \vec{r}_1' - \vec{r}_1 \quad \text{and} \quad \vec{z} = \frac{1}{2}(\vec{r}_1' + \vec{r}_1) - \vec{r}_2\,. \tag{70}$$

These variables are illustrated in Fig. 2.

$V^{NR}(z)$ is the nonrelativistic potential. Since $K_1(m_1 D)$ is a modified Bessel function, which decreases exponentially with increasing D, and with a range

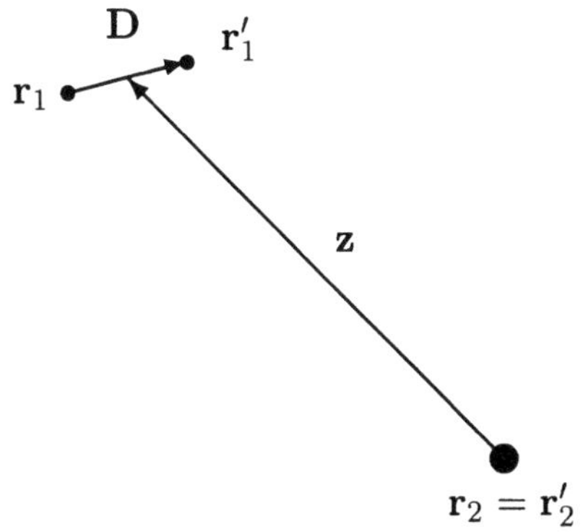

Fig. 2. Showing the nonlocality $D \approx m_1^{-1}$ of the interaction.

m^{-1}, the nonlocality of the potential is equal to the nonlocality of the light particle, which is given by its Compton wavelength.

In order to investigate the observable effects of this nonlocality, a simpel model for the scattering of two spinless particles will be considered, for which (3) can be solved exactly.

For that purpose the potential and the scattering amplitude are first expanded in partial waves:

$$V(s', s, \bar{t}) = \frac{4m_1 m_2}{\tau(s', s)} \sum_l (2l+1) V_l(s', s) P_l \left(\frac{\bar{t} - t_0}{\tau} \right) \tag{71}$$

and

$$M(s', s, \bar{t}) = \frac{4m_1 m_2}{\tau(s', s)} \sum_l (2l+1) M_l(s', s) P_l \left(\frac{\bar{t} - t_0}{\tau} \right) . \tag{72}$$

The argument of the Legendre functions is the cosine of the scattering angle θ^* in the centre of mass system. The quantities t_0 and τ are defined by

$$t_0(s', s) = -\frac{\lambda(\bar{s}, m_1^2, m_2^2)}{2\bar{s}} - \frac{1}{2}(\sqrt{s'} - \sqrt{s})^2 \tag{73}$$

and

$$\tau(s', s) = \frac{1}{2} \sqrt{\frac{\lambda(s', m_1^2, m_2^2)\lambda(s, m_1^2, m_2^2)}{s's}} = 2k'k \, , \tag{74}$$

where k' and k are the c.m.s. momenta.

It is often more convenient to use the K-matrix, which for each partial wave is defined by

$$K_l(s',s) = -\sqrt{\frac{c(s)}{c(s')}}\frac{M_l(s',s)}{c(s) - iM_l(s,s)} \qquad \text{with} \qquad c(s) = \frac{\sqrt{\lambda(s)}}{4\pi^2 m_1 m_2}\,. \tag{75}$$

With

$$W_l(s',s) \equiv -\frac{V_l(s',s)}{\sqrt{c(s')c(s)}} \tag{76}$$

the equation for this K-matrix finally becomes

$$K_l(s'',s) = W_l(s'',s) + \frac{1}{2\pi}\int_{s_+}^{\infty} \frac{W_l(s'',s')K_l(s',s)}{\sqrt{s'}(\sqrt{s'}-\sqrt{s})_P}\, ds' \tag{77}$$

with $s_+ = (m_1 + m_2)^2$. The on-shell solution of this equation gives the phase shifts according to

$$\tan\delta_l(s) = K_l(s,s)\,. \tag{78}$$

The on-shell scattering amplitude becomes

$$M_l(s,s) = \frac{\sqrt{\lambda(s)}}{8\pi^2 m_1 m_2}\,\eta_l(s) \tag{79}$$

with the phase function $\eta_l(s)$ defined by

$$\eta_l(s) = -2\sin\delta_l\, e^{i\delta_l}\,. \tag{80}$$

In terms of these phase functions the total cross section is

$$\sigma_{tot}(s) = \frac{4\pi s}{\lambda(s)}\sum_l (2l+1)|\eta_l(s)|^2\,. \tag{81}$$

In the framework as defined so far, it is now possible to formulate questions about the observability of nonlocal effects. For that purpose consider the scattering of two particles with a hard sphere interaction. In nonrelativistic quantum mechanics this interaction is formulated as a boundary condition: the wave function shall be zero when the distance between the particles has a certain value a. This method cannot be used in a relativistic theory, because the range of the hard sphere interaction cannot be defined with infinite precision, due to the nonlocal character of the states $|\vec{r}\rangle$, which were used for the definition of the position operator $\vec{Q}$.

It is, however, possible to construct a geometrical picture of the scattering process at high energies, by writing the on-shell scattering amplitude as a function $H(k,b)$ of the impact parameter b and the c.m.s. momentum $k = \frac{1}{2}\sqrt{\frac{\lambda(s)}{s}}$. It is defined (see, e.g., (Cottingham and Peierls 1965)) as the Fourier-Bessel transform of the scattering amplitude

$$H(k,b) = \frac{2\pi^2}{\sqrt{\lambda(s)}}\int_0^{2k} M(s,s,-x^2)J_0(bx)x\, dx \tag{82}$$

with

$$x^2 = \frac{\lambda(s)}{2s}(1-\cos\theta^*) = [2k\sin\frac{\theta^*}{2}]^2 = -t\,. \tag{83}$$

This is similar to the formula obtained for $M_l(s,s)$ by inverting (72) and using (79):

$$\eta_l(s) = \frac{\pi^2\sqrt{\lambda(s)}}{2s}\int_{-1}^{1} M(s,s,-x^2)P_l(\cos\theta^*)\,d\cos\theta^*\,. \tag{84}$$

The r.h.s.can be expressed in terms of $H(k,b)$, by first showing that the function

$$T(k,x) \equiv \frac{\sqrt{\lambda(s)}}{2\pi^2}\int_0^\infty H(k,b)J_0(bx)\,b\,db \tag{85}$$

is equal to

$$T(k,x) = \begin{cases} M(s,s,-x^2) & \text{if} \quad 0\le x\le 2k\,, \\ 0 & \text{if} \quad x>2k\,. \end{cases} \tag{86}$$

For the proof, (82) is substituted into (85) and use is made of the completeness of Bessel functions

$$\int_0^\infty J_0(bx)J_0(by)\,b\,db = \frac{1}{x}\delta(x-y)\,. \tag{87}$$

With the definition (85) and with

$$\int_0^1 P_l(1-2y^2)J_0(yz)\,y\,dy = \frac{J_{2l+1}(z)}{z}\,, \tag{88}$$

(84) then takes the form

$$\eta_l(k) = 2k\int_0^\infty H(k,b)J_{2l+1}(2kb)\,db\,. \tag{89}$$

Using the partial wave expansion (72) and the identity (88), (82) can be written as

$$H(k,b) = \frac{1}{kb}\sum_l (2l+1)\eta_l(k)J_{2l+1}(2kb)\,. \tag{90}$$

Notice that $H(k,b=0)=\eta_0(k)$.

The differential and total cross section can be written in terms of $H(k,b)$ and it is easy to show that

$$\frac{d\sigma}{dt} = \frac{4\pi^5}{\lambda(s)}|M(s,s,t)|^2 = \pi\left|\int_0^\infty H(k,b)J_0(bx)\,b\,db\right|^2 \tag{91}$$

and

$$\sigma_{tot}(k) = 2\pi \int_0^\infty |H(k,b)|^2 \, b \, db \,. \tag{92}$$

These expressions show that $H(k,b)$ can be interpreted as the on-shell scattering amplitude for a beam of particles passing through a ring of radius b and width db. The largest ring which still contributes to the scattering of a hard sphere, determines the size of this sphere. In order, however, to get a sharp definition of the edge, the momentum must be large and hence also the angular momentum. This leads to the classical limit

$$\text{Lim}: \qquad k \to \infty \quad l \to \infty \quad \rho = \frac{l}{k} \quad \text{fixed.} \tag{93}$$

In this limit the relation (89) between $\eta_l(k)$ and $H(k,b)$ becomes very simple. Replacing $J_{2l+1}(2kb)$ by the l.h.s. of (88), with $y \to \frac{x}{2k}$ and $z \to 2kb$, and using

$$\lim_{l \to \infty} P_l(1 - \frac{\rho^2 x^2}{2l^2}) = J_0(\rho x) \tag{94}$$

it turns out that

$$\text{Lim } \eta_l(k) = H(k = \infty, \rho) \equiv \overline{H}(\rho) \,. \tag{95}$$

If now for scattering by a hard sphere with radius a it turns out that

$$\overline{H}(\rho) \neq 0 \quad \text{for} \quad \rho < a \quad \text{and} \quad \overline{H}(\rho) = 0 \quad \text{for} \quad \rho > a \,, \tag{96}$$

then it will have been shown that the nonlocality of the potential, as exhibited in Fig. 2, has no observable effect.

In the remaining part of this paper it will be shown that (96) is indeed correct.

3.3 The Model

The potential to be used is the δ-shell potential, which in the nonrelativistic theory takes the form

$$V^{NR}(r) = aV_0\delta(r - a) \,. \tag{97}$$

The exact solution of the scattering- and the bound state problem has been extensively discussed by Gottfried (Gottfried 1966), Antoine et al. (Antoine et al. 1987) and Albeverio et al. (Albeverio et al. 1988).

Applying the recipe, described in the beginning of the previous section, for the construction of the relativistic potential, from its nonrelativistic limit, one obtains

$$V(s', s, \bar{t}) = \frac{2g}{\pi^2} \frac{\sin a\sqrt{-\bar{t}}}{a\sqrt{-\bar{t}}} \tag{98}$$

in which g is the dimensionless coupling constant

$$g = a^3 V_0 m_1 m_2 \,. \tag{99}$$

From the partial wave expansion (71) the functions $V_l(s', s)$ can be calculated. Via (76) one then finds

$$W_l(s', s) = -g \frac{[\lambda(s')\lambda(s)]^{1/4}}{\sqrt{s's}} j_l(k'a) j_l(ka) \,. \tag{100}$$

Like in the nonrelativistic case this is a separable potential, which is the reason why (77) can be solved exactly. The result is found to be

$$K_l(s', s) = \overline{K}_l \frac{[\lambda(s')\lambda(s)]^{1/4}}{\sqrt{s's}} j_l(k'a) j_l(ka) \tag{101}$$

with

$$\overline{K}_l = -\frac{g}{1 + g I_l} \tag{102}$$

and

$$I_l = \frac{1}{2\pi} \int_{s_+}^{\infty} \frac{\sqrt{\lambda(s')}\, j_l^2(k'a)}{s'^{3/2}(\sqrt{s'} - \sqrt{s}\,)_P} \, ds' \,. \tag{103}$$

The relation between s' and k' is

$$\sqrt{s'} = \sqrt{m_1^2 + k'^2} + \sqrt{m_2^2 + k'^2} \,. \tag{104}$$

For $V_0 \to \infty$, i.e., for $g \to \infty$, the shell becomes impenetrable, and is therefore equivalent to a hard sphere. In this case (78) for the phase shift becomes

$$\tan \delta_l(k) = -\frac{\pi}{\sqrt{m_1^2 + k^2} + \sqrt{m_2^2 + k^2}} \frac{J_{l+1/2}^2(ka)}{a I_l} \,. \tag{105}$$

A further simplification consists in taking the second particle to be infinitely heavy. Equation (105) then reads

$$\tan \delta_l(k) = -\frac{J_{l+1/2}^2(ka)}{F_l(ka)} \tag{106}$$

with

$$F_l(ka) = \frac{1}{\pi} \int_0^{\infty} \frac{J_{l+1/2}^2(k'a)\, k' \, dk'}{\sqrt{m_1^2 + k'^2}[\sqrt{m_1^2 + k'^2} - \sqrt{m_1^2 + k^2}]_P} \,. \tag{107}$$

For the calculation of $\overline{H}(\rho)$, defined in (95), the phase function $\eta_l(k)$, (80), must be determined. Using (106) it is found to be equal to

$$\eta_l(k) = 2[i + \frac{F_l(ka)}{J_{l+1/2}^2(ka)}]^{-1} \,. \tag{108}$$

In order to prove that $\overline{H}(\rho) = 0$ for $\rho > a$, Eq. (96), it must therefore be shown that

$$\frac{J^2_{l+1/2}(ka)}{F_l(ka)} \to 0 \quad \text{if} \quad l \to \infty,\ k \to \infty,\ \rho = \frac{l}{k} \quad \text{fixed and} \quad \rho > a\,. \tag{109}$$

In this limit (107) becomes

$$F_l(y) \simeq \frac{1}{\pi} \int_0^\infty \frac{J^2_{l+1/2}(x)}{(x-y)_P}\, dx\,. \tag{110}$$

Using the asymptotic behaviour of $J_{l+1/2}(cl)$ for $c > 1$ and for $c < 1$ (Abramowitz and Stegun (Eds.) 1965), it can be shown that this integral consists of three parts. The first part, stemming from $x < l$, goes to zero with $l \to \infty$. The integrand of the second part, for $x > l$, has a rapidly oscillating term, which can be omitted, and a finite term, which can be integrated. The final result is

$$\frac{J^2_{l+1/2}(l\frac{a}{\rho})}{F_l(l\frac{a}{\rho})} \simeq \frac{e^{-2l(\alpha_0 - \tanh\alpha_0)}}{4\arctan(\coth\frac{\alpha_0}{2})} \qquad \text{for} \quad \rho > a \tag{111}$$

and with $\rho = a\cosh\alpha_0$.

Since this ratio approaches zero for $l \to \infty$, it concludes the proof that $\overline{H}(\rho) = 0$ for $\rho > a$ (96).

4 Conclusions

In relativistic quantum mechanics practically the only acceptable operator for the position of a particle is the one proposed by Newton and Wigner (Newton and Wigner 1949). The eigenvectors of this operator,which are called *localised states,* however, have a wave function which makes the particle look as large as its Compton wavelength. Newton and Wigner pointed out that this has strange consequences, such as the loss of localisation after a boost or a time translation. Others have shown that as a result strict causality is violated (Hegerfeldt 1974, 1998), although measurable effects could not be found (Ruijsenaars 1981).

This is corroborated by the calculations in the present paper, of the probability to find an initially localised particle at a later time at a point which lies only a few Compton wavelengths beyond the causality limit. This probability turns out to decrease exponentially with the square root of the time (Eq. (65)).

In order to treat relativistic interacting particles, a Poincaré invariant theory was used, which is explained in the first two sections. With this theory it was then shown that the interaction potential in the coordinate representation, as defined by the eigenstates of the Newton-Wigner operator, becomes nonlocal. This nonlocality, however, does not spread over distances much larger than the Compton wavelengths of the particles involved (see Fig. 2).

Another problem of relativistic quantum mechanics is the question of how to formulate the interaction, that nonrelativistically is described by boundary conditions in coordinate space. In Section 3.2 it was shown how, in the present formalism, this problem is solved for scattering by a hard sphere. By using the impact-parameter representation, the classical formula (96) for the sharp definition of the edge, was reproduced. This proved that, by taking the limit $g \to \infty$, (98) can be used to describe a hard sphere. According to (106), the zeroes of $F_l(ka)$, (107), then give the resonance energies for relativistic hard sphere scattering.

In a similar fashion a box with hard walls can be defined for relativistic particles, and the energies of the bound states can be found in the usual way as the singularities on the positive imaginary axis of the phase functions $\eta_l(ka)$. At present, however, this will not be investigated any further.

References

Abramowitz, M. and Stegun, I.A., (Eds.), (1965): *Handbook of Mathematical Functions,* Dover Publ., New York, Eq. (9.3.2) and Eq. (9.3.3)

Albeverio, S., Gesztesy, F., Hoegh-Krohn, R. and Holden, H., (1988): *Solvable Models in Quantum Mechanics*, Springer

Antoine, J.-P., Gesztesy, F. and Shabani, J., (1987): J. Phys. **A20**, 3687

Cisło, J., Łopuszański, J. and Stichel, P.C., (1998): Fortschr. Phys. **40**, 45

Cottingham, W.N. and Peierls, R.F., (1965): Phys. Rev., **B137**, 147

Currie, D.G., (1963): J. Math. Phys., **4**, 1470

Currie, D.G., Jordan, T.F. and Sudarshan, E.C.G., (1963): Rev. Mod. Phys., **35**, 350

Currie, D.G., (1966): Phys. Rev., **142**, 817

Van Dam, H. and Wigner, E.P., (1965): Phys. Rev., **B138**, 1576

Van Dam, H. and Wigner, E.P., (1966): Phys. Rev., **B142**, 838

Dirac, P.A.M., (1949): Rev. Mod. Phys., **21**, 392

Gottfried, K., (1966): *Quantum Mechanics*, W.A. Benjamin, Inc, New York, Ch. III. 15

de Groot, E.H. and Ruijgrok, Th.W., (1975): *A unitary model for independent particle production*, Nucl. Phys., **B101**, 95

de Groot, E.H. and Ruijgrok, Th.W., (1986): *Relativistic Coulomb Scattering*, Acta Phys. Polon. **B17**, 365

Hegerfeldt, G.C., (1974): Phys. Rev., **D10**, 3320 and in Bohm, A., Doebner, H.-D., and Kielanowski, P., (Eds.), (1998): *Irreversibility and Causality in Quantum Theory-Semigroups and Rigged Hilbert Spaces*, Springer Lecture Notes in Physics, Vol. 504

Hersbach, H. and Ruijgrok, Th.W., (1991): *A quasipotential theory for the relativistic hydrogen atom*, Z. Phys., **D18**, 209

Hersbach, H., (1992): *A relativistic quantum theory in momentum space for weak and strong coupling*, Phys. Rev., **A46**, 3657

Hersbach, H., (1993): *Relativistic linear potential in momentum space*, Phys. Rev., **D47**, 3027
Hersbach, H., (1994): *Relativistic meson spectroscopy in momentum space*, Phys. Rev., **C50**, 2562
Hill, R.N., (1982): See Llosa (Ed.) 1982 p. 104, where also more references can be found
Kerner, E.H., (ed.), (1972): *The Theory of Action-at-a-Distance in Relativistic Particle Dynamics*, Gordon and Breach, Science Publishers, New York
Leutwyler, H., (1965): Nuovo Cimento, **37**, 556
Lippmann, B.A. and Schwinger, J., (1950): Phys. Rev., **79**, 469
Llosa, J., (Ed.), (1982): *Relativistic Action at a Distance: Classical and Quantum Aspects,* Proc. Barcelona Workshop, 1981, Springer, Berlin
van Meijgaard, E. and Ruijgrok, Th.W., (1985): *The Binding of Electrons to Heavy Ions*, Physica, **131A**, 570
Møller, C., (1952): *The Theory of Relativity,* Clarendon Press, Oxford, Chapter III, §31
Newton, T.D. and Wigner, E.P., (1949): Rev. Mod. Phys., **21**, 400
Prudnikov, A.R., Brychkov, Yu.A and Marichev, O.I., (1986): *Integrals and Series*, Vol. 1, Gordon and Breach, New York, Eq. 2.5.25.9 and Eq. 2.5.25.15
Ruijgrok, Th.W., (1998): Few-Body Systems, **25**, 5
Ruijgrok, Th.W., (1976): *On the Relativistic Hydrogen Atom*, Acta Phys. Polon., **B7**, 43
Ruijgrok, Th.W., (1983): B. Jancewicz and J. Lukierski, (Eds.), *Quantum Theories of Particles and Fields*, World Scientific, Singapore, pp. 117-138
Ruijgrok, Th.W. and Kirchner, E.J.J., (1990): A. Giovannini and W. Kittel, (Eds.), *Multiparticle Dynamics, Festschrift for Léon Van Hove and Proceedings, La Thuile, 1989*, World Scientific, Singapore, pp. 811–820
Ruijgrok, Th.W. and Kirchner, E.J.J., (1991): *The S-matrix and its pole structure for the relativistic Lee model* Acta Phys. Polon., **B22**, 47
Ruijgrok, Th.W., (1996): *Relativistic contributions to the electromagnetic binding and scattering of pions by atoms*, Acta Phys. Polon., **B27**, 1869
Ruijsenaars, S.N.M., (1981): Annals of Physics, **137**, 33
Sudarshan, E.C.G. and Mukunda, N., (1974): *Classical Dynamics: A Modern Perspective*, John Wiley and Sons, New York, Chapter 19

Dynamical Equivalence, Commutation Relations and Noncommutative Geometry*

P.C. Stichel

An der Krebskuhle 21, D-33619 Bielefeld,
Germany
e-mail: hanne@physik.uni-bielefeld.de

Dedicated to the 75th birthday of Prof. Jan Łopuszański

Abstract. We revisit Wigner's question about the admissible commutation relations for coordinate and velocity operators given their equations of motion (EOM). In more general terms we want to consider the question of how to quantize dynamically equivalent Hamiltonian structures. A unique answer can presumably be given in those cases, where we have a dynamical symmetry. In this case arbitrary deformations of the symmetry algebra should be dynamically equivalent. We illustrate this for the linear as well as the singular $1d$-oscillator. In the case of nonlinear EOM quantum corrections have to be taken into account. We present some examples thereof. New phenomena arise in case of more then one degree of freedom, where sometimes the interaction can be described either by the Hamiltonian or by nonstandard commutation relations. This may induce a noncommutative geometry (for example the $2d$-oscillator in a constant magnetic field). Also some related results from nonrelativistic quantum field theory applied to solid state physics are briefly discussed within this framework.

1 Introduction

It is well known, that the Lagrangean leading to a given description of a classical mechanical system is not unique. To be more specific, we have to ask for the set of all Lagrange functions, whose Euler-Lagrange equations have the same solutions in configuration space. Those Lagrangeans are called s-equivalent. The task of finding them has been solved completely for systems with one degree of freedom in terms of one arbitrary positive function (Currie and Saletan 1966; Cisło et al. 1998), while an extensive discussion for two degrees of freedom has been given by Douglas (Douglas 1941). As classical dynamics is described completely by trajectories in configurations space, s-equivalent Lagrangeans are dynamically equivalent. Dynamical equivalence my also be expressed in terms of a set of equivalent Hamiltonian structures $\{(\omega, H)\}$ where ω denotes a symplectic structure (fundamental Poisson brackets) and H a Hamilton function. In turning to quantum mechanics, a fundamental question arises:

* Supported by the Humboldt Foundation

How to quantize dynamically equivalent Hamiltonian structures?
One may ask also the subquestion: What are, given the equations of motion (EOM) for coordinate and velocity operators, the admissible commutation relations between them? This question was first asked by Wigner (Wigner 1950) for the $1d$-harmonic oscillator in the framework of the standard Hamiltonian. He got a whole set of solutions, which are characterized by one real parameter (Wigner 1950). His solutions are equivalent to the parabose algebra, as has been shown by Palev (Palev 1982).

Proceeding from s-equivalent Lagrangeans, Wigner's question was first considered by Okubo (Okubo 1980) for some unconventional examples. A general treatment of this question for quantum mechanical systems living on a finite dimensional Hilbert space as well as for the $1d$-oscillator (infinite dimensional Hilbert space) has been given very recently by Man'ko, Marmo, Sudarshan and Zaccaria (Man'ko et al. 1996). For reasons of completeness we will take up the discussion of the $1d$-oscillator again in the present paper. In particular we will discuss parabosons and different cases of q-deformations within a unique framework of nonlinear deformations of the oscillator algebra. In addition we examine representations for some simple nonlinear deformations and discuss the essential difference between the classical and quantum mechanical formalism for general deformations.

The harmonic oscillator is an exceptional case insofar, as the EOM are linear. They are identical in both classical mechanics and quantum mechanics. Therefore, in passing from classical Poisson brackets to commutators by means of Dirac's recipe we have no difficulties. But the situation becomes worse if nonlinear observables besides the Hamiltonian are involved. Then we are confronted with the inconsistency of Dirac's rule (Lledo and Gracia Sucre 1996; Gotay 1998). In those cases quantum corrections appear either in the EOM or in some observables. This will be demonstrated for the singular oscillator, for some power potentials and for spherically symmetric potentials.

In this context we will neither discuss modern treatments of the quantization or dequantization problem (cp. (Lledo and Gracia Sucre 1996; Gotay 1998), (Carinena et al. 1998), (Werner 1995)) nor their difficulties (Rieffel 1997). It is the aim of our paper, to consider our fundamental question by means of some simple but important physical examples, but not to discuss it in terms of a general mathematical framework.

Another very interesting question is the prescription of the interaction not in terms of a Hamiltonian but in terms of nonstandard commutation relations. We will demonstrate this for the example of a charged particle moving in a constant magnetic field in a plane. The resulting nonstandard commutation relations describe a noncommutative geometry. It is an exciting topic to extend this question to quantum field theory. We give a brief account of some related results in solid state physics.

The paper is organized as follows: In Sec. 2 we treat systems with one degree of freedom, divided into Hamiltonian mechanics with the most general symplectic form, quantum mechanics of the nonlinear deformed linear and singular oscillator and the case of a general potential. Sec. 3 is devoted to the movement of a charged particle in a constant magnetic field in a plane. In Sec. 4 spherically symmetric potentials are revisited. Sec. 5 contains some remarks on examples from nonrelativistic quantum field theory. In Sec. 6 we close with some final remarks including open questions.

2 Systems with One Degree of Freedom

In this section we study dynamical equivalence for the motion of either one particle in an external field or for the relative motion of two particles in $1d$-space within the framework of classical or quantum mechanics.

2.1 1d – Classical Mechanics

For reasons of simplicity we consider Newton's EOM for a conservative force only

$$\ddot{x} = -V'(x) \; . \tag{1}$$

The corresponding description in the standard formulation of the canonical formalism is given by the Hamilton function

$$H(u, x) = \frac{u^2}{2} + V(x) \, , \tag{2}$$

leading by means of the symplectic structure ω_0 for the independent variables $(y_1, y_2) = (x, u)$ with

$$(\omega_0)_{ij} = \epsilon_{ij} \tag{3}$$

to the canonical EOM

$$\dot{x} = \{x, H\}_{\omega_0} = u \tag{4}$$

$$\dot{u} = \{u, H\}_{\omega_0} = -V'(x) \tag{5}$$

and therefore to Newton's EOM (1) in x-space where we define the Poisson bracket $\{\cdot, \cdot\}_\omega$ for an arbitrary symplectic structure ω as usual[1]

$$\{A, B\}_\omega := \frac{\partial A}{\partial y_i} \omega_{ij} \frac{\partial B}{\partial y_j} \; . \tag{6}$$

Now we ask for the most general Hamiltonian structure $(\omega, \tilde{H})$ with

$$\frac{\partial \omega}{\partial t} = 0 \, , \tag{7}$$

which preserves the EOM (4), (5), i.e.,

[1] We use the summation convention for repeated indices.

$$\dot{x} = \{x, \tilde{H}\}_\omega = u\,, \tag{8}$$
$$\dot{u} = \{u, \tilde{H}\}_\omega = -V'(x)\,. \tag{9}$$

Such a structure $(\omega, \tilde{H})$ we call dynamically equivalent to (ω_0, H). According to Leubner and Marte (Leubner and Marte 1984) ω is a conserved quantity, which due to (7) has to be a function of H only. Therefore ω may be expressed in terms of an arbitrary nonvanishing function $\sigma(z)$ (which we choose to be positive), such that

$$\omega = \begin{pmatrix} 0 & 1/\sigma(H) \\ -1/\sigma(H) & 0 \end{pmatrix}\,. \tag{10}$$

It is easily seen, that for a given σ the Hamiltonian structure $(\omega, \tilde{H})$ with

$$\tilde{H} := \int^H dz\sigma(z) \tag{11}$$

satisfies the EOM (8), (9).

Remark: The foregoing results my be derived also within the Lagrangean framework (Currie and Saletan 1966; Cisło et al. 1998).

2.2 Quantum Mechanics of the Deformed 1d-Oscillator

The EOM in configuration space for the linear oscillator has the form

$$\ddot{x} + x = 0\,, \tag{12}$$

which may be written as a system of two first order equations

$$\dot{x} = u, \qquad \dot{u} = -x\,. \tag{13}$$

If we introduce raising and lowering operators as usual

$$a^\dagger := \frac{1}{\sqrt{2}}(x - iu)\,, \qquad a := \frac{1}{\sqrt{2}}(x + iu) \tag{14}$$

the EOM (13) takes the form

$$\dot{a} = -ia\,, \qquad \dot{a}^\dagger = ia^\dagger\,. \tag{15}$$

Due to (15) the operator N

$$N := a^\dagger a\,, \tag{16}$$

as well as the commutator $[a, a^\dagger]$ are conserved quantities.
We conclude, as in classical mechanics (cp. section 2.1), that the commutator may be expressed by a positive function σ which has to be a function of N only

$$[a, a^\dagger] = \frac{1}{\sigma(N)}\,. \tag{17}$$

For the standard description of the linear oscillator we take $\sigma = 1$. The generalization (17) defines an arbitrary deformation of the usual oscillator algebra. Now we ask for a new number operator K, related to the new Hamiltonian $\tilde{H}$ through $\tilde{H} = K + 1/2$, which preserves the EOM (15)

$$[K, a] = -a \,, \tag{18}$$

$$[K, a^\dagger] = a^\dagger \,. \tag{19}$$

Taken as a function of N our new number operator K satisfies the functional equation

$$K\left(N + \frac{1}{\sigma(N)}\right) = 1 + K(N) \,. \tag{20}$$

The solution of (20) is supposed to be unique up to a constant.
We prove (20) by starting with

$$Na^\dagger - a^\dagger\left(N + \frac{1}{\sigma(N)}\right) = 0 \,, \tag{21}$$

which follows from (16), (17). Therefore, for each holomorphic function f we obtain the relation

$$f(N)a^\dagger - a^\dagger f\left(N + \frac{1}{\sigma(N)}\right) = 0 \,. \tag{22}$$

By identifying $f(N)$ with $K(N)$ and using (19) we obtain immediately (20).
There exists an useful alternative formulation of (20): Suppose the commutator $[a, a^\dagger]$ is given instead of by (17) in terms of a positive function $\varphi(K)$

$$[a, a^\dagger] = \varphi(K) \tag{23}$$

and we have an implicit definition of K by means of a positive function F

$$N = F(K) \,. \tag{24}$$

Then, by applying F to (20) we obtain the relation

$$\varphi(K) = F(K+1) - F(K) \tag{25}$$

between the functions F and φ (cp. (Man'ko et al. 1997), (Katriel and Quesne 1996)).
There is a corollary to relation (25). Let us define the q-commutator (quommutator) $[a, a^\dagger]_q$ by[2]

$$[a, a^\dagger]_q := aa^\dagger - qa^\dagger a \,, \qquad q \in \mathbf{R}^1 \,. \tag{26}$$

Then the commutator (23) with φ given by (25) and the quommutator

$$[a, a^\dagger]_q = \psi(K) \tag{27}$$

[2] We don't consider complex valued q's in this paper.

are equivalent if

$$\psi(K) = F(K+1) - qF(K) \ . \tag{28}$$

We observe, that our quantum mechanical relation (25) looks different from the corresponding relation in classical mechanics, which, due to (11) has the form

$$\varphi(z) = F'(z) \tag{29}$$

with $z \in \mathbf{R}^1_+$. But as opposed to (29) our K in (25) is not a continuous variable but an operator taking discrete values $n \in \mathbf{N}$ on a $1d$-lattice space. In order to compare (25) with (29) we have to introduce a differential calculus on this lattice space (cp. (Dimakis and Müller-Hoissen 1997)). The simplest way to do this is to introduce an operator valued differential $dg(K)$ for an arbitrary holomorphic function $g(z)$. We define

$$dg(K) := [a, g(K)] \ . \tag{30}$$

Remarks on the definition (30):

1) It respects the Leibniz rule $d(gh) = (dg)h + gdh$.

2) By means of Dirac's rule the dequantization looks as follows

$$dg(K) \to i\hbar\{a, g(K)\}_\omega = g'(K)\hbar a \ .$$

If we specify $g(z) = z$ we get due to (18)

$$dK = a \ . \tag{31}$$

With that and using (18) again we finally obtain for (30)

$$dg(K) = (g(K+1) - g(K))dK \ . \tag{32}$$

This is a differential calculus in noncommutative geometry (cp. (Dimakis and Müller-Hoissen 1997)), because dK and K don't commute. On the contrary we find

$$[dK, K] = dK \ .$$

Let us define the left partial derivative of $g(K)$ by (Dimakis and Müller-Hoissen 1997)

$$dg(K) = (\partial_+ g)dK \ , \tag{33}$$

then we recognize $\varphi(K)$ in (25) as just this derivative of $F(K)$.

Remark: Instead of (30) we could have defined another operator valued differential

$$\tilde{d}g(K) := [g(K), a^\dagger]$$

leading finally to the identification of $\varphi(K)$ as the right partial derivative of $F(K)$ (with respect to the differential $\tilde{d}$).

We conclude, that agreement between the quantum mechanical and classical expressions (25) and (29) respectively may be obtained, if we use appropriate differential calculi in both cases.

The foregoing description of deformed oscillators yields a general and unique framework covering all the results known already. We want to illustrate this with some typical examples.

(1) The algebra of the Arik-Coon q-oscillator (Arik and Coon 1976)

$$[a, a^\dagger]_q = 1 \tag{34}$$

is equivalent to (cp. (Katriel and Quesne 1996))

$$[a, a^\dagger] = q^K \ . \tag{35}$$

This follows immediately from $\psi = 1$ and (28) leading to

$$F(K) = \frac{q^K - 1}{q - 1} \ . \tag{36}$$

If we take the inverse of (36) we get

$$K(N) = \frac{1}{lnq} \ln(1 + (q-1)N) \ , \tag{37}$$

which is obviously the solution of (20) with

$$\frac{1}{\sigma(N)} = 1 + (q-1)N$$

obtained from (34).

(2) The algebra of the Biedenharn-Macfarlane q-oscillator (Biedenharn 1989; Macfarlane 1989)

$$[a, a^\dagger]_q = q^{-K} \tag{38}$$

is equivalent to

$$[a, a^\dagger] = F(K+1) - F(K) \tag{39}$$

with

$$F(K) = \frac{q^K - q^{-K}}{q - q^{-1}} \ . \tag{40}$$

By inverting (40) we obtain with $q = e^\lambda$

$$K(N) = \frac{1}{\lambda} \ln(N \sinh\lambda + (N^2 \sin^2 h\lambda + 1)^{1/2}) \ . \tag{41}$$

The corresponding $\sigma(N)$ is given by

$$1/\sigma(N) = N(\cosh\lambda - 1) + (N^2 \sin^2 h\lambda + 1)^{1/2} .$$

Examples (1) and (2) show the equivalence between commutators and quommutators.

Systematic constructions of such an equivalence is given in (Katriel and Quesne 1996) by means of a recursive procedure in Fock-space.

(3) The parabose oscillator of order p is defined by[3] (cp. (Chaturvedi and Srinivasan 1991))

$$K = \frac{1}{2}\{a^\dagger, a\} - p/2 , \qquad p \in \mathbf{N} , \tag{42}$$

where the vacuum has to satisfy the relations

$$a|0\rangle = 0 \quad \text{and} \quad aa^\dagger|0\rangle = p|0\rangle . \tag{43}$$

In order to find the function φ we rewrite (42) as

$$K = F(K) + \frac{1}{2}\varphi(K) - p/2 . \tag{44}$$

If we eliminate $F(K)$ in (25) by means of (44) we obtain an equation for φ

$$\frac{1}{2}(\varphi(K) + \varphi(K+1)) = 1 ,$$

which has the solution

$$\varphi(K) = 1 + \alpha(-1)^K , \qquad \alpha \in \mathbf{R}^1 . \tag{45}$$

Finally by using (43) we obtain

$$\alpha = p - 1 . \tag{46}$$

This result has already been obtained in (Chaturvedi and Srinivasan 1991) by means of a more complicated procedure.

The representation of the deformed oscillator algebra (18), (19), (23), in particular the spectrum of the Hamiltonian, depends on the deformation given by the function φ (cp. (Quesne and Vansteenkiste 1996; Guichardet 1998)). In order to illustrate this statement, let us consider a very simple example of a family of deformations

[3] the symbol $\{\cdot,\cdot\}$ denotes the anticommutator

$$\varphi_q(K) = q + 2(q-1)K\,, \qquad q \in \mathbf{R}^1\,, \tag{47}$$

which interpolates between oscillator algebra ($q = 1$), $SU(2)$ ($q = 0$) and $SU(1,1)$ ($q = 2$). But if for $q \notin 1$ we shift the scale of K

$$K = K' - \frac{q}{2(q-1)} \tag{48}$$

and rescale a simultaneously

$$a = |q-1|^{1/2} a' \tag{49}$$

we arrive for $q > 1$ ($q < 1$) at the $SU(1,1)$ ($SU(2)$) algebra respectively

$$[K', a'] = -a'\,, \qquad [K', a'^\dagger] = a'^\dagger\,, \qquad [a', a'^\dagger] = \pm 2K'\,, \tag{50}$$

where the upper (lower) sign has to be taken for $q > 1 (q < 1)$.
We conclude that the limit $q \to 1$ in (47) is globally discontinuous.
But this is not the whole story. The deformed oscillator algebra has the Casimir operator (Roček 1991)

$$C = N - F(K)\,. \tag{51}$$

But in (24) we have identified $F(K)$ with N on our state space. Therefore, we can realize only those irreducible representations of our deformed algebra, which exhibit the value zero for C corresponding to the value zero for the Casimir operator C'

$$C' = N' \mp K'(K'-1) \tag{52}$$

of the algebra (50). The latter may be seen as follows: Define for $q \neq 1$

$$F_q(K) = K + (q-1)K^2 + \frac{1}{4(q-1)} - \frac{q-1}{4} \tag{53}$$

leading due to (25) to $\varphi_q(K)$ in (47). With the shift $K \to K'$ we obtain

$$F_q(K) = (q-1)K'(K'-1) \tag{54}$$

and therefore

$$C = |1-q|C'\,. \tag{55}$$

The value zero for C' corresponds for $SU(2)$ to a singlet state – the Hilbert space $\mathcal{H}$ consists of one state only!

Remark: For more general deformations, $\mathcal{H}$ might consist of a finite number of states (Roček 1991).

For the $SU(1,1)$ case, a vanishing Casimir C' is degenerate: $\mathcal{H}$ is the direct sum of a singlet state $|0\rangle$ and an infinite dimensional representation with $|1\rangle$ as the cyclic vector.
Here we denoted by $|h\rangle$ the eigenstates of K'

$$K'|h\rangle = h|h\rangle\ .$$

In both cases $F_q(K)$ is a nonnegative operator on $\mathcal{H}$ as required.

Only for reasons of completeness we note, that the eigenstates of K in configuration or momentum space differ from the undeformed case (cp. (Sicong Jing 1998), (Finkelstein and Marcus 1995)). In particular, deformations of the oscillator algebra change Heisenberg's uncertainty relation. In certain cases a minimal length uncertainty shows up (cp. (Kempf 1997; 1994)). An example for that is given by the Arik-Coon oscillator.

2.3 The Singular Oscillator

It is our aim to demonstrate in this subsection, that s-equivalence may be destroyed in the quantum case if we require the validity of a deformed symmetry algebra. As an example we consider the $1d$-singular oscillator, defined by the potential (cp. (Samsonov 1998))

$$V(x) = \frac{x^2}{2} + \frac{b}{x^2}\,, \qquad b > 0 \tag{56}$$

on the half-line $x \geq 0$.

Classical mechanics

The standard canonical description shows a dynamical $SU(1,1)$-symmetry. We describe this symmetry in terms of the standard Hamiltonian

$$H = \frac{u^2}{2} + V(x) \tag{57}$$

and

$$K_\pm := \frac{1}{2}\left(\frac{1}{2}(x \mp iu)^2 - \frac{b}{x^2}\right) \tag{58}$$

as follows

$$\{K_\pm, K_0\}_{\omega_0} = \pm\, iK_\pm\,, \tag{59}$$

$$\{K_-, K_+\}_{\omega_0} = -\,2iK_0\,, \tag{60}$$

with $K_0 := H/2$.
The EOM may be written either in the form (from (59))

$$\frac{d}{dt}K_\pm = \pm\, 2iK_\pm \tag{61}$$

or in the standard form

$$\dot{x} = u \,,$$
$$\dot{u} = -(x - \frac{2b}{x^3}) \,. \tag{62}$$

It can easily be seen, that (61) and (62) are s-equivalent, because they are connected by a nonsingular linear transformation.
According to the results presented in Sec. 2.1, the singular oscillator may be described in terms of a general Hamiltonian structure $(\omega, \tilde{H})$, which preserves the EOM in both forms. Accordingly we obtain a deformed $SU(1,1)$

$$\{K_\pm, K\}_\omega = \pm\, iK_\pm \,, \tag{63}$$
$$\{K_-, K_+\}_\omega = -\,\frac{iH}{\sigma(H)} \tag{64}$$

instead of (59), (60). We defined $K := \tilde{H}/2$. Because K is a function of H due to (11), the r.h.s. of (64) may be expressed also in terms of K

$$\{K_-, K_+\}_\omega = -i\psi(K) \,. \tag{65}$$

Quantum mechanics

In passing over from classical to quantum mechanics we want to keep the deformed $SU(1,1)$ algebra. Therefore, we don't apply Dirac's quantization recipe to the EOM of (62) but rather to (63). Our deformed $SU(1,1)$ then takes the form

$$[K, K_\pm] = \pm\, K_\pm \,, \tag{66}$$
$$[K_-, K_+] = \psi(K) \,. \tag{67}$$

But (66) is inconsistent with the quantization of the EOM (62) written in terms of Poisson brackets. By means of Dirac's rule we would obtain

$$i[\tilde{H}, x] = u \,, \qquad i[\tilde{H}, u] = -x + \frac{2b}{x^3} \,. \tag{68}$$

But let us now calculate the commutator $[K, K_\pm]$ by using (68) with (58). In this way we obtain

$$[K, K_\pm] = \pm\, K_\pm + \frac{ib}{2}\left[\frac{1}{x^2}, \frac{1}{x}[u,x]\frac{1}{x}\right] \tag{69}$$

instead of (66). If the commutator $[u, x]$ would have been a function of x only, both expressions would have coincided. But such an exclusive x-dependence of $[u, x]$ is for a generic deformation in disagreement with (67).
Because of $[u, x] = 0(\hbar)$ the additional term in (69) clearly is a quantum correction. It results from the well known fact, that the simultaneous application of Dirac's rule to different Poisson brackets is in general inconsistent (cp. (Lledo and Gracia Sucre 1996; Gotay 1998)).

Problem: What are the quantum corrections to (68) if we start with (66)?

2.4 Arbitrary Potentials $V(x)$

Not much can be said for an arbitrary potential $V(x)$ which differs from the harmonic one.
Suppose we have the quantum analogon to the classical Hamiltonian structure $(\omega, \tilde{H})$ such , that corresponding to (8), (9) our EOM look as follows

$$\frac{i}{\hbar}\left[\tilde{H}, x\right] = u\,, \tag{70}$$

$$\frac{i}{\hbar}\left[\tilde{H}, u\right] = -V'(x)\ . \tag{71}$$

By means of Jacobi's identity we conclude, that the commutator $[u, x]$ is conserved

$$[\tilde{H}, [u, x]] = 0\ .$$

But we are not able to express $[u, x]$ in terms of H as in the case of the classical Poisson bracket. With the exception of the oscillator or a constant force H is not a conserved quantity if $[u, x]$ differs from a c-number. Therefore we must describe $[u, x]$ in terms of $\tilde{H}$ from the very beginning

$$[u, x] = \frac{\hbar}{i}\varphi(\tilde{H})\ . \tag{72}$$

Remark: In simple cases for φ and $V(x)$ we may construct a conserved extension $\hat{H}$ of H. For example consider the deformation

$$\varphi(\tilde{H}) = 1 + \alpha\tilde{H}, \qquad \alpha \in \mathbf{R}^1\,, \tag{73}$$

where α has the dimension of inverse energy, together with a power potential

$$V(x) = \lambda x^n\ , \qquad n = 3, 4\ .$$

Then we obtain by straightforward calculation

$$\hat{H}_3 = H_3 - \frac{\alpha\lambda}{2}\hbar^2 x \tag{74}$$

and

$$\hat{H}_4 = H_4 - \alpha\lambda\hbar^2 x^2\,, \tag{75}$$

where we defined $H_n := \frac{u^2}{2} + \lambda x^n$.
In the limit $\alpha \to 0$ or $\hbar \to 0$ respectively, we obtain the standard result. The additional terms in (74), (75) are again quantum corrections.

3 Systems with Two Degrees of Freedom

We are not going to generalize the consideration of Sec. 2 to two dimensions, but concentrate on those phenomena, which are typical for the $2d$-case. In particular we will consider a charged particle (charge e) moving in a constant magnetic field perpendicular to the plane of motion with (or without) an additional harmonic potential.
In the presence of a harmonic potential the EOM in configuration space looks as follows

$$\ddot{x}_i = -\omega^2 x_i + \kappa \epsilon_{ij} \dot{x}_j \tag{76}$$

with $\kappa := \frac{e}{c} B$.
The EOM (76) are linear equations. Therefore no problems arise with the quantization procedure and we may consider the quantum mechanical formulae from the beginning. First we consider the standard description, which is given by the Hamiltonian

$$H_1 = \frac{p_i^2}{2} + \frac{\tilde{\omega}^2}{2} x_i^2 - \frac{\kappa}{2} J \tag{77}$$

with angular momentum

$$J := \epsilon_{ij} x_i p_j \tag{78}$$

and shifted frequency

$$\tilde{\omega}^2 := \omega^2 + (\kappa/2)^2 \; . \tag{79}$$

By means of the canonical commutation relations

$$[p_i, x_j] = \frac{1}{i} \delta_{ij} \; , \tag{80}$$

$$[x_i, x_j] = 0 \; , \tag{81}$$

$$[p_i, p_j] = 0 \; , \tag{82}$$

the EOM in phase space lead as usual to (76).
As a first alternative we study the case of noncommuting space variables

$$[x_i, x_j] = i \frac{\kappa}{\omega^2} \epsilon_{ij} \tag{83}$$

leaving the other commutators (80) and (82) unchanged.
Such a noncommutative space has been studied recently by Lukierski, Zakrzewski and the present author (Lukierski et al. 1997) in connection with Galilean symmetry in $(2+1)$-dimensions including a second central charge of the extended algebra. By means of the commutation relations (80), (82) and (83) the Hamiltonian

$$H_2 = \frac{p_i^2}{2} + \frac{\omega^2}{2} x_i^2 \tag{84}$$

leads to the EOM in phase space

$$\dot{x}_i = p_i + \kappa\epsilon_{ij}x_j \tag{85}$$
$$\dot{p}_i = -\omega^2 x_i \tag{86}$$

and, by combining these, we arrive at the EOM (76) again.
It is an essential point of the latter approach, that the interaction with the external B-field has been shifted from the Hamiltonian to the commutator (83) inducing a noncommutative geometric structure.

As a second alternative we write the Hamiltonian again in the form (84)

$$H_3 = \frac{u_i^2}{2} + \frac{\omega^2}{2}x_i^2\,, \tag{87}$$

where the u_i are the velocities now related to the canonical momenta p_i as usual

$$u_i = p_i - \kappa A_i \tag{88}$$

and the vector potential A_i describes a B-field of unit strength

$$A_i = -\frac{1}{2}\epsilon_{ij}x_j\ . \tag{89}$$

With (87) and (88) H_3 is identical with H_1, but we consider the commutation relations of the velocities as the primary objects now. We obtain for them

$$[u_i, u_j] = i\epsilon_{ij}\kappa\,, \tag{90}$$

i.e., we have a noncommutative structure of velocity space now.
This approach may be generalized. As (90) is independent of the potential term in (87), we may consider the potential free case. With new variables

$$b := \frac{1}{\sqrt{2}}(u_1 + iu_2)\,, \qquad b^\dagger := \frac{1}{\sqrt{2}}(u_1 - iu_2)\,, \tag{91}$$

the commutator (90) together with the Hamiltonian leads, as is well known, to the oscillator algebra (we put $\kappa = 1$)

$$[b, b^\dagger] = 1\,, \tag{92}$$
$$[H, b] = -b\,, \qquad [H, b^\dagger] = b^\dagger\,. \tag{93}$$

Now we may again consider an arbitrary, nonlinear deformation

$$[b, b^\dagger] = \varphi(K)\,, \tag{94}$$
$$[K, b] = -b\,, \qquad [K, b^\dagger] = b^\dagger\,. \tag{95}$$

For a discussion of this algebra we refer to Sec. 2.2 of this paper. The particular case of a $SU_q(2)$ deformation has been considered by Hojman recently (Hojman 1991).

4 The s-Equivalence for Spherically Symmetric Potentials Revisited

For a particle moving in an arbitrary spherically symmetric potential $V(r)$ a whole set of dynamically equivalent Hamiltonian structures exists (Henneaux and Shepley 1982), (Cisło et al 1995). This set is characterized by the symplectic structure

$$\{x_i, u_j\}_\omega = \delta_{ij} - \frac{GL_iL_j}{1+G\mathbf{L}^2}\,, \tag{96}$$

where L_i denotes the i-th component of the angular momentum and $G(\mathbf{L})$ is a homogeneous function of degree (-3).

For the particular case

$$G(\mathbf{L}) = \frac{\gamma}{L^3}\,, \quad \gamma \in \mathbf{R}^1 \tag{97}$$

with

$$L := \sqrt{\mathbf{L}^2}$$

the new Hamiltonian $\tilde{H}$ may be expressed explicitly in terms of canonical variables $(\mathbf{x}, \mathbf{p})$ as follows (Henneaux and Shepley 1982)

$$\tilde{H} = \frac{p^2}{2} + \frac{1}{2r^2}\left(\gamma^2 - 2\gamma J\right) + V(r)\,, \tag{98}$$

where $\mathbf{J}$ is the canonical angular momentum and $J := \sqrt{\mathbf{J}^2}$.

If quantized, the spectrum of the Hamilton operator (98) differs from the standard one due to the additional second term. But in quantum mechanics, the EOM derived from (98) is not s-equivalent to the standard form

$$\dot{u}_i = -\frac{x_i}{r}V'(r)\,. \tag{99}$$

Let us show this by an explicit calculation:

We obtain by means of (98)

$$u_i := \dot{x}_i = p_i - \frac{i\gamma}{\hbar r^2}[J, x_i] \tag{100}$$

and therefore

$$\dot{u}_i = -\frac{x_i}{r}V'(r) + \gamma A_{1,i} + \gamma^2 A_{2,i} \tag{101}$$

with the quantum corrections A_i defined by

$$A_{1,i} := -\frac{i}{\hbar}\left[\frac{J}{r^2}, p_i\right] - \frac{1}{2\hbar^2}\left[J, \left[\frac{x_i}{r^2}, p^2\right]\right]\,, \tag{102}$$

$$A_{2,i} := \frac{1}{r^4}\left(x_i - \frac{1}{\hbar^2}\left[J,\left[J,x_i\right]\right]\right) . \tag{103}$$

If we dequantize these $A_{j,i}$ by means of Dirac's rule we obtain zero as required. It is easy to show that $A_{2,i}$ e.g., is nonvanishing. By taking matrix elements of $A_{2,i}$ between angular momentum eigenstates, we obtain

$$\langle \ell+1|A_{2,i}|\ell\rangle = \langle \ell+1|\frac{x_i}{r^4}|\ell\rangle B(\ell) \tag{104}$$

with

$$B(\ell) := 1 - \left(\sqrt{(\ell+1)(\ell+2)} - \sqrt{\ell(\ell+1)}\right)^2 . \tag{105}$$

For finite ℓ we have a nonvanishing $B(\ell)$ but it vanishes for large ℓ (classical limit) as $0(\ell^{-1})$ in agreement with dequantization.

5 Interactions as Modified Commutators in Quantum Field Theory

In Sec. 3 we observed that for a simple example the interaction can be expressed either in terms of a Hamiltonian or in terms of modified commutators. The application of this idea to quantum field theory is a highly exciting matter. At present we are far away from a systematic treatment of such an idea. In this section it is our aim to give a brief account of some existing examples pointing in this direction. As a matter of convenience we will limit ourselves to nonrelativistic Fermi systems as they appear in solid state physics.

A well known example of an integrable model is the $1d$-Luttinger model (Luttinger 1963). It has been shown by Komori and Wadati (Komori and Wadati 1996) that this model can be expressed equivalently by two Fermi fields ψ_j ($j = 1,2$) satisfying free field EOM but anyon-like commutation relations for $j \neq k$

$$\begin{aligned} \psi_j(x)\psi_k^\dagger(y) + \exp(i(-1)^j\lambda)\psi_k^\dagger(y)\psi_j(x) &= 0 , \\ \psi_j(x)\psi_k(y) + \exp(i(-1)^{j+1}\lambda)\psi_k(y)\psi_j(x) &= 0 , \end{aligned} \tag{106}$$

where λ is proportional to the coupling strength between the two fields.

A quite similar situation arises in two space dimensions if a charged matter field couples minimally to an abelian gauge field described by a Chern-Simons term. This coupling can be removed by a gauge transformation such, that the new matter field will be described by a free Hamiltonian but anyonic commutation relations (Jackiw and Pi 1990; Lerda 1992). This theory is of importance in relation to the fractional quantum Hall effect.

In a recent paper P.W. Anderson et al. (Anderson and Khveshchenko 1995) described $2d$-Fermions by first bosonizing them and then modifying the bosonic commutation relations. They didn't succeed in finding the corresponding Fermion representation. But this can be achieved for electrons with an on-site repulsive interaction of infinite strength in the Hubbard model (Hubbard 1963). In this case at most one electron can occupy a lattice site i. Usually this will be achieved by means of the Gutzwiller projector

$$C^{\dagger}_{i,\sigma} \to C'^{\dagger}_{i\sigma} := (1 - N_{i,-\sigma})C^{\dagger}_{\sigma,i}\,, \tag{107}$$

where now the $C'_{i,\sigma}$ obey complicated commutation relations. But it is easier to introduce field operators satisfying a new on-site algebra

$$\begin{aligned} d_{i,\sigma}d_{i,\sigma'} &= 0\,, \\ d_{i,\sigma}d^{\dagger}_{i,\sigma'} &= \delta_{\sigma\sigma'}(1 - N_{i,-\sigma})\,. \end{aligned} \tag{108}$$

It may easily be seen, that by operating on state vectors the algebra of the $C'_{i,\sigma}$ or $d_{i,\sigma}$ respectively are the same. But (108) has the advantage to hold as an operator relation.

The algebra (108) or modifications of it may be generalized to the continuum without any difficulty.

It is an open question how to formulate on-site interaction of finite strength in terms of a new Fermionic algebra. For Bosons a first step in this direction has been done quite recently by Flores (Flores 1997).

6 Conclusions

We have seen from the examples given for one- and two-space dimensions, that our question "how to quantize dynamically equivalent Hamiltonian structures" has presumably a unique answer in those cases where we have a underlying dynamical symmetry. It remains to be shown that this is also true for the $2d$-Coulomb problem as well as for $3d$-examples (hydrogen atom, harmonic oscillator). A general framework is also missing.
The example of a noncommutative $2d$-space inducing the interaction of an oscillator with a constant magnetic fields also calls for generalizations.
Finally, in field theory, we have to answer the question in which cases the interaction can be described by nonstandard commutation relations instead of an interaction term in the Hamiltonian.

References

Anderson, P.W., Khveshchenko, D., (1995): Phys. Rev. **B52**, 16415
Arik, M., Coon, D.D., (1976): J. Math. Phys. **17**, 524
Biedenharn, L.C., (1989): J. Phys. **A22**, L873;
Macfarlane, A.J., (1989): J. Phys. **A23**, 4581

Carinena, J.F., Clemente-Gallardo, J., Follana, E., Gracia-Bondia, J.M., Rivero, A., Varilly, J.C., (1998): Connes tangent groupoid and deformation quantization, J. Geom. and Phys. (math.DG/9802102)
Chaturvedi, S., Srinivasan, V., (1991): Phys. Rev. **A44**, 8024
Cisło, J., Łopuszanski, J., Stichel, P.C., (1995): Fortschr. Phys. **43**, 745
Currie, D.G., Saletan, E.J., (1966): J. Math. Phys. **7**, 967;
Cisło, J., Łopuszański, J., Stichel, P.C., (1998): Fortschr. Phys. **46**, 45
Dimakis, A., Müller-Hoissen, F., (1997): Some aspects of noncommutative geometry and physics, preprint physics/9712004
Douglas, J., (1941): Trans. Am. Math. Soc. **50**, 71
Finkelstein, R., Marcus, E., (1995): J. Math. Phys. **36**, 2652
Flores, J.C., (1997): Bose-Hubbard Hamiltonian from generalized commutation relations, preprint cond-mat/9712006
Henneaux, M., Shepley, L.C., (1982): J. Math. Phys. **23**, 2101
Hojman, P.A., (1991): J. Phys. A. Math. Gen. **24**, 249
Hubbard, J., (1963): Proc. Roy. Soc. **A276**, 238
Jackiw, R., Pi, S.Y., (1990): Phys. Rev. **D42**, 3500;
Lerda, A., (1992): *Anyons-quantum mechanics of particles with fractional statistics*, Lecture Notes in Physics m14, Springer
Katriel, J., Quesne, C., (1996): J. Math. Phys. **37**, 1650
Kempf, A., (1997) and (1994): Phys. Rev. **D55**, 7909 and J. Math. Phys. **35**, 4483
Komori, Y., Wadati, M., (1996): Phys. Lett. **A218**, 42
Leubner, C., Marte, M.A.M., (1984): Phys. Lett. **101A**, 179
cp. Lledo, M.A., Gracia-Sucre, M., (1996): J. Math. Phys. **37**, 160; Gotay, M.J., (1998): Obstructions to quantization, preprint math-ph/9809011 and the literature quoted there
Lukierski, J., Stichel, P.C., Zakrzewski, W., (1997): Ann. Phys. **260**, 224
Luttinger, J.M., (1963): J. Math. Phys. **4**, 1154
Man'ko, V.I., Marmo, G., Sudarshan, E.C.G., Zaccaria, F., (1996): Int. J. Mod. Phys. (quant-ph/9612007)
Man'ko, V.I., Marmo, G., Sudarshan, E.C.G., Zaccaria, F., (1997): Phys. Scripta **55**, 528
Palev, T.D., (1982): J. Math. Phys. **23**, 1778
Okubo, P., (1980): Phys. Rev. **D22**, 919
Quesne, B., Vansteenkiste, N., (1996): Helv. Phys. Acta **69**, 141;
Guichardet, A., (1998): J. Math. Phys. **39**, 4965
Rieffel, M.A., (1997): Questions on Quantization, preprint quant-ph/9712009 v 2.
Roček, M., (1991): Phys. Lett. **B255**, 554
Samsonov, B.F., (1998): J. Math. Phys. **39**, 967
Sicong Jing, (1998): J. Phys. A. Math. Gen. **31**, 6347
Werner, R.F., (1995): The classical limit of quantum theory, preprint quant-th/9504016
Wigner, E.P., (1950): Phys. Rev. **77**, 711

Quantum Channels of the Einstein–Podolsky–Rosen Kind

A. Uhlmann

Institut für Theoretische Physik
Universität Leipzig, Germany

Dedicated to Jan Łopuszański,
friend and colleague

Abstract. An EPR-channel consists of two Hilbert spaces, H^A and H^B, and of a density operator sitting on the their direct product space H^{AB}. The channel is triggered by a von Neumann measurement on H^A, resulting in a state (density operator) ω^A. Because a measurement in H^A can be considered as a measurement in H^{AB} equally well, it induces a new state in H^{AB} and, hence, a new state, ω^B, in H^B. The map $\omega^A \to \omega^B$ depends only on the channel's original density operator, and not on the chosen complete von Neumann measurement. This map is referred to as "channel map".

The construction of the channel map is described together with various of its properties, including an elementary link to the modular conjugation and to some related questions.

The (noisy) quantum teleportation channel is treated as an example. Its channel map can be decomposed into two EPR channel maps.

1 Introduction

In 1935 A. Einstein, B. Podolsky, and N. Rosen posed an intelligent and far reaching question (Einstein et al. 1935), albeit suggesting a misleading answer. The abbreviation "EPR" in several of the following notations is pointing to these authors. Further early contributions to the EPR-effect are due to Schrödinger, (Schrödinger 1935a), (Schrödinger 1935b). Since then a wealth of papers had appeared on the subject, mostly discussing "non-locality" and similar aspects which seem to contradict our causal feelings. They touch the question whether and how space and time can live with the very axioms of quantum physics, axioms which, possibly, are prior to space and time.

Quantum information theory considers the EPR-effect not as a paradox but as a "channel", as (part of) a protocol to transfer "quantum information" from one system to another one, see for example (Ekert 1991; Bennett and Wiesner 1992).

The use of the word "protocol" may well be compared with the way it is used in, say, ordinary telecommunication: It is independent with what physical device the bits of its commands are stored and processed. The implementation independence in quantum information theory is guaranteed by the use of Hilbert spaces, states (density operators), and operations between and on them. It is *not* said, what they physically describe, whether we are dealing with spins, polarizations, energy levels, particle numbers, or whatever you can imagine. Because of this, the elements of quantum information theory, to which the EPR-channel belongs, are of rather abstract nature. Their physical realization is generally much much more difficult and often not visible yet.

The abstract setting of an EPR-channel starts with two Hilbert spaces, $\mathcal{H}^A$ and $\mathcal{H}^B$, and their direct product

$$\mathcal{H}^A \otimes \mathcal{H}^B = \mathcal{H}^{AB} \tag{1}$$

and is completed with a density operator, ϱ^{AB}, on it.

The subsystems given by $\mathcal{H}^A$ and $\mathcal{H}^B$, respectively, are referred to as A– or B–system, or by its "owners", Alice and Bob, who are responsible for local actions. A local action of Alice is by definition a measurement which can be performed by an observable of the A–system, or by an operation which can be expressed by operators of the form $X^A \otimes 1^B$, $X^A \in \mathcal{B}(\mathcal{H}^A)$. Similarly one defines Bob's local actions.

This definition of locality is compatible, but not equivalent, with a possible localization of the two subsystems in space. They may be sit even macroscopically space-like one to another, but they need not do so.

Neither Alice nor Bob have access to all the operators (observables) of the total system. Hence they can see the state ϱ^{AB} only partially. For instance, the state ϱ^A, induced in the A–system, is gained by partial tracing over the B–system, and is defined by

$$\mathrm{Tr}_A \varrho^A X = \mathrm{Tr}_{AB} \varrho^{AB} (X \otimes 1^B) \,, \quad \forall X \in \mathcal{B}(\mathcal{H}^A) \,.$$

A classical message is a sequence of letters from an alphabet or, equivalently, a sequence of positions in a set which is called "alphabet". A quantum message is a sequence of positions in a state space of a quantum system, hence a sequence of states. A quantum channel is supposed to transfer the quantum messages. It maps the state space of the sender into that of the receiver. Without some knowledge of the possible positions used as letters, and of the channel's action, attempts to encode the quantum message are hopeless. To be useful one needs some additional classical message, transported through a classical channel, and some conventions between sender and receiver.

The sender in a general EPR-channel is a measuring apparatus in the A–system which performs a von Neumann measurement, (von Neumann 1932), (Lüders 1951). Let X be the observable describing its action. The duty of X is to *prepare* one of the eigenstates of X, and to *distinguish* it from the other ones by pointing to its eigenvalue. The physical meaning of the eigenvalues of X is *not* relevant for the purpose in question.

The state ϱ^{AB} correlates the Alice' system with that of Bob. These correlations constitute the channel. Measuring X destroys these correlations, thereby creating a new state in the A–system and inducing a new one in Bob's system. Repeating this procedure, which includes the regeneration of

ϱ^{AB}, Alice creates a random quantum message. The EPR-channel transmits the quantum message to Bob, who receives, generally, a deformed version of it. How much the message will be deformed depends on the strength of the correlations provided by ϱ^{AB}, but also on the choice of X relative to ϱ^{A}.

Now we define the channel map. Let

$$\pi^{A} = |\phi^{A}\rangle\langle\phi^{A}|, \quad \phi^{A} \in \mathcal{H}^{A} \tag{2}$$

be a rank one projection operator. Let us assume π^{A} is a non-degenerate eigenstate of X, and Alice's measuring apparatus points to the eigenvalue associated with π^{A}. (This happens with probability $\langle\phi^{A}|\varrho^{A}|\phi^{A}\rangle$.) Now, a measurement in a system is always a measurement in every larger system, in our case in the AB–system. Lüder's rule provides us with the new state prepared by that measurement:

$$\varrho^{AB} \mapsto \omega^{AB} = (\pi^{A} \otimes 1^{B})\varrho^{AB}(\pi^{A} \otimes 1^{B}) = \pi^{A} \otimes \omega^{B} . \tag{3}$$

Obviously, ω^{B} is, up to normalization, the state prepared in Bob's system by Alice' measurement.

The *channel map* is the map

$$\pi^{A} \mapsto \omega^{B} := \Phi_{\varrho}^{AB}(\pi^{A}), \quad \varrho = \varrho^{AB} . \tag{4}$$

We shall see that this map exists, i. e. it does not depend on Alice's action.

2 The Maps s^{BA} and s^{AB}

We start with EPR-channel maps *for vectors*, ψ, so that

$$\varrho^{AB} = |\psi\rangle\langle\psi|, \quad \psi \in \mathcal{H}^{AB} .$$

Lemma 1

Let $\psi \in \mathcal{H}^{AB}$ be written as a sum

$$\psi = \sum \tilde{\phi}_{j}^{A} \otimes \tilde{\phi}_{j}^{B} \tag{5}$$

with vectors $\tilde{\phi}_{i}^{A}$ and $\tilde{\phi}_{k}^{B}$ from $\mathcal{H}^{A}$ and $\mathcal{H}^{B}$ respectively.

Then the map

$$\phi^{A} \mapsto s^{BA}\phi^{A}, \quad \phi^{A} \in \mathcal{H}^{A}, \tag{6}$$

given by

$$s^{BA}\phi^{A} = \sum_{j}\langle\phi^{A}|\tilde{\phi}_{j}^{A}\rangle\,\tilde{\phi}_{j}^{B} \tag{7}$$

is uniquely defined by ψ. Hence it can be denoted by s_{ψ}^{BA}.

Proof: The idea is in assuming a von Neumann measurement by Alice to check whether her system is in the state given by ϕ^{A}. If the answer is "YES", the vector

$$\varphi := \{\, |\phi^A\rangle\langle\phi^A| \otimes 1^B \,\}\, \psi$$

is prepared in the AB–system which can depend on ϕ^A and ψ only. By the help of (5) this vector is written

$$\varphi = \phi^A \otimes \sum \langle \phi^A | \tilde{\phi}_j^A \rangle \tilde{\phi}_j^B \, .$$

Comparing this expression with the definition (7) we obtain the important relation

$$\{\, |\phi^A\rangle\langle\phi^A| \otimes 1^B \,\}\, \psi = \phi^A \otimes s_\psi^{BA} \phi^A \, , \quad \phi^A \in \mathcal{H}^A \, . \tag{8}$$

This shows that the map (7) does not depend on the way the vector ψ is represented as a sum (5).

Corollary 2 If Alice is successful in preparing ϕ^A by a von Neumann measurement, the prepared vector of the AB-System is given by (8).

Corollary 3 Let ϕ_1^A, $\phi_2^A, \ldots$ be any collection of vectors satisfying

$$1^A = \sum |\phi_k^A\rangle\langle\phi_k^A| \tag{9}$$

then

$$\psi = \sum \phi_k^A \otimes s_\psi^{BA} \phi_k^A \, . \tag{10}$$

Indeed, this follows easily from (8). The mathematical content of lemma 1 is nothing than the well know theorem, stating that the Hilbert space (1) is isomorphic to the linear Hilbert–Schmidt maps from $\mathcal{H}^A$ into the dual of $\mathcal{H}^B$. One can map the latter by an antiunitary map onto $\mathcal{H}^B$. In this way we see, how quantum measurements provide a randomly pointwise realization of Hilbert–Schmidt maps:

Corollary 4 Every antilinear Hilbert–Schmidt map from $\mathcal{H}^A$ into $\mathcal{H}^B$ can be identified with exactly one of the maps s_ψ^{BA}.

An interesting observation, (Fivel 1995), appendix[1], is the *antilinearity* of (6), clearly seen from its definition (7). Such a map cannot be tensored with a linear one, for instance with the identity map of another (complex!) Hilbert space. The sign of the imaginary unit cannot be fixed in such a construct. On the physical side this is very good: An antilinear map can be represented by a linear one followed by time reversal. A direct product of an antilinear and a linear map would be equivalent, up to a linear operation, to reversing time in the first but not in the other system, and this is forbidden. Instead we have to apply (equivalents of) time reversal simultaneously to *all* quantum systems which can share entanglement.

[1] Thanks to D. DiVincenzo for the hint to Fivel's paper

A further important fact is the possibility to exchange the roles of Alice and of Bob. There is *no preferred direction* A → B or B → A in the game, but a complete symmetry with respect to the exchange A ↔ B in all equations and relations.

In particular the map

$$\phi^B \mapsto s_\psi^{AB}\, \phi^B = \sum_j \langle \phi^B | \tilde{\phi}_j^B \rangle\, \tilde{\phi}_j^A, \quad \phi^B \in \mathcal{H}^B \tag{11}$$

is well defined and there are counterparts to all the conclusions above. In particular

$$\{\, 1^A \otimes |\phi^B\rangle\langle\phi^B|\,\}\, \psi = s_\psi^{AB} \phi^B \otimes \phi^B, \quad \phi^B \in \mathcal{H}^B\,. \tag{12}$$

We mention some of the cross-relations between these maps. With two arbitrary vectors ψ and φ from $\mathcal{H}^{AB}$ one has

$$\mathrm{Tr}_A s_\varphi^{AB} s_\psi^{BA} = \mathrm{Tr}_B s_\varphi^{BA} s_\psi^{AB} = \langle \psi, \varphi \rangle\,. \tag{13}$$

To derive these equations, represents the vectors by any decompositions

$$\psi = \sum \tilde{\phi}_j^A \otimes \tilde{\phi}_j^B, \quad \varphi = \sum \hat{\phi}_j^A \otimes \hat{\phi}_j^B\,.$$

Then

$$s_\varphi^{BA} s_\psi^{AB} = \sum |\tilde{\phi}_j^B\rangle\langle\tilde{\phi}_k^A, |\hat{\phi}_j^A\rangle\langle\hat{\phi}_k^B|\,,$$

$$s_\varphi^{AB} s_\psi^{BA} = \sum |\hat{\phi}_k^A\rangle\langle\tilde{\phi}_j^B, |\hat{\phi}_k^B\rangle\langle\tilde{\phi}_j^A|\,.$$

Taking the relevant trace one gets (13).

$$\langle \phi^B | s^{BA} \phi^A \rangle = \langle \phi^A | s^{AB} \phi^B \rangle \;\forall\, \phi^A \in \mathcal{H}^A,\, \phi^B \in \mathcal{H}^B\,, \tag{14}$$

which is shortly rewritten as

$$(\, s_\psi^{BA}\,)^* = (\, s_\psi^{AB}\,)\,. \tag{15}$$

3 EPR Channel Maps for States and Observables

Knowing a convenient description of EPR-channel maps for vectors, we extend the formalism to Einstein-Podolsky-Rosen channels based on an arbitrary density operator. To start with, the density operator, ϱ^{AB}, may be in any decomposition

$$\varrho^{AB} = \sum |\psi_i\rangle\langle\psi_i|, \quad \psi_i \in \mathcal{H}^{AB}\,, \tag{16}$$

so that the trace of ϱ^{AB} equals $\sum\langle\psi_i|\psi_i\rangle$.

According to lemma 1 and (11), every one of the vectors ψ_i gives rise to two antilinear maps

$$\psi_i \leftrightarrow s_i^{BA} \leftrightarrow s_i^{AB} \,. \tag{17}$$

Again we ask for the state change if Alice' von Neumann measurement confirms the state $|\phi^A\rangle\langle\phi^A|$ characterized by the vector ϕ^A. The preparation causes the change

$$\varrho^{AB} \mapsto \left(|\phi^A\rangle\langle\phi^A| \otimes 1^B\right) \varrho^{AB} \left(|\phi^A\rangle\langle\phi^A| \otimes 1^B\right) . \tag{18}$$

There is no difficulty at all to insert the decomposition (16) and to arrive, for all i, to a problem solved by lemma 1. We use (8) for the kets and, after respecting (14, 15), also for the bras. In this way the right hand side of (18) is converted into

$$(|\phi^A\rangle\langle\phi^A|) \otimes \sum_i s_i^{BA}|\phi^A\rangle\langle\phi^A|\, s_i^{AB} \,.$$

This expression depends only on ϱ^{AB} and on $|\phi^A\rangle\langle\phi^A|$, the latter dependence can be extended to arbitrary finite sums of positive rank one operators and, for infinite dimensional Hilbert spaces, to all trace-class operators. By the same argument as in the proof of lemma 1 we get, therefore,

Lemma 5
There is a channel map

$$\omega^A \mapsto \Phi_\varrho^{BA}(\omega^A), \quad \varrho \equiv \varrho^{AB} , \tag{19}$$

such that for all

$$\pi^A = |\phi^A\rangle\langle\phi^A|, \quad \phi^A \in \mathcal{H}^A$$

one gets

$$\left(\pi^A \otimes 1^B\right) \varrho^{AB} \left(\pi^A \otimes 1^B\right) = \pi^A \otimes \Phi_\varrho^{BA}(\pi^A) . \tag{20}$$

This defines a map map from the trace-class operators on $\mathcal{H}^A$ into those of $\mathcal{H}^B$. Every decomposition (16) yields

$$\omega^A \mapsto \Phi_\varrho^{BA}(\omega^A) = \sum_i s_i^{BA}\omega^A s_i^{AB} \,. \tag{21}$$

Φ_ϱ^{BA} is called the *EPR channel map from Alice to Bob, based on* ϱ.

We see the following:

a) To be a genuine channel map, the definition of Φ^{BA} should not depend on the actions of Alice. This is obviously true.

b) The maps Φ^{BA}, though *not* completely positive themselves, become such after sandwiching them between antiunitaries. Using for the latter a conjugation, we see that their action on density operators is that of a complete copositive operator, see (Woronowicz 1976). Hence we may call the maps (21) *completely *-copositive.*

c) We may exchange the roles of Alice and Bob getting Φ^{BA}, the channel map from Bob to Alice.

d) There is a one-to-one correspondence

$$\varrho \Longleftrightarrow \Phi_{\varrho}^{BA} \Longleftrightarrow \Phi_{\varrho}^{AB}\,, \quad \varrho \in \mathcal{H}^{AB}\,.$$

e) There is a virtual "dictionary" translating any property of density operators of the AB–system into a property of the associated channel map, and vice versa.

The dual of a channel map as described above is a map from Bob's observables (operators) into those of Alice. Its duty is to look at what is going on in Alice'system by transporting her expectation values to Bob. (In case of infinitely many degrees of freedom there are circumstances, where it is advisable just to start by mapping observables.)

To construct the dual we need for every operator Y, acting on $\mathcal{H}^B$, an operator X, acting on Alice' Hilbert space with the following property:

If von Neumann measurement of Alice results in a pure state density operator, π^A, and if Φ^{BA} is the channel map introduced by lemma 5, then

$$\mathrm{Tr}_A \pi^A X = \mathrm{Tr}_B \Phi^{BA}(\pi^A) Y\,. \tag{22}$$

That is, according to (21),

$$\langle \phi^A | X | \phi^A \rangle = \sum \langle s_i^{BA} \phi^A | Y | s_i^{BA} \phi^A \rangle\,.$$

Let us denote by Φ_{AB} the wanted operator,

$$X = \Phi_{AB}(Y), \quad \Phi_{AB} \equiv \Phi_{AB}^{\varrho}\,.$$

From (14 or 15) one gets

$$\Phi_{AB}(Y) = \sum_i s_i^{BA} Y^* s_i^{BA} = \sum_i \left(s_i^{BA} Y s_i^{BA} \right)^*\,. \tag{23}$$

Again we see the antilinearity. At the first instant one may think it irrelevant: Should we not restrict ourselves to selfadjoint (Hermitian) oberservables? However, such a practice is an oversimplification. Indeed, any normal operator, Y, $Y^*Y = YY^*$, is a perfect observable. Its complex eigenvalues may be read off as points from a screen. The crux with the antilinearity is that: A sequence of points, appearing on Bob's screen, is an affine deformation of those seen by Alice *together with a reflection on a certain line.* The latter comes from the complex conjugation of the eigenvalues, enforced by the Hermitian conjugation $Y \to Y^*$. In other words, the orientations of Alice' and Bob's screen are mutually opposite. The determinant of the mapping between the screens, if not degenerated, is negative.

The antilinearity of the channel map produces similar effects on geometric Berry phases.

4 An Excursion to Quantum Teleportation

The teleportation protocol (Bennett et al. 1993) of Bennett, Brassard, Crepeau, Jozsa, Peres, and Wootters, needs a quantum and a classical information channel. Here we are concerned with the quantum one and its channel maps.

Quantum teleportation lives on the direct product of three Hilbert spaces,

$$\mathcal{H}^A \otimes \mathcal{H}^B \otimes \mathcal{H}^C = \mathcal{H}^{ABC} . \tag{24}$$

At first, as in the EPR case, one has to have control on the transfer of vectors. The input of the teleportation channel consists of an (unknown) unit vector to be teleported, multiplied by an auxiliary one, the *ancilla*, carrying "entanglement",

$$\psi^{ABC} = \phi^A \otimes \psi^{BC} , \quad \phi^A \in \mathcal{H}^A , \quad \psi^{BC} \in \mathcal{H}^B \otimes \mathcal{H}^C . \tag{25}$$

The channel is triggered by a complete measurement of the AB–system with respect to an orthonormal basis

$$\psi_j^{AB} \in \mathcal{H}^A \otimes \mathcal{H}^B , \quad j = 1, 2, \dots . \tag{26}$$

It seems natural to consider the maps

$$s^{CB} , \quad s_j^{BA} , \quad j = 1, 2, \dots , \tag{27}$$

where the first one is defined by lemma 1 with respect of ψ^{BC}. The other ones are associated to the members ψ_j^{AB} of the basis (26) accordingly.

It is tempting to compose these maps to transporters $s^{CB} s_j^{BA}$ from $\mathcal{H}^A$ to $\mathcal{H}^C$. And, indeed, that is the essence of the quantum part of the famous teleportation protocol. One needs no further assumption on the nature of the vectors (25) and (26) to run the protocol, though its effectiveness (or failure) depends critically on them. This is the content of the following lemma.

Lemma 6

If, with the notations above, the measurement device acting on $\mathcal{H}^{AB}$ is pointing to the i-th vector of (26), the teleportation is providing in $\mathcal{H}^C$ the vector

$$\phi_i^C := t_i^{CA} \phi^A , \quad t_i^{CA} := s^{CB} s_i^{BA} . \tag{28}$$

Proof: In (24) the vector

$$\varphi_i = \left(|\psi_i^{AB}\rangle\langle\psi_i^{AB}| \otimes 1^C \right) \psi^{ABC} \tag{29}$$

is prepared by the measurement. Choosing in $\mathcal{H}^B$ an orthonormal basis $\{\phi_j^B\}$ gives the opportunity to write

$$\psi^{BC} = \sum \phi_j^B \otimes s^{CB} \phi_j^B$$

and hence

$$\varphi_i = |\psi_i^{AB}\rangle\langle\psi_i^{AB}| \sum_j |\phi^A \otimes \phi_j^B\rangle \otimes s^{CB}\phi_j^B .$$

Now we choose in $\mathcal{H}^A$ an orthonormal basis $\{\phi_k^A\}$ to resolve the scalar product in the last equation:

$$\varphi_i = \psi_i^{AB} \otimes \sum_{jk} \langle\phi_k^A|\phi^A\rangle\langle s_i^{BA}\phi_k^A|\phi_j^B\rangle s^{CB}\phi_j^B .$$

Using antilinearity,

$$\varphi_i = \psi_i^{AB} \otimes s^{CB} \sum_k \langle\phi^A|\phi_k^A\rangle \sum_j \langle\phi_j^B|s_i^{BA}\phi_k^A\rangle\phi_j^B .$$

The summation over j results in $s_i^{BA}\phi_k^A$. Now, again by antilinearity, the sum over k comes down to

$$s_i^{BA} \sum_k \langle\phi_k^A|\phi^A\rangle\phi_k^A .$$

Thus, we finally get assertion of the lemma:

$$\Big(|\psi_i^{AB}\rangle\langle\psi_i^{AB}| \otimes 1^C\Big)\psi^{ABC} = \psi_i^{AB} \otimes s^{CB} s_i^{BA}\phi^A . \tag{30}$$

We have seen that the i-th teleportation channel map is composed of two antilinear Hilbert-Schmidt maps. Hence t_i^{CA} is *linear* and of *trace class.* In estimating its magnitude by a norm, an adequate one is certainly the *trace norm*

$$\| t_i^{CA} \|_1 := \mathrm{Tr}\,\sqrt{t_i^{AC} t_i^{CA}}\,, \qquad t_i^{AC} := (t_i^{CA})^* . \tag{31}$$

This norm depends as follows on the reduced density operators

$$\varrho_i^B = \mathrm{Tr}_A|\psi_i^{AB}\rangle\langle\psi_i^{AB}|, \quad \varrho^B = \mathrm{Tr}_C|\psi^{BC}\rangle\langle\psi^{BC}| . \tag{32}$$

Lemma 7

The trace norm $\| t_i^{CA} \|_1$ is the square root of the transition probability (fidelity) between ϱ_i^B and ϱ^B.

There are two more or less straightforward generalizations of lemma 6. At first, we can convert the channel maps for vectors to ones for density operators (states). In doing so we consider the preparation $|\varphi_i\rangle\langle\varphi_i|$ with φ_i from (29). In this expression we vary the vector ϕ^A of (25) and add them up. This tells us how to teleport, through the i-th channel, an arbitrary density operator, say ω^A, to the C–system. The transport is done by

$$\omega^A \Longrightarrow t_i^{CA}\omega^A t_i^{AC} . \tag{33}$$

Resolving the map according to lemma 6, and applying lemma 5, we can replace the pure vector ψ^{BC} of (24) by an arbitrary density operator ϱ^{BC}. Then, with such an arbitrary ancilla, the i-th teleportation channel map reads

$$\omega^A \Longrightarrow \Phi_\varrho^{CB}\Big(s_i^{BA}\omega^A s_i^{AB}\Big) \tag{34}$$

with $\varrho \equiv \varrho^{BC}$.

5 Something More About EPR Channel Maps

To prepare the next section, and for its own sake, we return to section 2 and add some further relations. Let ψ be a vector from (1) and ϱ^A and ϱ^B its partial traces. Then

$$s^{BA}\, s^{AB} = \varrho^B\,, \quad s^{AB}\, s^{BA} = \varrho^A\,. \tag{35}$$

From this one deduces the polar decompositions

$$\begin{aligned} s^{BA} &= j^{BA}\sqrt{\varrho^A} = \sqrt{\varrho^B} j^{BA}\,, \\ s^{AB} &= j^{AB}\sqrt{\varrho^B} = \sqrt{\varrho^A} j^{AB}\,. \end{aligned} \tag{36}$$

The partial antiunitaries $j_\psi^{AB} \equiv j^{AB}$ and $j_\psi^{BA} \equiv j^{BA}$ are uniquely fixed by demanding one or both of the conditions

$$j^{AB} j^{BA} = \operatorname{supp} \varrho^A\,, \quad j^{BA} j^{AB} = \operatorname{supp} \varrho^B\,, \tag{37}$$

ψ is called *completely entangled* (with respect to Alice) iff the support of ϱ^A is the identity map of $\mathcal{H}^A$. This is equivalent with calling ϱ^A *faithful*, or with calling ψ *separating* with respect to $\mathcal{B}(\mathcal{H}^A) \otimes 1^B$. If this occurs, the dimension of $\mathcal{H}^B$ cannot be smaller than that of $\mathcal{H}^A$. In case the dimensions are finite and equal, ψ is completely entangled with respect to Alice iff it does so to Bob. If $\dim \mathcal{H}^A$ is finite, ψ can be *maximally entangled* with respect to Alice. That means, ϱ^A is proportional to 1^A. If both dimension are finite and equal, maximal entanglement with respect to Alice implies the same to Bob, and the reference to one of them is not necessary. Indeed, in calling a vector of a bipartite system *maximally entangled*, one supposes finiteness and equality of the dimensions by implication.

The equations above can be established by the help of a Gram–Schmidt decomposition of ψ. Denote by ϕ_j^A the vectors of a complete orthonormal system of eigenvectors of ϱ^A, and p_j the corresponding eigenvectors. Then there are orthonormal eigenvectors ϕ_j^B such that

$$\psi = \sum \sqrt{p_j} \phi_j^A \otimes \phi_j^B\,. \tag{38}$$

One immediately infers from its definitions

$$s_\psi^{BA}(\phi_j^A) = \sqrt{p_j} \phi_j^B\,, \quad s_\psi^{AB}(\phi_j^B) = \sqrt{p_j} \phi_j^A \tag{39}$$

and the Gram-Schmidt form of the s-maps,

$$\begin{aligned} s_\psi^{BA}(\phi^A) &= \sum_1^m \sqrt{p_j}\, \langle \phi^A | \phi_j^A \rangle\, \phi_j^B\,, \\ s_\psi^{AB}(\phi^B) &= \sum_1^m \sqrt{p_j}\, \langle \phi^B | \phi_j^B \rangle\, \phi_j^A\,. \end{aligned} \tag{40}$$

Some of the eigenvalues of p_k may be zero, and then the k–th term in the Gram-Schmidt representation will vanish automatically. For the j-maps we had to exclude them explicitly. Hence, assuming (38), we should write

$$
\begin{aligned}
j_\psi^{BA}(\phi^A) &= \sum_{p_j \neq 0} \langle \phi^A | \phi_j^A \rangle \, \phi^B \, , \\
j_\psi^{AB}(\phi^B) &= \sum_{p_j \neq 0} \langle \phi^B | \phi_j^B \rangle \, \phi^A \, .
\end{aligned} \tag{41}
$$

Remarks:

(a) For any given partial antiunitary map j^{BA} there are vectors $\psi \in \mathcal{H}^{AB}$ such that $j^{BA} = j_\psi^{BA}$.

(b) j_ψ^{BA} is itself Hilbert-Schmidt iff there are only finitely many terms in the sums (41). In other words: If and only if s^{BA} is of finite rank, there is ψ' such that $j^{BA} = s_{\psi'}^{BA}$.

(c) Obviously, we need not care of finite rank and Hilbert-Schmidt conditions in dealing with finite dimensional Hilbert spaces.

(d) ψ is completely entangled with respect to Alice iff j^{AB} is an antiunitary map from $\mathcal{H}^A$ into $\mathcal{H}^B$.

Finally, let us mention the action of an operator of the form $X^A \otimes Y^B$,

$$
\varphi = (X^A \otimes Y^B)\,\psi \, . \tag{42}
$$

A look at (7) and (11) shows

$$
s_\varphi^{BA} = Y^B s_\psi^{BA} (X^A)^* \, , \quad s_\varphi^{AB} = X^A s_\psi^{AB} (Y^B)^* \, . \tag{43}
$$

We get from this and (36) with unitary factors, U^A and U^B,

$$
j_\varphi^{BA} = U^B j_\psi^{BA} (U^A)^* \, , \quad j_\varphi^{AB} = U^U j_\psi^{AB} (U^B)^* \, , \tag{44}
$$

where now $\varphi = (U^A \otimes U^B)\psi$. Let us assume ψ completely entangled with respect to Alice and to Bob. Then the j are antiunitaries, and from (44) the implication

$$
U^B = j_\psi^{BA} U^A j_\psi^{AB} \quad \Longrightarrow (U^A \otimes U^B)\psi = \psi \tag{45}
$$

follows. The unitaries $U^A \otimes j_\psi^{BA} U^A j_\psi^{AB}$ form the *local stabilizer group of* ψ.

6 Operator Lifts to $\mathcal{H}^{AB}$

With a pair of maps, one from Alice to Bob and and one from Bob to Alice, one can compose an antilinear (linear) map in $\mathcal{H}^{AB}$ provided both are antilinear (or both linear). Here we are concerned with the antilinear case only. Let us denote the "twisting" operation doing this by $\tilde{\otimes}$. Requiring antilinearity, the maps are characterized completely by their actions on product vectors:

$$
\begin{aligned}
j_\psi \tilde{\otimes} j_\psi \, (\phi^A \otimes \phi^B) &= j_\psi^{AB} \phi^B \otimes j_\psi^{BA} \phi^A \,, \\
j_\psi \tilde{\otimes} s_\psi \, (\phi^A \otimes \phi^B) &= j_\psi^{AB} \phi^B \otimes s_\psi^{BA} \phi^A \,, \\
s_\psi \tilde{\otimes} j_\psi \, (\phi^A \otimes \phi^B) &= s_\psi^{AB} \phi^B \otimes j_\psi^{BA} \phi^A \,, \\
s_\psi \tilde{\otimes} s_\psi \, (\phi^A \otimes \phi^B) &= s_\psi^{AB} \phi^B \otimes s_\psi^{BA} \phi^A \,.
\end{aligned} \tag{46}
$$

The first of these operators is a standard one: It is the *modular conjugation* of Tomita-Takesaki's theory,

$$J_\psi = j_\psi \tilde{\otimes} j_\psi \,.$$

In this theory one considers $\mathcal{H}^{AB}$ as representation space of a *–representation of $\mathcal{B}(\mathcal{H}^A) \otimes 1^B$, (see, for example, (Haag 1993, sections III.2 and V.2)). The *modular operator*, Δ_ψ, and the operator S_ψ,

$$\Delta_\psi = \varrho^A \otimes (\varrho^B)^{-1} \,, \quad S_\psi = J_\psi \sqrt{\Delta_\psi} \,,$$

if they exist, satisfy

$$\sqrt{\Delta}(j \tilde{\otimes} s) = s \tilde{\otimes} j \,, \quad S\,(1^A \otimes \sqrt{\varrho^B}) = s \tilde{\otimes} j \,. \tag{47}$$

If we can rely on a Gram-Schmidt decomposition (38), then

$$
\begin{aligned}
J_\psi \phi_j^A \otimes \phi_k^B &= \phi_k^A \otimes \phi_j^B \,, \quad p_j p_k \neq 0 \,, \\
J_\psi \phi_j^A \otimes \phi_k^B &= 0 \,, \quad p_j p_k = 0 \,.
\end{aligned} \tag{48}
$$

Similarly explicit expressions one obtains for the other operators.

There is a lot more to say, including the discussion of examples. But, hopefully, also this sketchy paper is of use.

Acknowledgment

Part of this work was completed during the 1998 Elsag-Bailey – I.S.I. Foundation research meeting on quantum computation. I thank the organizers of the XII Max Born Symposium for the privilege to speak in honour of Jan Łopuszański. Thanks to B. Crell for discussions.

References

Bennett, C., Brassard, G., Crepeau, C., Jozsa, R., Peres, A., Wootters, W., (1993): Teleporting an unknown quantum state via dual classical and Einstein-Podolsky-Rosen channels, Phys. Rev. Lett., **70**, 1895–1898

Ekert, A.K., (1991): Quantum Cryptography Based on Bell's Theorem, Phys. Rev. Lett., **67** 661–663; Bennett, Ch.H., Wiesner, S.J., (1992): Communication via One- and Two-Particle Operators on Einstein-Podolsky-Rosen States, Phys. Rev. Lett., **69**, 2881–2884

Einstein, A., Podolsky, B., Rosen, N., (1935): Can quantum-mechanical description of physical reality be considered complete ? Phys. Rev. **47**, 777–780

Fivel, D.I., (1995): Remarkable Phase Oscillations Appearing in the Lattice Dynamics of Einstein-Podolsky-Rosen States, Phys. Rev. Lett. **74**, 835–838

Haag, R., (1993): Local Quantum Physics. Springer Verlag, Berlin, Heidelberg, New York

Lüders, L., (1951): Über die Zustandsänderung durch den Meßprozeß, Ann. d. Physik, **8**, 322–328

von Neumann, J. (1932): *Mathematische Grundlagen der Quantenmechanik*, Springer Verlag, Berlin

Schrödinger, E., (1935a) Die gegenwärtige Situation in der Quantenmechanik, Naturwissenschaften, **35**, 807–812, 823–828, 844–849

Schrödinger, E., (1935b): Discussion of probability relations between separated systems, Proc. Cambr. Phil. Soc., **31**, 555–563

Woronowicz, S.L., (1976): Positive maps of low dimensional matrix algebras, Rep. Math. Phys. **10**, 165–183

Part II

Quantum Physics: Different Approach

Quantum Dynamics in the Proper Time

Z. Haba

Institute of Theoretical Physics
University of Wrocław
50–205 Wrocław, Pl. Maxa Borna 9, Poland

Abstract. We discuss a relativistic quantization scheme which treats particle's wave function and quantum field fluctuations at the same footing. We introduce a stochastic wave function whose dynamics is determined by a non-linear Schrödinger-type evolution equation in an additional time parameter.

Keywords: proper time, stochastic quantization, quantum fields

1 Introduction

Quantum mechanics consists of two sets of rules. The first rule describes an intrinsic time evolution of state vectors. The second one concerns the relation to an observation. These rules should have a classical limit when a quantum system approximates a classical one. There has been a substantial progress in recent years (Zurek 1991) in understanding the classical limit of non-relativistic quantum systems. In particular, the role of an infinite environment has been emphasized. In non-relativistic quantum mechanics the relation to an observation is fixed by the principle of an instantaneous wave function reduction. It leads to non-local phenomena which are hard to reconcile with relativity. It seems that we must reconsider the basic principles of quantum mechanics in relativistic theories.

The conventional way out is to abandon the relativistic quantum mechanics of interacting particles in favor of the relativistic quantum field theory (QFT). We need QFT in order to describe the processes of particle creation and annihilation. However, these processes make the notion of a particle difficult to sustain. Only free particles in an infinite past and in an infinite future have a meaning. On the other hand there should be a well-defined classical limit of QFT. From the conventional QFT we can obtain the limit $\hbar \to 0$. However, such a limit leads to the classical field theory rather than to the classical mechanics. We suggest here an explicitly relativistic invariant quantization scheme which treats simultaneously the wave function ψ_τ and the quantum fields A. In the classical limit $|\psi_\tau(x)|^2 \simeq |\psi(x_\tau)|^2$, where x_τ is the evolution (described by the proper time τ) of the classical particle interacting with the classical field A (the classical limit of the quantum field A). We can obtain the conventional quantum field theory by an averaging over the proper time.

2 Relativistic Wave Equation

We consider a relativistic particle in an electromagnetic field A moving on a general pseudoriemannian manifold $\mathcal{M}$ (see earlier papers on such wave equations (Stueckelberg 1941), (Horwitz and Piron 1973)). We introduce a universal time τ which is independent of the choice of coordinates and (in the quantum theory) it is independent of the space-time point $x \in \mathcal{M}$ (in particular τ is independent of the coordinate x_0)

$$i\hbar\partial_\tau\psi = \frac{1}{2M}g^{\mu\nu}\left(-i\hbar\partial_\mu + \frac{1}{c}A_\mu + \hbar g^{\alpha\beta}\Gamma_{\mu\alpha\beta}\right)\left(-i\hbar\partial_\nu + \frac{1}{c}A_\nu\right)\psi\,, \qquad (1)$$

where $g^{\mu\nu}$ is the Riemannian metric (we choose the signature (1,1,1,-1)), c is the velocity of light and Γ is the Christoffel symbol.

Equation (1) can be justified by the conventional quantization rules of the classical mechanics. Let us recall that if the relativistic Lagrangian is chosen in the form invariant under the reparametrization $x(\gamma) \to x(f(\gamma))$

$$L = \int d\gamma\sqrt{\left(\frac{dq}{d\gamma}\right)^2} + \int A dq\,,$$

then the canonical Hamiltonian

$$H = \frac{1}{2M}\left(p_\mu + \frac{1}{c}A_\mu\right)\left(p^\mu + \frac{1}{c}A^\mu\right), \qquad (2)$$

is identically equal to zero. The constraint $H = 0$ (Hansson et al. 1976) generates correct equations of motion if the time parameter is interpreted as the proper time. If from the beginning we choose γ as the proper time then $H \neq 0$. Then, we look for a canonical change of coordinates (determined by the generating function W) such that in the new coordinates $H \to H + \partial_\tau W = 0$. The generating function W is defined by the solution of the Hamilton-Jacobi equation

$$\partial_\tau W_\tau + \frac{1}{2M}g^{\mu\nu}(\partial_\mu W_\tau + A_\mu)(\partial_\nu W_\tau + A_\nu) = 0\,. \qquad (3)$$

Equation (1) can be considered as a quantization of (3) (we associate to W the wave function $\psi = \exp(iW/\hbar)$). In the standard quantization scheme of constrained systems (Hansson et al. 1976) one argues that the quantum theory should be invariant under the choice of the parameter γ (reparametrization invariance). This invariance leads to the Klein-Gordon equation $H\psi = 0$. However, in our interpretation the proper time has a physical meaning. Hence, we make this preferred choice of γ in the Lagrangian. In such a case the canonical Hamiltonian (2) is different from zero and it generates the time evolution (1). Conversely, the classical dynamics (3) results as a limit $\hbar \to 0$ of the quantum dynamics (1) if τ is identified with the classical proper time.

In fact, consider (1) with the initial condition $\phi = \exp(iW/\hbar)\Phi$. Then, ϕ_τ is a solution of (1) if and only if Φ_τ is the solution of the equation (the Lorentz gauge for A_μ is assumed)

$$i\hbar\partial_\tau\Phi = -\frac{\hbar^2}{2M}\Box_g\Phi - \frac{i\hbar}{M}g^{\mu\nu}(\partial_\mu W_\tau + A_\mu)\partial_\nu\Phi_\tau - \frac{i\hbar}{2M}\Box_g\Phi_\tau\,, \tag{4}$$

where $\Box_g$ is the wave operator on the pseudoriemannian manifold $\mathcal{M}$

$$\Box_g = g^{\mu\nu}\partial_\mu\partial_\nu + \frac{1}{2}g^{\nu\rho}\Gamma^\mu_{\nu\rho}\partial_\mu\,.$$

In the formal limit $\hbar \to 0$ of (4) we obtain $\Phi_\tau(x) \approx \Phi(\xi(\tau))$ where ξ is the solution of the equation $(0 \leq s \leq \tau)$

$$\frac{d\xi^\mu}{ds} = -\frac{1}{M}g^{\mu\nu}(\xi(s))\Big(\partial_\nu W(\tau - s, \xi(s)) + A_\nu(\xi(s))\Big)\,. \tag{5}$$

Differentiating (5) once more and using the Hamilton-Jacobi equation (3) we obtain the equation

$$\frac{d^2\xi^\mu}{ds^2} + \Gamma^\mu_{\nu\rho}\frac{d\xi^\nu}{ds}\frac{d\xi^\rho}{ds} = \frac{1}{M}F^{\mu\nu}(\xi(s))\frac{d\xi_\nu}{ds}\,, \tag{6}$$

where

$$F_{\mu\nu} = \partial_\mu A_\nu - \partial_\nu A_\mu\,.$$

Equation (6) shows that the correct classical limit (as a geodesic equation) of a motion of the quantum particle on a general pseudoriemannian manifold results if and only if τ is the proper time. The interpretation of τ as the classical proper time remains true not only in the leading order in $\hbar$ but also in all subsequent terms because the leading order determines the subsequent interpretation of τ.

The relativistic measurement problem was one of our motivations for a search of a new quantization method. The measurement can be considered as an interaction of a quantum particle with a macroscopic environment. Such an interaction can be described by a random wave function. The resulting quantum state is defined by a density matrix

$$\rho(x,y) = E[\psi(x)\overline{\psi}(y)]\,.$$

If we require that the trace and positivity of ρ are preserved by the time evolution then we obtain the Lindblad equation (Lindblad 1976) (Lindblad equation for the relativistic quantum mechanics has been discussed earlier in (Blanchard and Jadczyk 1996))

$$\partial_\tau\rho = -\frac{i}{\hbar}[H,\rho] - \frac{1}{2}\sum_k R_k^+R_k\rho - \frac{1}{2}\sum_k \rho R_k^+R_k + \sum_k R_k\rho R_k^+\,, \tag{7}$$

where R describes a (phenomenological) dissipation.

We can represent the dissipative dynamics (7) by the Ito stochastic wave equation (i.e., a random perturbation of (1))

$$d\psi = -\frac{i}{\hbar} H\psi d\tau + i\sum_k R_k \psi d\mathcal{B}_k - \frac{1}{2}\sum_k R_k^+ R_k \psi d\tau\,, \tag{8}$$

where $\mathcal{B}_k$ are independent complex Brownian motions

$$E[\mathcal{B}_k\mathcal{B}_r] = 0\,,$$

$$E[\overline{\mathcal{B}_l}(s,x)\mathcal{B}_k(\tau,y)] = \delta_{kl}\min(\tau,s)\delta(x-y)\,.$$

If the operators R are Hermitian then (8) can be expressed in a more compact form

$$d\psi = -\frac{i}{\hbar} H\psi d\tau + i\sum_k R_k\psi \circ d\tilde{B}_k\,, \tag{9}$$

where the circle denotes the Stratonovitch differential (Ikeda and Watanabe 1981) and $\tilde{B}_k$ are independent real Brownian motions.

At the end of this section let us note that (1) has the correct non-relativistic limit. The non-relativistic energy ϵ is related to the relativistic energy p_0, mass M and the momentum $\mathbf{p}$ by the formula

$$\epsilon = c\left(\mathbf{p}^2 + M^2c^2\right)^{1/2} - Mc^2 = \frac{\mathbf{p}^2}{2M} + o\left(\frac{1}{c}\right)\,.$$

Let (we write now $x_0 = ct$)

$$\psi_\tau = \exp\left(\frac{-iMc^2(\tau - 2t)}{2\hbar}\right)\tilde{\psi_\tau}\,.$$

Then, in the limit $c \to \infty$ we obtain (we write $A = (\mathbf{A}, V)$, i.e., $A_0 = V$)

$$i\hbar\partial_\tau\tilde{\psi_\tau} = -i\hbar\partial_t\tilde{\psi_\tau} + \frac{1}{2M}\left(\left(-i\hbar\nabla + \frac{1}{c}\mathbf{A}\right)^2\tilde{\psi_\tau} + V(\mathbf{x},t)\right)\tilde{\psi_\tau}\,. \tag{10}$$

If the potentials A and V are t-independent then we can express (10) as the Schrödinger equation with a new time $\hat{t} = t + \tau$. In such a case τ is just a global shift of time in the non-relativistic quantum mechanics. If the potentials are time-dependent then we obtain Howland's description of the time evolution in time-dependent potentials. In such a case t is treated as a coordinate on an equal footing with $\mathbf{x}$.

3 Quantum Fields

A relativistic wave equation does not describe the processes of particle creation and annihilation. A way to include such processes goes through a non-linear generalization of (1) and its quantization. We could approach the problem in a way similar to the standard quantization of the scalar field with the potential ϕ^4 when we first add a non-linear term to the Klein-Gordon equation and subsequently treat the fields as operators in the Fock space. Then, the evolution equation reads

$$\partial_\tau \phi = i\hbar(\Box - M^2)\phi - ig\hbar\phi^3 . \tag{11}$$

We are looking for a solution of this equation with an initial condition at $-\infty$

$$\lim_{\tau \to -\infty} \phi_\tau(x) = \phi_{in}(x) ,$$

where $(\Box - M^2)\phi_{in}=0$ is the wave equation for the quantum scalar free field of mass M (in the Fock space). We rewrite (11) as an integral equation

$$\phi_\tau(x) = \phi_{in}(x) - ig\hbar \int_{-\infty}^{\tau} \exp\Big(- i\hbar(\Box - M^2)(\tau - s)\Big)\phi_s^3 ds . \tag{12}$$

We solve (12) perturbatively. We assume that ϕ^3 is defined by the Wick normal product. Then, in the vacuum state $|0\rangle$ the Wick theorem allows us to calculate $\langle 0|T(\phi_\tau(x)\phi_{\tau'}(x'))|0\rangle$ where the T-ordering is understood in x_0 rather than in τ. In order to relate the τ-dependent quantum field theory to the conventional one we average the vacuum expectation values of the time-ordered products of ϕ_τ fields over τ. We have checked in the lowest order of perturbation theory that such averaging results in the standard time-ordered vacuum expectation values (if $(\Box - M^2)^{-1}$ is interpreted as the causal propagator).

We suggest here another method of quantization which resembles an interaction with the environment of (9). We introduce stochastic fluctuations corresponding to the Heisenberg uncertainty principle. We represent these fluctuations by the real independent Gaussian fields B_a with the covariance

$$E[B_a(\tau, x)B_c(s, y)] = \delta_{ac} \min(\tau, s)\delta(x - y) . \tag{13}$$

We consider an interaction among relativistic fields ϕ_a described by a general Lagrangian $\mathcal{L}(\phi_a)$. Its action integral is denoted $L(\phi_a)$. Then, a stochastic counterpart of (11) reads

$$d\phi_a(\tau, x) = i\hbar\frac{\delta L}{\delta\phi_a(\tau, x)}d\tau + \sqrt{2}\,\hbar dB_a(\tau, x) . \tag{14}$$

The solution of (14) depends on the proper time parameter τ. We suggest that correlations which are observed in experiments result from an average over the rapid oscillations in the proper time, i.e.,

$$\langle\phi(x_1)....\phi(x_k)\rangle = \lim_{T\to\infty}(T-\tau_0)^{-1}\int_{\tau_0}^{T} d\tau E[\phi(\tau,x_1).....\phi(\tau,x_k)]\,. \tag{15}$$

We can prove that (15) leads to the conventional time-ordered vacuum expectation values if (14) is solved with zero as the initial condition.

Let us discuss the quantization method first for the free field. If L corresponds to a free scalar field of mass M then (14) takes the form

$$d\phi(\tau,x) = \frac{i}{\hbar}(\Box - M^2)\phi(\tau,x)d\tau + \sqrt{2}\,\hbar dB(\tau,x)\,. \tag{16}$$

The solution of (16) is a sum of two pieces: the first part is the wave function (1) and the second one is a noise. We describe the quantum field by means of the random part. Its covariance is

$$\begin{aligned} &E[((\phi_\tau,f) - E[(\phi_\tau,f)])((\phi_{\tau'},f') - E[(\phi_{\tau'},f')])] \\ &\quad = \frac{\hbar^2}{2i}\Big(f,\mathcal{A}^{-1}\Big(\exp(-i\mathcal{A}|\tau-\tau'|) - \exp(-i(\tau+\tau'-2\tau_0)\mathcal{A})\Big)f'\Big)\,. \end{aligned} \tag{17}$$

where $\mathcal{A} = \hbar(\Box - M^2)$. Now, the average (15) can be calculated

$$\begin{aligned} &\lim_{T\to\infty}(T-\tau_0)^{-1}\int_{\tau_0}^{T} d\tau E[((\phi_\tau,f) - E[(\phi_\tau,f)])((\phi_\tau,f') - E[(\phi_\tau,f')])] \\ &\quad = \frac{\hbar^2}{2i}(f,\mathcal{A}^{-1}f') - \frac{\hbar^2}{4}\lim_{T\to\infty}(T-\tau_0)^{-1} \\ &\qquad \times (\mathcal{A}^{-1}f,\Big(\exp(-2i(T-\tau_0)\mathcal{A}) - 1\Big)\mathcal{A}^{-1}f')\,. \end{aligned} \tag{18}$$

The definition of the limit $T\to\infty$ is equivalent to a definition of $\mathcal{A}^{-1}$. We define the inverse by ($\epsilon > 0$)

$$\begin{aligned} &(\Box - M^2)^{-1}(x,y) \\ &\quad = \lim_{\epsilon\to 0}(2\pi)^{-d}\int dp\exp(-ip(x-y))(-p^2 - M^2 - i\epsilon)^{-1} \\ &\quad = \triangle_F(x-y)\,, \end{aligned} \tag{19}$$

where $\triangle_F$ denotes Feynman's causal function which is equal to the time-ordered vacuum expectation value of the real scalar free field

$$\langle 0|T(\phi(x)\phi(y))|0\rangle = i\hbar\triangle_F(x-y)\,. \tag{20}$$

It is important to note that if $(x-y)^2 \neq 0$ then the limit $T\to\infty$ in (15) holds true not only in a distributional sense but also pointwise. In general, for n-point correlation functions we can show that

$$\langle((\phi_\tau, f_1) - E[(\phi_\tau, f_1)])((\phi_\tau, f_2) - E[(\phi_\tau, f_2)]) \ldots$$

$$((\phi_\tau, f_{2n}) - E[(\phi_\tau, f_{2n})])\rangle = \hbar^{2n} \sum_{\text{pairs }(j,k)} \prod \frac{1}{2i}(f_j, \mathcal{A}^{-1} f_k)\,, \tag{21}$$

where the sum is over the product of all pairs (j, k) in agreement with the Gaussian integral combinatorics.

We can directly generalize (16) to an arbitrary pseudoriemannian manifold $\mathcal{M}$. In order to take the average over the proper time we need to define $(\Box_g - M^2)^{-1}$ (one usually adds the scalar curvature term $\frac{1}{6}R$ to $\Box_g$ but this is not necessary at this early stage). The kernel of this operator can be defined on a globally hyperbolic manifold (DeWitt 1975). In such a case there exists a complete set of solutions of the wave equation

$$(\Box_g - M^2)u_j = 0\,. \tag{22}$$

Then, we can define the two-point function of quantum fields

$$\Delta_g^{(+)}(x, y) = \sum_j \overline{u}_j(x) u_j(y)\,. \tag{23}$$

On a globally hyperbolic manifold there exists a choice of the x_0-coordinate (Kay 1980) such that

$$(\Box_g - M^2)^{-1}(x, y) \equiv \Delta_F(x, y)$$

$$= \theta(x_0 - y_0)\Delta_g^{(+)}(x, y) + \theta(y_0 - x_0)\Delta_g^{(+)}(y, x)\,, \tag{24}$$

is independent of the choice of coordinates. In such a way we can define quantum fields by the time-ordered expectation values starting from the stochastic equation (14).

Let us note that the averaged value (15) is formally expressed by the Feynman integral

$$\int d\phi \exp\left(\frac{i}{\hbar} L^{\text{free}}(\phi)\right) \phi(x_1) \ldots \phi(x_{2n}) = \langle\phi(x_1) \ldots \phi(x_{2n})\rangle^{\text{free}}\,,$$

here

$$L^{\text{free}}(\phi) = -\frac{1}{2} \int \phi(\Box - M^2)\phi\,.$$

We can prove such a result for a general Lagrangian L. We average over the proper time τ (as in (15)). Then, we express this average in terms of a measure ν. We can show in each order of the perturbation expansion that the corresponding (formal) measure is given by the Feynman formula

$$d\nu(\phi) = d\phi \exp\left(\frac{i}{\hbar} L(\phi)\right)\,,$$

where

$$L(\phi) = \int dx\mathcal{L}(\phi(x)) = -\int dx\Big(\frac{1}{2}\phi(\Box - M^2)\phi + gV(\phi(x))\Big)\,, \tag{25}$$

where V is an interaction potential.

We can add a dissipation to (14) in a way similar to that in (9) but now we admit a general non-linear function $R(\phi)$. We obtain a non-linear dissipative wave equation

$$d\phi = -i\hbar(\Box - M^2)\phi d\tau + ig\hbar V'(\phi) + iR(\phi) \circ d\tilde{B} + \sqrt{2}\,\hbar dB\,. \tag{26}$$

In such a case the corresponding Feynman measure reads

$$d\nu(\phi) = d\phi \,\exp\left(\frac{i}{\hbar}L_\gamma(\phi)\right)\,,$$

where

$$L_\gamma(\phi) = \int dx\mathcal{L}(\phi(x)) = \int dx\Big(-\frac{1}{2}\phi\Box\phi - gV(\phi(x)) + \frac{\gamma^2}{2}\mathcal{F}(\phi)\Big)\,, \tag{27}$$

where $\mathcal{F}$ is complex with a positive imaginary part.

The quantization method discussed here may be especially useful for a quantization of the Einstein gravity when we would like to treat the background metric together with the gravitons. The conventional Feynman integral is a formal tool which needs a proper definition. Methods developed so far fail when applied to Einstein gravity which has the action unbounded from below. An application of the stochastic quantization of Parisi and Wu (Parisi and Yong-Shi Wu 1981) (but with a complex action and real time) to the Einstein gravity has been suggested first by Rumpf (Rumpf 1986). We interpret the fictitious time of Parisi and Wu as the proper time with a physical meaning. Such an interpretation should have experimental consequences for a time evolution of the graviton wave function.

Acknowledgment

This contribution is dedicated to Professor Jan Łopuszański on the occasion of his 75th birthday. His mimeographed lecture notes on axiomatic quantum field theory happened to be my first encounter with the research of Wrocław physicists (before I became a student).

References

Blanchard, Ph., Jadczyk, A.Z., (1996): Found. Phys. **26**, 1669
Hansson, A.J., Regge, T., Teitelboim, C., (1976): Constrained Hamiltonian Systems, Accademia dei Lincei, Roma
Horwitz, L.P., Piron, C., (1973): Helv. Phys. Acta **46**, 316

Ikeda, N., Watanabe, S., (1981): Stochastic Differential Equations and Diffusion Processes, North Holland, Amsterdam
Kay, B.S., (1980): Comm. Math. Phys. **71**, 29
Lindblad, G., (1976): Comm. Math. Phys. **48**, 119
Parisi, G., Yong-Shi Wu, (1981): Sci. Sin. **24**, 483
Rumpf, H., (1986): in: "Fields and Geometry", XXII Winter School in Karpacz, Ed. A.Z. Jadczyk, World Scientific
Stueckelberg, E.C.G., (1941): Helv. Phys. Acta, **14**, 372; **14**, 588
DeWitt, B.S., (1975): Phys. Rep. **19C**, 295
Zurek, W.H., (1991): Physics Today **44**, No. 10, 36

Is It Possible to Construct Exactly Solvable Models?

O. Haschke and W. Rühl

Department of Physics,
University of Kaiserslautern, P.O.Box 3049
67653 Kaiserslautern, Germany

Dedicated to Professor Jan Łopuszański on the occasion of his 75-th birthday

Abstract. We develop a constructive method to derive exactly solvable quantum mechanical models of rational (Calogero) and trigonometric (Sutherland) type. This method starts from a linear algebra problem: finding eigenvectors of triangular finite matrices. These eigenvectors are transcribed into eigenfunctions of a selfadjoint Schrödinger operator. We prove the feasibility of our method by constructing an "AG_3 model" of trigonometric type (the rational case was known before from Wolfes 1975). Applying a Coxeter group analysis we prove its equivalence with the B_3 model. In order to better understand features of our construction we exhibit the F_4 rational model with our method.

1 Introduction

The completely integrable models are traditionally characterized by their relation with simple Lie algebras A_n, B_n, C_n, D_n, G_2, F_4, E_6, E_7, E_8. This relation is the starting point of the Hamiltonian reduction method exploited by Olshanetsky and Perelomov (Olshanetsky and Perelomov 1977, 1983). These models possess as limiting cases the trigonometric (Sutherland) and rational (Calogero) models that are exactly soluble, i.e., their eigenvalues and eigenvectors can be derived by elementary methods.

This exact solvability has been shown to follow from the fact that the Schrödinger operators can, after a "gauge transformation", be rewritten as a quadratic form of Lie algebra operators. These Lie algebra operators are represented as differential operators acting on polynomial spaces. This program was formulated in (Turbiner 1994, 1995) and successfully applied first to the A_n series in (Rühl and Turbiner 1995). Then it was carried over to the other sequences B_n, C_n, D_n and G_2 and even to corresponding supersymmetric models (Rosenbaum et al. 1997, Brink et al. 1998).

Our aim was to turn the arguments around and to develop an algorithm which may allow us to construct new exactly soluble models. First investigations were presented in (Haschke and Rühl 1999). The program contains two major and separate issues, to render a second order differential operator curvature free and to find a first order differential operator satisfying an integrability constraint. In this paper we present our algorithm in the following version. We start from a standard flat Laplacian and introduce Coxeter (or Weyl) group invariants as new coordinates. If the Coxeter group contains a symmetric group as subgroup, these invariants are built from elementary symmetric polynomials. The second order differential operators obtained this way are curvature free by construction, and act on polynomial spaces of these Coxeter invariants that form a flag. This flag is defined by means of a characteristic vector (**p**-vector).

Then we solve the integrability constraints by constructing "prepotentials" with a fixed algorithm. These prepotentials define the gauge transformation alluded to above which renders the differential operator the form of a standard Schrödinger operator of N particles in 1-dimensional space with a potential. Each prepotential contributes an additive term to this potential with a free (real) coupling constant. Finally the prepotentials define the ground state wave function of the Schrödinger operator which originates from the trivial polynomial in the flag and thus contains no further information. Except a possible oscillator prepotential in the translation invariant cases, the prepotentials are in one-to-one relation with the orbits of the Coxeter group.

We show that all known exactly soluble models can be obtained this way (at present we have to make an exemption with respect to E_6, E_7, E_8, but this will soon be overcome). Applying the method of constructing the Coxeter invariants of A_2 (Rosenbaum et al. 1997) to A_3, we obtain an "AG_3 model". Its Coxeter diagram is that of the affine Coxeter group $\hat{B}_3$, which possesses the same invariants as the Coxeter group B_3. This leads to an explicit proof of the equivalence of the AG_3 model with the B_3 model. Thus a translation invariant four–particle model after separation of the c.m. motion is shown to be equivalent with a translation non–invariant three-particle model. In this paper we also discuss F_4 from the view point of our algorithm. The Schrödinger operator obtained (only the rational case) deviates slightly from the one given in (Olshanetsky and Perelomov 1977, 1983) (probably due to a simple printing error in (Olshanetsky and Perelomov 1977, 1983)).

Thus our method shifts the centre of interest from the simple Lie algebras and their homogeneous spaces to the corresponding Weyl groups and by generalization to the Coxeter groups. On the other hand, the differential operators acting on polynomial spaces of Coxeter invariants define Lie algebras of their own, but at present these algebras are only of marginal interest.

2 The Constructive Program

We are interested here in the bound state spectrum of Schrödinger operators. The whole analysis is therefore performed in real spaces. Consider a flag of polynomial spaces $V_N(\mathbf{p})$, $N \in \mathbb{Z}_{\geq}$, $\mathbf{p} \in \mathbb{N}^n$

$$V_N(\mathbf{p}) = \text{span}\,\{z_1^{r_1} z_2^{r_2} ... z_n^{r_n} | r_1 p_1 + r_2 p_2 + ... + r_n p_n \leq N\} \,, \qquad (2.1)$$
$$(p_i \in \mathbb{N}) \,.$$

We consider differential operators of first order

$$D^{(1)}_{[\boldsymbol{\alpha};a]} = z^{[\boldsymbol{\alpha}]}\frac{\partial}{\partial z_a}\,, \tag{2.2}$$

($\boldsymbol{\alpha}$ a multi-exponent)
and of second order

$$D^{(2)}_{[\boldsymbol{\alpha};a,b]} = z^{[\boldsymbol{\alpha}]}\frac{\partial^2}{\partial z_a \partial z_b}\,, \tag{2.3}$$

that leave each space $V_N(\mathbf{p})$ invariant. If

$$\mathbf{p} = (1, 1, ..., 1)\,, \tag{2.4}$$

then the operators (2.2) generate the full linear (inhomogeneous) group of $\mathbb{R}_n$ and the operators of second order (2.3) can be obtained as products from the first order operators, i.e., in (2.2)

$$\boldsymbol{\alpha} = e^{(c)},\ e^{(c)}_b = \delta^c_b \quad \text{or} \quad \boldsymbol{\alpha} = 0\,, \tag{2.5}$$

and in (2.3)

$$\boldsymbol{\alpha} = e^{(c)} + e^{(d)} \quad \text{or} \quad \boldsymbol{\alpha} = \mathrm{e}^{(c)} \quad \text{or} \quad \boldsymbol{\alpha} = 0\,. \tag{2.6}$$

Now we consider a candidate for a future Schrödinger operator

$$D = -\sum_{\boldsymbol{\alpha},a,b} g_{[\boldsymbol{\alpha};a,b]} D^{(2)}_{[\boldsymbol{\alpha};a,b]} + \sum_{\boldsymbol{\beta},c} h_{[\boldsymbol{\beta};c]} D^{(1)}_{[\boldsymbol{\beta};c]}\,. \tag{2.7}$$

The eigenvectors and values of D in V_N can be calculated easily by finite linear algebra methods. Let

$$U_N = V_N / V_{N-1}\,, \tag{2.8}$$

and the diagonal part of D on U_N be defined as D_N

$$D_N U_N = D U_N \cap U_N\,. \tag{2.9}$$

If the eigenvalues of D_N are all different, the number of eigenvectors equals $\dim U_N$. But if some eigenvalues coincide (this is true in the generic case!) the number of eigenvectors is smaller. Then the Hilbert space on which the final selfadjoint Schrödinger operator is acting is not an L^2 -space. The missing eigenfunctions can be described. For more details see (Haschke and Rühl 1999).

If we want completely integrable models we must make sure that a complete set of involutive differential operators exists. For this task Lie algebraic methods may be very helpful.

Given a differential operator (2.7) one can characterize the vector $\mathbf{p}$ in (2.1) by inequalities

$$g_{[\boldsymbol{\alpha};a,b]} \neq 0 \Rightarrow \mathbf{p}\boldsymbol{\alpha} - p_a - p_b \leq 0\,, \tag{2.10}$$
$$h_{[\boldsymbol{\beta};c]} \neq 0 \Rightarrow \mathbf{p}\boldsymbol{\beta} - p_c \leq 0\,. \tag{2.11}$$

There should be enough equality signs in (2.10),(2.11) for a chosen $\mathbf{p}$ so that $D_N \neq 0$. It turns out that there exists a minimal $\mathbf{p}$-vector $\mathbf{p}_{\min}$ so that the $V_N(\mathbf{p}_{\min})$ spaces are maximal: For each $N, \mathbf{p}$ there is N' so that

$$V_N(\mathbf{p}) \subset V_{N'}(\mathbf{p}_{\min})\,. \tag{2.12}$$

It is convenient to work only with this minimal $\mathbf{p}$-vector.

The first step in transforming D into a Schrödinger operator is to write it symmetrically

$$D = -\sum_{a,b} \frac{\partial}{\partial z_a} g_{ab}^{-1}(z) \frac{\partial}{\partial z_b} + \sum_a r_a(z) \frac{\partial}{\partial z_a}\,, \tag{2.13}$$

where

$$g_{ab}^{-1} = \sum_{\boldsymbol{\alpha}} g_{[\boldsymbol{\alpha};a,b]} z^{[\boldsymbol{\alpha}]}\,. \tag{2.14}$$

We write g_{ab}^{-1} because this is the inverse of a Riemann tensor. The Riemann tensor g_{ab} is assumed to be curvature free. The task to make it so will not arise in this work. But we mention that we developed a minimal algorithm to solve this issue.

Following the notations of (Haschke and Rühl 1999) we "gauge" the polynomial eigenfunctions φ of D by

$$\psi(z) = e^{-\chi(z)} \varphi(z)\,, \tag{2.15}$$

so that

$$e^{-\chi} D e^{+\chi} = -\frac{1}{\sqrt{g}} \sum_{a,b} \frac{\partial}{\partial z_a} (\sqrt{g}\, g_{ab}^{-1}) \frac{\partial}{\partial z_b} + W(z)\,, \tag{2.16}$$

$(g = (\det g^{-1})^{-1})$.
This is possible if and only if

$$\sum_b g_{ab}^{-1}(z) \frac{\partial}{\partial z_b} \left[2\chi - \ln \sqrt{g}\right] = r_a(z)\,, \tag{2.17}$$

which implies integrability constraints on the functions $\{r_a(z)\}$. If they are fulfilled we obtain a "prepotential"

$$\rho = \ln P\,, \tag{2.18}$$

so that

$$\rho = 2\chi - \ln \sqrt{g}\,. \tag{2.19}$$

In most cases studied, we found solutions for ρ as follows. Let

$$\det g^{-1}(z) = \prod_{i=1}^{r} P_i(z)\,, \tag{2.20}$$

where $\{P_i(z)\}$ are different real polynomials. Then

$$\rho(z) = \sum_{i=1}^{r} \gamma_i \ln P_i(z) \tag{2.21}$$

with free parameters γ_i solves the requirement that $\{r_a(z)\}$ (2.17) belong to differential operators leaving each V_N invariant. In particular

$$r_a^{(i)}(z) = \frac{1}{P_i(z)} \sum_b g_{ab}^{-1}(z) \frac{\partial P_i}{\partial z_b} \tag{2.22}$$

are polynomials. Inserting (2.20), (2.21) in (2.19) we obtain finally

$$\chi = \frac{1}{2} \sum_{i=1}^{r} \left(\gamma_i - \frac{1}{2}\right) \ln P_i\,. \tag{2.23}$$

We will later see that in the case of the models of Calogero type a term

$$\gamma_0 \ln P_0 \tag{2.24}$$

can be added to ρ, where

$$P_0(z) = e^{z_1} \tag{2.25}$$

is not contained in $\det g^{-1}$ as a factor. This prepotential gives rise to the oscillator potential.

Finally we mention that $e^{-\chi}$ is the ground state wave function of the Schrödinger operator, as follows from (2.15).

The expression (Haschke and Rühl 1999), (6.17) for the potential $W(z)$ contains a term linear in χ

$$-\sum_{a,b} \frac{\partial}{\partial z_a} \left(g_{ab}^{-1} \frac{\partial \chi}{\partial z_b} \right) = -\frac{1}{2} \sum_{i=1}^{r} \left(\gamma_i - \frac{1}{2}\right) \sum_a \frac{\partial}{\partial z_a} r_a^{(i)}\,. \tag{2.26}$$

Each divergence

$$\sum_a \frac{\partial}{\partial z_a} r_a^{(i)}(z) = C^{(i)} \tag{2.27}$$

ought to be a constant. From now on we shall dismiss all constant terms in $W(z)$.

We can then write the potential as

$$W(z) = \sum_{i,j} \gamma_{ij} R_{ij}(z)\,, \tag{2.28}$$

$$R_{ij} = \sum_{a,b} g_{ab}^{-1} \frac{\partial \ln P_i}{\partial z_a} \frac{\partial \ln P_j}{\partial z_b}, \tag{2.29}$$

$$\gamma_{ij} = \frac{1}{4}\left(\gamma_i \gamma_j - \frac{1}{4}\right) \quad (i, j \neq 0). \tag{2.30}$$

In the cases of this article

$$R_{ij} = \text{const if } i \neq j. \tag{2.31}$$

If we then set

$$\gamma_i = -\nu_i + \frac{1}{2} \quad (i \neq 0), \tag{2.32}$$

we obtain

$$W(z) = \sum_{i=1}^{r} \gamma_{ii} R_{ii}(z) \tag{2.33}$$

with

$$\gamma_{ii} = \frac{1}{4}\nu_i(\nu_i - 1). \tag{2.34}$$

As stated in the Introduction the variables $\{z_i\}$ appearing in this section are identified with Coxeter invariants formed from root space coordinates $\{x_n\}$ or $\{y_n\}$. These invariants are either polynomial or trigonometric. Finally we return from the invariant coordinates $\{z_i\}$ to the root space coordinates $\{x_n\}$ in the Schrödinger operator (2.16). Each contribution

$$R_{ii} = \frac{Q_{ii}}{P_i} \tag{2.35}$$

admits a partial fraction decomposition due to the factorization of the prepotentials P_i (Section 5). The label $i = 1$ is always reserved to a "Vandermonde prepotential", i.e.,

$$P_1 \sim \prod_{i<j}(x_i - x_j)^2 \quad \text{or} \quad \prod_{i<j}(\sin(x_i - x_j))^2, \tag{2.36}$$

or alike.

3 Translation Invariant Models

3.1 Relative Coordinates

The Laplacian for an Euclidean space $\mathbb{R}_N$

$$\Delta = \sum_{i=1}^{N} \frac{\partial^2}{\partial x_i^2} \tag{3.1}$$

is translation invariant. We introduce relative coordinates by

$$y_i = x_i - \frac{1}{N} X \, , \tag{3.2}$$

$$X = \sum_{i=1}^{N} x_i \, . \tag{3.3}$$

They separate the Laplacian such that

$$\Delta = N \frac{\partial^2}{\partial X^2} + \sum_{i=1}^{N} \frac{\partial^2}{\partial y_i^2} - \frac{1}{N} \left(\sum_{i=1}^{N} \frac{\partial}{\partial y_i} \right)^2 . \tag{3.4}$$

We use all $\{y_i\}_{i=1}^N$ as coordinates on the plane

$$\sum_{i=1}^{N} y_i = 0 \, , \tag{3.5}$$

in order to maintain permutation symmetry.

3.2 Elementary Symmetric Polynomials

Elementary symmetric polynomials of N variables $\{q_i\}_{i=1}^N$ are defined by a generating function

$$\sum_{n=0}^{N} p_n(q) t^n = \prod_{i=1}^{N} (1 + q_i t) \, . \tag{3.6}$$

They are invariant under the symmetric group S_N. For each $g \in S_N$ we have a sector (simplex) $E_g \subset \mathbb{R}_N$

$$E_g = \{ q_{i_1} < q_{i_2} < \ldots < q_{i_N} ; \quad i_n = g(n) \} \, , \tag{3.7}$$

so that

$$\mathbb{R}_N = \bigcup_{g \in S_N} \bar{E}_g \, . \tag{3.8}$$

Inside E_g we can use the $\{p_n\}_{n=1}^N$ as coordinates since

$$\mathcal{M}_{ni} = \frac{\partial p_n}{\partial q_i} \, , \tag{3.9}$$

$$\det \mathcal{M} = (-1)^{[\frac{N}{2}]} V(q_1, q_2, \ldots q_N) \, , \tag{3.10}$$

where V is the Vandermonde determinant.

3.3 The A_{N-1} Series

The root system of A_{N-1} and the corresponding Weyl group possess elementary symmetric polynomials as invariants. We express the Laplacian in each sector E_g (3.7) intersected with the plane (3.5) in terms of these polynomials

$$\tau_n(y_1, ..., y_N) = p_n(q)|_{q_i = y_i \text{ all } i} \,. \tag{3.11}$$

The dynamics will be bounded to such sectors by corresponding potential walls automatically.

Then (see (Rühl and Turbiner 1995)) it results

$$\begin{aligned} &\sum_{i=1}^{N} \frac{\partial^2}{\partial y_i^2} - \frac{1}{N}\left(\sum_{i=1}^{N} \frac{\partial}{\partial y_i}\right)^2 \\ &= \sum_{n,m=2}^{N} g_{nm}^{-1} \frac{\partial^2}{\partial \tau_n \partial \tau_m} + \sum_{n=2}^{N} h_n \frac{\partial}{\partial \tau_n} \end{aligned} \tag{3.12}$$

with

$$g_{nm}^{-1}(\tau) = \frac{1}{N}(m-1)(N-n+1)\tau_n\tau_m - T_{n-1,m-1}(\tau) \,, \tag{3.13}$$

and

$$T_{nm}(\tau) = \sum_{l \geq 1}(2l + n - m)\tau_{n+l}\tau_{m-l} \,. \tag{3.14}$$

Here it is understood that

$$\begin{aligned} &\tau_0 = 1 \,, \\ &\tau_1 = 0 \,, \\ &\tau_n = 0 \text{ for } n < 0,\, n > N \,. \end{aligned} \tag{3.15}$$

In this case $\det g^{-1}$ is indecomposable as a polynomial, so we set

$$P_0 = e^{\omega \tau_2} \,, \tag{3.16}$$

$$P_1 = \det g^{-1} = C_N V(y_1, ..., y_N)^2 \,. \tag{3.17}$$

The resulting vectors $\{r_a\}_2^N$ are

$$r^{(0)} = (-2\tau_2, -3\tau_3, ..., -N\tau_N) \,, \tag{3.18}$$

$$r^{(1)} : \text{ explicit formulas known only for } N \leq 4 \,, \tag{3.19}$$

and the potential is

$$\frac{1}{2}W(x) = \frac{1}{2}\omega^2 \sum_{i=1}^{N} x_i^2 + g \sum_{1 \leq i < j \leq N} (x_i - x_j)^{-2} \,. \tag{3.20}$$

The corresponding Sutherland models are obtained as follows. We use as coordinates a system $\{\sigma_n\}_{n=2}^{N}$ defined by (these differ from those in (Rühl and Turbiner 1995))

$$\sigma_0 = \prod_{i=1}^{N} \cos y_i \,, \tag{3.21}$$

and

$$\sigma_n = \sigma_0 \cdot p_n(q)|_{q_i=\tan y_i} \,. \tag{3.22}$$

The identity

$$\begin{aligned} 1 &= \exp\left(i \sum_{j=1}^{N} y_j\right) \\ &= \prod_{j=1}^{N} (\cos y_j + i \sin y_j) \\ &= \sum_{n=0}^{N} i^n \sigma_n(y) \,, \end{aligned} \tag{3.23}$$

allows us to eliminate σ_0 and σ_1 in terms of the remaining $\{\sigma_n\}_{n=2}^{N}$ so that polynomials go into polynomials.

The Laplacian is expressed correspondingly as

$$\begin{aligned} &\sum_{i=1}^{N} \frac{\partial^2}{\partial y_i^2} - \frac{1}{N}\left(\sum_{i=1}^{N} \frac{\partial}{\partial y_i}\right)^2 = \\ &= \sum_{n,m=2}^{N} g_{nm}^{-1} \frac{\partial^2}{\partial \sigma_n \partial \sigma_m} + \sum_{n=2}^{N} h_n \frac{\partial}{\partial \sigma_n} \,, \end{aligned} \tag{3.24}$$

$$\begin{aligned} g_{nm}^{-1}(\sigma) = &-T_{n+1,m+1}(\sigma) - T_{n+1,m-1}(\sigma) \\ &-T_{n-1,m+1}(\sigma) - T_{n-1,m-1}(\sigma) \\ &+\frac{1}{N}[(m+1)\sigma_{m+1} + (m-1)\sigma_{m-1}] \\ &\times[(N-n-1)\sigma_{n+1} + (N-n+1)\sigma_{n-1}] \,, \end{aligned} \tag{3.25}$$

with T_{nm} as in (3.14).

Once again $\det g^{-1}$ is indecomposable, so we set

$$P_1 = \det g^{-1} = C_N' \tilde{V}(y_1, ..., y_N)^2 \,, \tag{3.26}$$

where

$$\tilde{V}(y_1, ..., y_N) = \prod_{i<j} \sin(y_i - y_j) \,, \tag{3.27}$$

has the symmetry of the Vandermonde determinant (translations and permutations). The vector $r^{(1)}$ is known only up to $N = 4$. Finally we obtain as potential

$$\frac{1}{2}W(x) = g \sum_{1 \le i < j \le N} \sin(x_i - x_j)^{-2} . \tag{3.28}$$

In each case A_{N-1} the minimal p-vector is $(1, 1, ..., 1) \in \mathbb{N}^{N-1}$.

3.4 The G_2 and AG_3 Models

The models G_2 and AG_3 belong also to the domain of translation invariant models (Rosenbaum et al. 1997). For G_2 we start from A_2 and extend its Weyl group by a $\mathbb{Z}_2$ group

$$y_i \to -y_i .$$

As invariant variables we use (Rosenbaum et al. 1997)

$$\lambda_2 = \tau_2 , \tag{3.29}$$

$$\lambda_3 = \tau_3^2 . \tag{3.30}$$

In these variables

$$\sum_{i=1}^{3} \frac{\partial^2}{\partial y_i^2} - \frac{1}{3}\left(\sum_{i=1}^{3} \frac{\partial}{\partial y_i}\right)^2 =$$

$$= \sum_{a,b=2}^{3} g_{ab}^{-1} \frac{\partial^2}{\partial \lambda_a \partial \lambda_b} + \sum_{a=2}^{3} h_a \frac{\partial}{\partial \lambda_a} . \tag{3.31}$$

We find

$$g^{-1}(\lambda) = \begin{pmatrix} -2\lambda_2, & -6\lambda_3 \\ -6\lambda_3, & +\frac{8}{3}\lambda_2^2\lambda_3 \end{pmatrix} , \tag{3.32}$$

so that

$$\det g^{-1} = -\frac{4}{3}\lambda_3(4\lambda_2^3 + 27\lambda_3) . \tag{3.33}$$

Thus as ansatz for the prepotentials we use

$$P_0 = e^{\omega \lambda_2} , \tag{3.34}$$

$$P_1 = 4\lambda_2^3 + 27\lambda_3 , \tag{3.35}$$

$$P_2 = \lambda_3 . \tag{3.36}$$

The r-vectors (justifying this ansatz) are

$$r^{(0)} = (-2\lambda_2, -6\lambda_3) , \tag{3.37}$$

$$r^{(1)} = (-6, 0) , \tag{3.38}$$

$$r^{(2)} = (-6, +\frac{8}{3}\lambda_2^2) . \tag{3.39}$$

The minimal **p**-vector is

$$\mathbf{p} = (1,2)\,. \tag{3.40}$$

The potential is

$$\frac{1}{2}W(x) = \frac{1}{2}\omega^2 \sum_{i=1}^{3} x_i^2 + g_1 \sum_{1\le i<j\le 3} (x_i - x_j)^{-2} + g_2 \sum_{i<j,k\notin(i,j)} (x_i + x_j - 2x_k)^{-2}\,, \tag{3.41}$$

with

$$\begin{aligned} g_1 &= \nu_1(\nu_1 - 1)\,,\\ g_2 &= 3\nu_2(\nu_2 - 1)\,. \end{aligned} \tag{3.42}$$

If

$$\nu_2 = 0 \text{ or } \nu_2 = 1\,, \tag{3.43}$$

we return to the A_2 model.

In the Sutherland case we use as variables

$$\mu_2 = \sigma_2\,, \tag{3.44}$$

$$\mu_3 = \sigma_3^2\,. \tag{3.45}$$

leading to the inverse Riemann tensor

$$g^{-1} = \begin{pmatrix} -2\mu_2 - 2\mu_2^2 + \frac{2}{3}\mu_3, & -\mu_3(6 + \frac{16}{3}\mu_2) \\ -\mu_3(6 + \frac{16}{3}\mu_2), & \frac{8}{3}\mu_2^2\mu_3 - 8\mu_3^2 \end{pmatrix}\,. \tag{3.46}$$

Now $\det g^{-1}$ is decomposable with

$$\det g^{-1} = -\frac{4}{3}\mu_3 P_1(\mu)\,, \tag{3.47}$$

and

$$P_1(\mu) = 4\mu_3^2 + \mu_3(8\mu_2^2 + 36\mu_2 + 27) + 4\mu_2^3(1 + \mu_2)\,, \tag{3.48}$$

$$P_2(\mu) = \mu_3\,. \tag{3.49}$$

The r-vectors are

$$r^{(1)} = (-6 - 8\mu_2, -16\mu_3)\,, \tag{3.50}$$

$$r^{(2)} = (-6 - \frac{16}{3}\mu_2, \frac{8}{3}\mu_2^2 - 16\mu_3)\,. \tag{3.51}$$

The resulting potential is

$$\frac{1}{2}W(x) = g_1 \sum_{1\le i<j\le 3} \sin(x_i - x_j)^{-2}$$
$$+\frac{1}{9}g_2 \sum_{i<j,k\notin(i,j)} \sin\frac{1}{3}(x_i + x_j - 2x_k)^{-2}\,. \qquad (3.52)$$

In the case of the A_2 models the spaces V_N decompose into even and odd subspaces in τ_3 (or σ_3) which are left invariant separately under action of the Laplacian. In the case of the odd spaces we can factor $\tau_3(\sigma_3)$ and leave an even space as well. In each case we obtain a polynomial space in the variables $\lambda_2, \lambda_3 = \tau_3^2$ $(\mu_2, \mu_3 = \sigma_3^2)$. Thus starting from such polynomial space and multiplying with $\tau_3^{\nu_2}$ $(\sigma_3^{\nu_2})$ we obtain the A_2 model if $\nu_2 = 0$ or $\nu_2 = 1$ but a new potential in all other cases.

It is plausible that a similar procedure works for A_3 but not for A_{N-1}, $N \geq 5$. In the latter models we have two or more odd variables $\tau_3, \tau_5, ...(\sigma_3, \sigma_5, ...)$ and there is no factorization of the odd invariant subspaces. Let us sketch the A_3 model whose extension leads to the AG_3 model (Haschke and Rühl 1998).

In this case the variables are chosen as in (3.29), (3.30), (3.44), (3.45)

$$\lambda_2 = \tau_2,\ \lambda_3 = \tau_3^2,\ \lambda_4 = \tau_4\,. \qquad (3.53)$$

The inverse Riemann tensor is

$$g^{-1} = \begin{pmatrix} -2\lambda_2, & -6\lambda_3, & -4\lambda_4 \\ -6\lambda_3, & 4\lambda_3(\lambda_2^2 - 4\lambda_4), & \lambda_2\lambda_3 \\ -4\lambda_4, & +\lambda_2\lambda_3, & -2\lambda_2\lambda_4 + \frac{3}{4}\lambda_3 \end{pmatrix}\,. \qquad (3.54)$$

The determinant is decomposable as

$$\det g^{-1} = \lambda_3 P_1(\lambda)\,, \qquad (3.55)$$

and the ansatz for the prepotentials is

$$P_0(\lambda) = e^{\omega\lambda_2}\,, \qquad (3.56)$$
$$P_1(\lambda) = 27\lambda_3^2 - 256\lambda_4^3 + 128\lambda_2^2\lambda_4^2 - 16\lambda_2^4\lambda_4 + 4\lambda_2^3\lambda_3 - 144\lambda_2\lambda_3\lambda_4\,, \qquad (3.57)$$
$$P_2(\lambda) = \lambda_3\,. \qquad (3.58)$$

The r-vectors come out as

$$r^{(0)} = (-2\lambda_2, -6\lambda_3, -4\lambda_4)\,, \qquad (3.59)$$
$$r^{(1)} = (-12, 0, -2\lambda_2)\,, \qquad (3.60)$$
$$r^{(2)} = (-6, 4(\lambda_2^2 - 4\lambda_4)\lambda_2)\,. \qquad (3.61)$$

The potential for this Calogero type model is

$$\frac{1}{2}W(x) = \frac{1}{2}\omega^2 \sum_{i=1}^{4} x_i^2 \qquad (3.62)$$
$$+g_1 \sum_{1\le i<j\le 4} (x_i - x_j)^{-2} + g_2 \sum_{3\text{ terms}} (x_i + x_j - x_k - x_l)^{-2},$$

with

$$g_1 = \nu_1(\nu_1 - 1),\ g_2 = 2\nu_2(\nu_2 - 1). \qquad (3.63)$$

It was discovered first by Wolfes, (Wolfes 1974).

The Sutherland model is obtained in the same fashion. With

$$\mu_2 = \sigma_2,\ \mu_3 = \sigma_3^2,\ \mu_4 = \sigma_4, \qquad (3.64)$$

the inverse Riemann tensor is

$$g_{22}^{-1} = -2\mu_2 - 2\mu_2^2 - 8\mu_4 + 2\mu_3 + 8\mu_2\mu_4 + 8\mu_4^2, \qquad (3.65)$$
$$g_{23}^{-1} = -6\mu_3 - 4\mu_2\mu_3, \qquad (3.66)$$
$$g_{24}^{-1} = -4\mu_4 - 6\mu_2\mu_4 + \mu_3 + 4\mu_4^2, \qquad (3.67)$$
$$g_{33}^{-1} = 4\mu_3[-4\mu_4 + \mu_2^2 - 4\mu_2\mu_4 + 4\mu_4^2 - 2\mu_3], \qquad (3.68)$$
$$g_{34}^{-1} = \mu_2\mu_3 - 6\mu_3\mu_4, \qquad (3.69)$$
$$g_{44}^{-1} = -2\mu_2\mu_4 + \frac{3}{4}\mu_3. \qquad (3.70)$$

Its determinant decomposes

$$\det g^{-1} = -\mu_3 P_1(\mu), \qquad (3.71)$$
$$P_1(\mu) = 256\mu_4^6 + 32 \text{ further terms}, \qquad (3.72)$$

(Eq. (A.2) from (Haschke and Rühl 1998))

$$P_2(\mu) = \mu_3 \qquad (3.73)$$

and the r-vectors are

$$r^{(1)} = (-16\mu_2 - 12, -24\mu_3, -12\mu_4 - 2\mu_2), \qquad (3.74)$$
$$r^{(2)} = (-4\mu_2 - 8, 16\mu_4^2 - 16\mu_4\mu_2 + 4\mu_2^2 - 8\mu_3 - 16\mu_4, -6\mu_4 + \mu_2). \qquad (3.75)$$

The factorization of σ_3 which is necessary in this case is

$$\sigma_3 = -\prod_{1\le i<j\le 3} \sin(y_i + y_j), \qquad (3.76)$$

implying

$$\frac{Q_{22}}{P_2} = 4 \sum_{1\le i<j\le 3} (\sin(y_i + y_j))^{-2}. \qquad (3.77)$$

This gives the potential

$$\frac{1}{2}W(x) = g_1 \sum_{1\le i<j\le 4} (\sin(x_i - x_j))^{-2}$$
$$+\frac{1}{4}g_2 \sum_{3\text{ cases}} (\sin\frac{1}{2}(x_i + x_j - x_k - x_l))^{-2}. \qquad (3.78)$$

The discussion of this AG_3 model is resumed in Section 5.

4 Translation Non-Invariant Models

4.1 The BC_N and D_N Models

As we shall see there is only one series with two (Calogero) and three (Sutherland) independent coupling constants. For any such model we use as Cartesian coordinates $\{x_i\}_{i=1}^N$ and require permutation symmetry S_N and reflection symmetry $(\mathbb{Z}_2)^N$ $x_i \to -x_i$ for each i separately. Then the natural coordinates invariant under these group actions are (Brink et al. 1998)

$$\lambda_n(x) = p_n(q)|_{q_i=x_i^2,\,\text{all}\,i}\,. \tag{4.1}$$

There is a bilinear relation with the $\{p_n(x)\}_{n=1}^N$

$$\lambda_n(x) = \sum_{k=0}^{2n}(-1)^{n-k}p_{2n-k}(x)p_k(x)\,. \tag{4.2}$$

The inverse Riemann tensor for the full Laplacian (3.1) is then

$$g^{-1}_{nm}(\lambda) = 4M_{nm}(\lambda)\,, \tag{4.3}$$

where we introduce the shorthand

$$M_{nm}(\lambda) = \sum_{l\geq 0}(2l+n-m+1)\lambda_{n+l}\lambda_{m-1-l}\,. \tag{4.4}$$

Its determinant factorizes

$$\det g^{-1} = (-1)^{[\frac{N}{2}]}4^N\lambda_N P_1(\lambda)\,, \tag{4.5}$$

where

$$\begin{aligned} P_1(\lambda) &= N^N\lambda_N^{N-1} + \ldots \\ &= D_N V(x_1^2, x_2^2, \ldots x_N^2)^2 \end{aligned} \tag{4.6}$$

and

$$P_2(\lambda) = \lambda_N\,. \tag{4.7}$$

Both functions P_1, P_2 factorize in a trivial way. In the general case there is no explicit expression for $r^{(1)}$ but

$$r_a^{(2)} = 4(N-a+1)\lambda_{a-1}\,. \tag{4.8}$$

If follows

$$R_{22} = 4\frac{\lambda_{N-1}}{\lambda_N} = 4\sum_{i=1}^{4}x_i^{-2}\,. \tag{4.9}$$

The resulting potential is, including an oscillator potential

$$\frac{1}{2}W(x) = \frac{1}{2}\omega^2 \sum_{i=1}^{N} x_i^2 + g_1 \sum_{1\le i<j\le N} [(x_i - x_j)^{-2} + (x_i + x_j)^{-2}]$$
$$+ g_2 \sum_{i=1}^{N} x_i^{-2} , \tag{4.10}$$
$$g_1 = \nu_1(\nu_1 - 1) , \tag{4.11}$$
$$g_2 = \frac{1}{2}\nu_2(\nu_2 - 1) . \tag{4.12}$$

In the Sutherland case we use coordinates

$$\mu_0 = \prod_{i=1}^{N} \cos^2 x_i , \tag{4.13}$$

$$\mu_n(x) = \mu_0(x) p_n(q)|_{q_i=\tan^2 x_i,\ \text{all}\ i} \qquad n \in \{1, 2, ...N\} . \tag{4.14}$$

From the identity

$$1 = \prod_{i=1}^{N} (\cos^2 x_i + \sin^2 x_i) = \sum_{n=0}^{N} \mu_n(x) \tag{4.15}$$

we learn how to eliminate μ_0 in facour of $\{\mu_n\}_{n_1}^N$ so that a polynomial of $\{\mu_n\}_{n=0}^N$ remains a polynomial.

In this case the inverse Riemannian is

$$g^{-1}_{nm} = 4\{M_{n+1,m+1}(\mu) + M_{n,m}(\mu) - M_{n,m+1}(\mu) - M_{n+1,m}(\mu)\} \tag{4.16}$$

and the determinant decomposes as

$$\det g^{-1} = 4^N (-1)^{[\frac{N}{2}]} \mu_0 \mu_N P_1(\mu) . \tag{4.17}$$

Now the factorization of $P_1(\mu)$ is

$$P_1(\mu) = D'_N \prod_{1\le i<j\le N} (\cos^2 x_i \sin^2 x_j - \sin^2 x_i \cos^2 x_j)^2 \tag{4.18}$$

and we choose

$$P_2(\mu) = \mu_N , \tag{4.19}$$
$$P_3(\mu) = \mu_0 . \tag{4.20}$$

Again we have no general explicit expression for $r^{(1)}$ but

$$r_a^{(2)} = 4[(N-a+1)\mu_{a-1} - (N-a)\mu_a]\,, \tag{4.21}$$
$$r_a^{(3)} = 4[(a+1)\mu_{a+1} - a\mu_a]\,, \tag{4.22}$$

so that

$$R_{22} = \frac{\mu_{N-1}}{\mu_N} = 4\sum_{i=1}^{N} \cot^2 x_i\,, \tag{4.23}$$

$$R_{33} = \frac{\mu_1}{\mu_0} = 4\sum_{i=1}^{N} \tan^2 x_i\,. \tag{4.24}$$

Thus we end up with a potential

$$\begin{aligned}\frac{1}{2}W(x) = g_1 &\sum_{1\le i<j\le N} [(\sin(x_i - x_j))^{-2} + (\sin(x_i + x_j))^{-2}] \\ &+ g_2 \sum_{i=1}^{N} (\sin x_i)^{-2} \\ &+ g_3 \sum_{i=1}^{N} (\cos x_i)^{-2}\,,\end{aligned} \tag{4.25}$$

where $g_{1,2}$ are as in (4.11),(4.12) and

$$g_3 = \frac{1}{2}\nu_3(\nu_3 - 1)\,. \tag{4.26}$$

An alternative form of the potential is obtained from

$$\frac{g_2}{\sin^2 x} + \frac{g_3}{\cos^2 x} = \frac{g_2 - g_3}{\sin^2 x} + \frac{4g_3}{\sin^2 2x}\,. \tag{4.27}$$

If we set $g_2 = g_3$ or $g_3 = 0$ we obtain different samples of the BC_N or D_N series. We mention finally that the minimal p-vector is in all cases

$$\mathbf{p} = (1, 1, \ldots 1) \in \mathbb{N}^N\,. \tag{4.28}$$

4.2 The F_4 Model

The F_4 model belongs also to the translation noninvariant class. The Weyl group of F_4 possesses four basic polynomial invariants

$$I_1(x),\ I_3(x),\ I_4(x), I_6(x)\,, \tag{4.29}$$

(I_n of degree $2n$) which can be expressed as polynomials in the $\{\lambda_n\}_{n=1}^4$ as follows

$$I_1 = \lambda_1 , \tag{4.30}$$

$$I_3 = \lambda_3 - \frac{1}{6}\lambda_1\lambda_2 , \tag{4.31}$$

$$I_4 = \lambda_4 - \frac{1}{4}\lambda_1\lambda_3 + \frac{1}{12}\lambda_2^2 , \tag{4.32}$$

$$I_6 = \lambda_4\lambda_2 - \frac{1}{36}\lambda_2^3 + \frac{1}{24}\lambda_2^2\lambda_1^2 - \frac{1}{64}\lambda_2\lambda_1^4 . \tag{4.33}$$

In these coordinates the inverse Riemannian can be given as

$$g_{1m}^{-1} = 4mI_m , \tag{4.34}$$

$$g_{33}^{-1} = \frac{20}{3}I_4I_1 - \frac{2}{3}I_3I_1^2 , \tag{4.35}$$

$$g_{34}^{-1} = 8I_6 - 3I_3^2 - \frac{13}{3}I_4I_1^2 - \frac{3}{4}I_3I_1^3 , \tag{4.36}$$

$$g_{36}^{-1} = 16I_4^2 + I_6I_1^2 + 14I_4I_3I_1 + \frac{5}{2}I_3^2I_1^2 - \frac{1}{4}I_4I_1^4 - \frac{5}{32}I_3I_1^5 , \tag{4.37}$$

$$g_{44}^{-1} = -4I_4I_3 - 2I_6I_1 + \frac{3}{4}I_4I_1^3 + \frac{3}{4}I_3^2I_1 + \frac{3}{16}I_3I_1^4 , \tag{4.38}$$

$$g_{46}^{-1} = 8I_4^2I_1 + 2I_4I_3I_1^2 - \frac{1}{8}I_4I_1^5 , \tag{4.39}$$

$$\begin{aligned} g_{66}^{-1} &= 30I_6I_4I_1 + \frac{21}{2}I_6I_3I_1^2 - \frac{3}{32}I_6I_1^5 + 12I_4^2I_3 + 6I_4I_3^2I_1 \\ &\quad -\frac{3}{8}I_4I_3I_1^4 + \frac{3}{4}I_3^3I_1^2 + \frac{3}{1024}I_3I_1^8 - \frac{3}{32}I_3^2I_1^5 . \end{aligned} \tag{4.40}$$

The determinant decomposes into two factors

$$\det g^{-1} = \frac{1}{3072}P_1(I)P_2(I) , \tag{4.41}$$

where $P_1(I)$ is connected with the Vandermonde determinant squared as usual

$$\begin{aligned} P_1(I) = &-4096I_4^3 + 432I_3^4 + 3072I_6^2 - 2304I_6I_4I_1^2 \\ &-576I_6I_3I_1^3 + 864I_4I_3^2I_1^2 + 216I_4I_3I_1^5 \\ &+432I_4^2I_1^4 + 27I_3^2I_1^6 - 2304I_6I_3^2 + 216I_3^3I_1^3 , \end{aligned} \tag{4.42}$$

or in factorized form

$$P_1(I) = -16 \prod_{1\le i<j\le 4} (x_i^2 - x_j^2)^2 , \tag{4.43}$$

and $P_2(I)$

$$\begin{aligned} P_2(I) = &\,36864I_6^2 - 18432I_6I_4I_1^2 - 4608I_6I_3I_1^3 + 32I_6I_1^6 \\ &-49152I_4^3 - 36864I_4^2I_3I_1 + 1536I_4^2I_1^4 \\ &+768I_4I_3I_1^5 - 12I_4I_1^8 - 9216I_4I_3^2I_1^2 \\ &-768I_3^3I_1^3 + 96I_3^2I_1^6 - 3I_3I_1^9 , \end{aligned} \tag{4.44}$$

which factorizes as

$$P_2(I) = -12\lambda_4(64\lambda_4 - 16\lambda_2^2 + 8\lambda_2\lambda_1^2 - \lambda_1^4)^2$$
$$= -12x_1^2x_2^2x_3^2x_4^2 \prod_{\nu_2,\nu_3\nu_4\in\{1,0\}} (x_1 - \sum_{i=2}^{4}(-1)^{\nu_i}x_i)^2 . \qquad (4.45)$$

The r-vectors are

$$r^{(1)} = (48, -2I_1^2, 0, 36I_4I_1 + 12I_3I_1^2 - \frac{3}{16}I_1^5) , \qquad (4.46)$$

$$r^{(2)} = (48, -4I_1^2, -12I_3, 24I_4I_1 + 6I_1^2I_3 - \frac{3}{8}I_1^5) . \qquad (4.47)$$

The potential resulting is

$$\frac{1}{2}W(x) = \frac{1}{2}\omega^2 \sum_{1\leq i\leq 4} x_i^2 + g_1 \sum_{1\leq i<j\leq 4} [(x_i - x_j)^{-2} + (x_i + x_j)^{-2}]$$
$$+g_2\{ \sum_{\substack{\nu_2,\nu_3,\nu_4\\ \in\{+1,0\}}} 4\left(x_1 - \sum_{i=2}^{4}\nu_i x_i\right)^{-2} + \sum_{i=1}^{4} x_i^{-2}\} , \qquad (4.48)$$

where $g_{1,2}$ are as in (4.11),(4.12). The minimal p-vector is

$$\mathbf{p} = (1, 2, 3, 5) . \qquad (4.49)$$

5 Coxeter Groups, Orbits and Prepotentials

The prepotentials used in the empirical constructions of sections 3 and 4 necessitate a mathematical interpretation. Let W be a Coxeter group generated by the reflections

$$\{s_\alpha\} , \qquad (5.1)$$

where α are roots running over a set

$$\Phi = \{\alpha\}_1^M . \qquad (5.2)$$

The roots span an Euclidean space V. In this space the reflections $\{s_\alpha\}$ act by

$$x \in V : s_\alpha x = x - 2\frac{(\alpha, x)}{(\alpha, \alpha)}\alpha . \qquad (5.3)$$

If the Coxeter group W is "crystallographic", it is a Weyl group (for more details see (Humphreys 1990)).

We denote a set of basic polynomial invariants of W by

$$\{z_1(x), \ldots, z_n(x)\}, \quad n = \dim V. \tag{5.4}$$

Invariance means

$$\begin{aligned} z_i(w^{-1}x) &= z_i(x) \\ &= wz_i(x), \end{aligned} \tag{5.5}$$

for all $w \in W$. The Jacobian for the transition $\{x_j\} \to \{z_i\}$

$$J = \det \left\{ \frac{\partial z_i}{\partial x_j} \right\} \tag{5.6}$$

can be factorized as follows ((Humphreys 1990), Proposition 3.13).

Each reflection s_α leaves a hyperplane H_α in V pointwise fixed, let H_α be given by a linear function l_α

$$l_\alpha(x) = 0. \tag{5.7}$$

Then due to the proposition

$$J = C \prod_{\alpha \in \Phi^+} l_\alpha(x) \tag{5.8}$$

with Φ^+ the set of positive roots. The proof of this proposition is rather elementary.

For any inverse Riemann tensor $\{g^{-1}\}$ of Sections 3 and 4 we obtain this way

$$\det g_{ab}^{-1} = C^2 \prod_{\alpha \in \Phi^+} l_\alpha(x)^2. \tag{5.9}$$

If Φ decomposes into orbits under W

$$\Phi = \bigcup_i \Phi_i, \tag{5.10}$$

then

$$P_i = \prod_{\alpha \in \Phi_i^+} l_\alpha(x)^2 \tag{5.11}$$

is an invariant polynomial under action of W and therefore a polynomial in the basic invariants

$$P_i = P_i(z_1, \ldots, z_n). \tag{5.12}$$

These polynomials are the prepotentials constructed in Sections 3 and 4. The factorization of these prepotentials as quoted at the end of Section 2 (eqs. (2.35),(2.36)) and used throughout in Sections 3 and 4 is based on (5.11).

We emphasize that our empirical results of Sections 3 and 4 indicate the validity of further mathematical propositions which could not be traced in the literature:

1. an analogous factorization theorem for the trigonometric invariants;
2. the polynomial properties ("integrability") of the functions $r^{(i)}(z)$ (2.22).

Now we return to the AG_3 model of Section 3. We identify the roots involved in a model using (5.7),(5.9)

$$l_\alpha(x) = (\alpha^\vee, x)$$
$$(\alpha^\vee = \frac{2\alpha}{(\alpha,\alpha)}, \text{the"dual"of}\alpha) \tag{5.13}$$

and the Sutherland version whose potential is

$$\frac{1}{2}W(x) = \sum_{\text{orbits}i} g_i \sum_{\alpha\in\Phi_i^+} [\sin l_\alpha(x)]^{-2} \,. \tag{5.14}$$

Thus the simple roots of A_3

$$\begin{aligned} \alpha_1 &= e_1 - e_2 \,, \\ \alpha_2 &= e_2 - e_3 \,, \\ \alpha_3 &= e_3 - e_4 \,, \end{aligned} \tag{5.15}$$

are completed by a fourth root in AG_3

$$\alpha_4 = e_3 + e_4 - e_1 - e_2 \,. \tag{5.16}$$

The corresponding Coxeter-diagram is shown in Fig. 1. It belongs to the affine Coxeter group $\hat{B}_3$ ((Humphreys 1990), Figure 1 in Section 2.4).

The coordinates of the $\hat{B}_3$ root space with respect to the standard basis $\{f_i\}_{i=1}^3$ are denoted $\{\xi_i\}_{i=1}^3$, those of AG_3 with respect to the standard basis $\{e_i\}_{i=1}^4$ by $\{x_i\}_{i=1}^4$ as before. The simple roots of B_3 are

$$\beta_1 = f_1 - f_2, \quad \beta_1 = f_2 - f_3, \quad \beta_3 = f_3 \,, \tag{5.17}$$

and $\hat{B}_3$ is obtained by adjoining

$$\beta_4 = -f_1 - f_2 \,. \tag{5.18}$$

It follows that

$$s_4 \begin{pmatrix} \xi_1 \\ \xi_2 \\ \xi_3 \end{pmatrix} = \begin{pmatrix} -\xi_2 \\ -\xi_1 \\ \xi_3 \end{pmatrix} \tag{5.19}$$

leaves the Coxeter invariants of B_3

$$\lambda_1(\xi) = \sum_{1\le i\le 3} \xi_i^2 \,, \tag{5.20}$$

$$\lambda_2(\xi) = \sum_{1\le i<j\le 3} \xi_i^2\xi_j^2 \,, \tag{5.21}$$

$$\lambda_3(\xi) = \xi_1^2\xi_2^2\xi_3^2 \,, \tag{5.22}$$

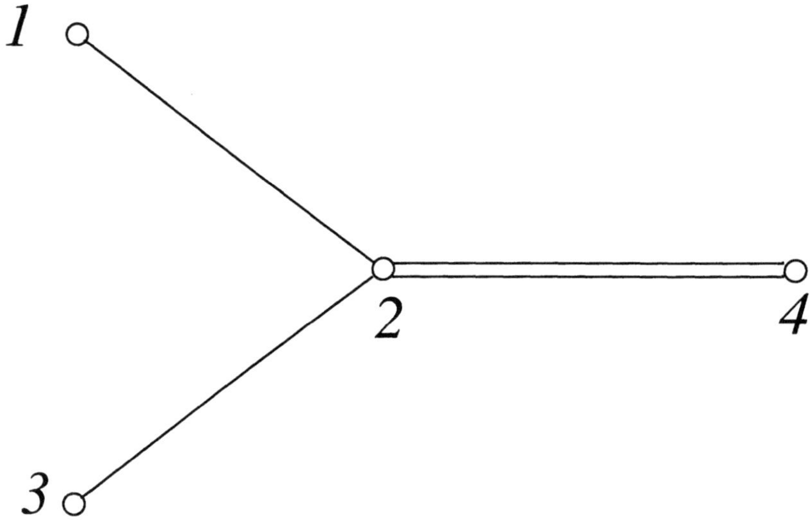

Fig. 1. Coxeter diagram of $\hat{B}_3$

invariant, too. This suggests the equivalence of the AG_3 and the B_3 models.

An explicit identification of the simple roots

$$f_1 = \frac{1}{2}(e_1 - e_2 - e_3 + e_4)\,, \tag{5.23}$$

$$f_2 = \frac{1}{2}(-e_1 + e_2 - e_3 + e_4)\,, \tag{5.24}$$

$$f_3 = \frac{1}{2}(-e_1 - e_2 + e_3 + e_4)\,, \tag{5.25}$$

gives $(i, j \in \{1, 2, 3\})$

$$x_i - x_j = \xi_i - \xi_j\,, \tag{5.26}$$

$$x_4 - x_j = \sum_{i(\neq j)} \xi_i\,. \tag{5.27}$$

It follows

$$g_1 \sum_{1 \le i < j \le 4} [\sin(x_i - x_j)]^{-2} + \frac{1}{4} g_2 \sum_{\text{3 cases}} [\sin \frac{1}{2}(x_i + x_j - x_k - x_l)]^{-2}$$
$$= g_1 \sum_{1 \le i < j \le 3} \{[\sin(\xi_i - \xi_j)]^{-2} + [\sin(\xi_i + \xi_j)]^{-2}\} + \frac{1}{4} g_2 \sum_{i=1}^{3} [\sin \xi_i]^{-2}\,. \tag{5.28}$$

Moreover the rational invariants (3.64) can be identified with the invariants (5.20)–(5.22)

$$\mu_2(x) = -\frac{1}{2}\lambda_1(\xi)\,, \tag{5.29}$$

$$\mu_3(x) = +\frac{1}{4}\lambda_3(\xi)\,, \tag{5.30}$$

$$\mu_4(x) = -\frac{1}{4}\lambda_2(\xi) + \frac{1}{16}\lambda_1(\xi)^2\,. \tag{5.31}$$

This establishes the equivalence between the two models.

Our method involves a reduction of the affine Coxeter group $\hat{B}_3$ to the Coxeter group B_3 having the same invariants. It may therefore be of interest that the construction performed in (Rosenbaum et al. 1997) is analogous (see Fig. 2).

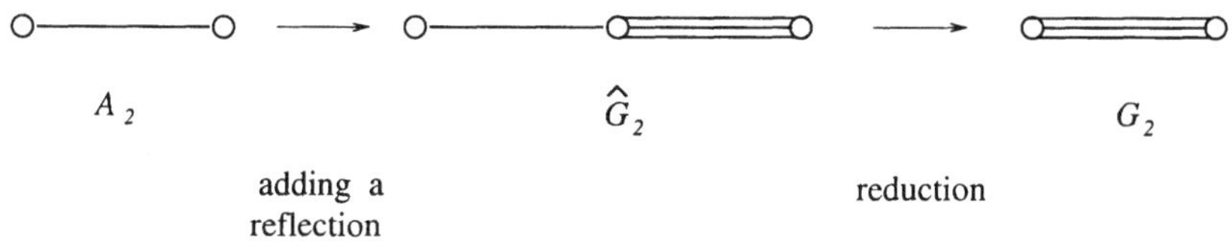

Fig. 2. Extending the Coxeter diagram of A_2 to $\hat{G}_2$ and reduction to G_2

References

Brink, L., Turbiner, A., Wyllard, N., (1998): J. Math. Phys. **39**, 1285 1285

Haschke, O., Rühl, W., (1998): Construction of exactly solvable quantum models of Calogero and Sutherland type with translation invariant four-particle interactions, (hep-th 9807194)

Haschke, O., Rühl, W., (1999): Exactly solvable dynamical systems in the neighborhood of the Calogero model, J. Mod. Phys. **A14**, 387; in Eq. (7.14) a_7 should read $a_7 - 3$

Humphreys, J.E., (1990): Reflection groups and Coxeter groups, Cambridge University Press 1990

Olshanetsky, M.A., Perelomov, A.M., (1977): Lett. Math. Phys. **2**, 7–13 and Phys. Reports **94**, (1983) 313

Rosenbaum, M., Turbiner, A., Capella, A., (1997): Solvability of the G_2 Integrable System; (sol-int 9707005)

Rühl, W., Turbiner, A., (1995): Mod. Phys. Letters **A10**, 2213 and (hep-th 9506105)

Turbiner, A.V., (1994): "Lie algebras and linear operators with invariant subspace", in "Lie algebras, cohomologies and new findings in quantum mechanics" (N.

Kamran and P.J. Olver eds.) AMS, Vol 160, pp. 263–310, 1994;
"Lie-algebras and quasi-exactly-solvable differential equations", Vol 3: New Trends in Theoretical Developments and Computational Methods, Chapter 12, CRC Press (N. Ibragimov ed.) (1995): pp. 331–366, (hep-th 9409068)
Wolfes, J., (1974): Ann. Phys. (N.Y.) **85**, 454

Summational Invariants

H. Reeh

Institut für Theoretische Physik
der Universität Göttingen
Bunsenstraße 9, D 37073 Göttingen

For Jan Łopuszański, scientist and friend,
on his 75th birthday

1 Introduction

In 1872 Ludwig Boltzmann derived the Maxwell distribution for the momenta of particles in a gas starting from the Boltzmann equation. The distribution function $F(\mathbf{p}), \mathbf{p} \in \mathbf{R}^3$, is stationary if and only if

$$F(\mathbf{p}_1)F(\mathbf{p}_2) = F(\mathbf{p}_3)F(\mathbf{p}_4)\,. \tag{1'}$$

Here $\mathbf{p}_1, \mathbf{p}_2$ are the momenta of two particles before a collision, $\mathbf{p}_3, \mathbf{p}_4$ the momenta of these after the collision, for which of course

$$\mathbf{p}_1 + \mathbf{p}_2 = \mathbf{p}_3 + \mathbf{p}_4, \qquad \text{and} \qquad \omega_1(p_1) + \omega_2(p_2) = \omega_3(p_3) + \omega_4(p_4)\,, \tag{2}$$

with $\omega_i(p) = p^2/2\mathbf{M}_j$ (or $\sqrt{p^2 + \mathbf{M}_j^2}$). Putting

$$f = \log F\,,$$

we have instead of (1')

$$f(\mathbf{p}_1) + f(\mathbf{p}_2) = f(\mathbf{p}_3) + f(\mathbf{p}_4)\,. \tag{1}$$

Functions f fulfilling (1) for scattering processes are called "summational invariants" or "additive collision invariants". The task of Boltzmann then was: *Find all solutions of eq.* (1) holding for all $\mathbf{p}_i$, *realizable in scattering processes.*

Related - but different - is to look for all solutions of (1) on the manifold given by (2).

If the restriction by the energy conservation is released, one has

$$f(\mathbf{p}_1) + f(\mathbf{p}_2) = f(\mathbf{p}_3) + f(\mathbf{p}_4)\,,$$

for

$$\mathbf{p}_1 + \mathbf{p}_2 = \mathbf{p}_3 + \mathbf{p}_4\,.$$

In the rest system of particle 2, $\mathbf{p}_2 = 0$, and putting

$$g(\mathbf{p}) = f(\mathbf{p}) - f(o), \quad \mathbf{p}_3 = \mathbf{p}, \quad \mathbf{p}_4 = \mathbf{q},$$

one gets

$$g(\mathbf{p}+\mathbf{q}) = g(\mathbf{p}) + g(\mathbf{q}), \qquad \forall \mathbf{p}, \mathbf{q} \in \mathbf{R}^3, \tag{3}$$

which is Cauchy's functional equation on $\mathbf{R}^3$.

Under rather mild smoothness conditions (e.g., g being continuous, or only Lebesgue measurable, or having a measurable majorant) it is well known that all solutions are of the form

$$g(\mathbf{p}) = \mathbf{ap},$$

with 3 constants $\mathbf{a}$.

There are, however, other solutions of (3). They were given by Georg Hamel in the same short paper (Hamel 1905) in which he introduced what since then is called a Hamel basis. He considers $\mathbf{R}$ as vector space over the field $\mathbf{Q}$ of the rationals: Every $x, x' \in \mathbf{R}$ may he expressed uniquely as

$$x = \alpha a + \beta b + \gamma c + \ldots, ; \quad \alpha, \beta, \gamma, \ldots \in \mathbf{Q}, \qquad a, b, c, \ldots \in \mathbf{B},$$

$$x' = \alpha' a + \beta' b + \gamma' c + \ldots,$$

where the sum is only over a finite number of elements of the basis B (The elements of B are linearly independent over $\mathbf{Q}$). Hamel defines $g(x)$ by

$$g(x) = \alpha g(a) + \beta g(b) + \gamma g(c) + \ldots,$$

where $g(a), g(b), \ldots$ are to be given (also allowed to be 0 or ∞). Then

$$g(x+x') = (\alpha + \alpha')g(a) + \ldots = g(x) + g(x').$$

Unless the quotients $g(a)/a$, $g(b)/b$, $g(c)/c, \ldots$ are all equal, we have $g(x) \neq$ const. x and $g(x)$ can not be continuous (in fact it is very discontinuous!)

After this digression we return to (1). There are investigations of (1) on the whole of the manifold of solutions of (2) or on a suitable submanifold (see (Amigo and Reeh 1983) for references) showing that

$$f(\mathbf{p}) = \mathbf{a} + \mathbf{bp} + \mathbf{c}_\omega(p),$$

under some smoothness assumptions (without the latter we might e.g., add an arbitrary Hamel type function $g(\omega(p))$!)

There is, however, the following problem: *If there is besides of energy momentum conservation another still unknown conservation law further restricting the manifold on which (1) is given, the result again might be different.* Therefore, from the point of view of a physicist an (unpublished) result of E. Wichmann for the relativistic case (Łopuszański 1991) is rather welcome:

Given (1) for one non-trivial scattering process and for all Poincaré transforms of this process, then $f = a + \mathbf{bp} + c\omega(p)$ (under suitable smoothness assumptions).

For proving this assertion, it is convenient to realize that there are reference systems for which $\mathbf{p}_1, \ldots, \mathbf{p}_4$ are in a plane (In the center of mass system in which $\mathbf{p}_1 = -\mathbf{p}_2$, $\mathbf{p}_3 = -\mathbf{p}_4$, $\mathbf{p}_1$, and $\mathbf{p}_3$ define a plane.)

2 Generalizations

In classical mechanics, an observable is a function on phase space, possibly with an explicit dependence on time, $F(\mathbf{x}, \mathbf{p}, t)$. It is "conserved" if

$$\frac{d}{dt}F(\mathbf{x}, \mathbf{p}, t) = 0,$$

on trajectories in phase space. The natural generalization of the notion of collision invariants therefore is:

Look for all asymptotically conserved observables (scattering invariants) which are additive on asymptotic particle configurations in all scattering process, i.e., look at

$$F(\mathbf{x}_1^{\text{in}}, \mathbf{p}_1^{\text{in}}, \mathbf{x}_2^{\text{in}}, \mathbf{p}_2^{\text{in}}, \ldots, t)$$
$$= \sum_i f_i(\mathbf{x}_i^{\text{in}}, \mathbf{p}_i^{\text{in}}, t) = \sum_i f_i(\mathbf{x}_i^{\text{out}}, \mathbf{p}_i^{\text{out}}, t) = F(\mathbf{x}_1^{\text{out}}, \mathbf{p}_1^{\text{out}}, \ldots, t),$$

with

$$\frac{d}{dt}f_i(\mathbf{x}(t), \mathbf{p}(t), t) = 0,$$

on asymptotic orbits.

From the last equation it follows that f_i may be written as

$$f_i(\mathbf{x}, \mathbf{p}, t) = g_i(\mathbf{s}, \mathbf{p}),$$

with

$$\mathbf{s} = \mathbf{M}_i\mathbf{x} - \mathbf{p}t, \quad \text{and} \qquad \mathbf{s} = \sqrt{\mathbf{p}^2 + \mathbf{M}_i^2}\,\mathbf{x} - \mathbf{p}t,$$

for Galilei and Lorentz invariance respectively. We therefore study the equation

$$\sum_{i=1,\ldots,4} f_i(\mathbf{p}_i, \mathbf{s}_i) = 0, \tag{4}$$

(with a temporary change of sign of f_3, f_4), where $\mathbf{p}_i$, $\mathbf{s}_i \in \mathbf{R}^3$ are the data of two particles before (i =1, 2) and after (i = 3, 4) a non trivial scattering process (i.e., the momenta change during collision!) or all the Galilei (or Poincaré) transformed of that process. In view of particle physics we allow that the particles change their identity and mass during collision. Our aim is to determine the general solution of (4).

As far as the smoothness of the f_i is concerned, it can be shown (by convolution with Galilei or Poincaré transformations[1] (Amigo and Reeh 1983) that it suffices to study C^∞ functions for getting the result for functions being locally integrable. Hence we assume that the f_i are C^∞ in $\mathbf{p}$ and $\mathbf{s}$ and polynomially bounded in $\mathbf{s}$ for all finite $\mathbf{p}$. Assume $\mathbf{M}_i > 0$. We consider

[1] For $M = 0, C^\infty$ follows only for $\mathbf{p} \neq 0$.

pointlike particles without intrinsic angular momentum. Because of conservation of angular momentum, we assume that the two particle scattering occurs in a plane. The result is

Theorem: Assume (4) and the assumptions just listed. Then $f_i(\mathbf{p},\mathbf{s}) = a^0\omega_i(p) + \mathbf{ap} + \mathbf{b}(\mathbf{s}\times\mathbf{p})/M_i + \mathbf{cs} + d(\mathbf{ps})/\mathbf{M}_i + e\mathbf{s}^2/M_i + k_i$ with $M_1 + M_2 = M_3 + M_4$ and with 12 constant numbers $a^0, \mathbf{a}, \ldots, e$ (which may vanish !) independent of i, and a number k_i $(k_1 + k_2 = k_3 + k_4)$ in the Galilei invariant case. Whereas $f_i = a^0\omega_i(p) + \mathbf{ap} + \mathbf{b}(\mathbf{s}\times\mathbf{p})/(p^2 + M_i)^{1/2} + \mathbf{cs} + k_i$ in the Lorentz invariant case with 10 + 1 constants $a^0, \ldots, \mathbf{c}$ and k_i.

With other words: The f_i are linear combinations of the 12 generators of the Schrödinger group (Niederer 1972, 1974) or the Poincaré group respectively.

The content of the theorem is a classical version of Coleman and Mandula's theorem of particle physics. t have studied the classical version partly because it might be easier to find counterexamples.

To prove the theorem, it is useful to show first (using the assumption of non trivial scattering) that the dependence of f_i on $\mathbf{s}$ is only polynomial (Here the assumption on polynomial boundedness is used. I do not actually know whether this is needed for the technique of proof only). Having that, one reduces the general case to Wichmann's result by applying small Galilei or Poincaré transformations on f_i (A small time translation, for instance, amounts to a differentiation of f_i with respect to the explicit t in $\mathbf{s}$. This reduces the degree in $\mathbf{s}$ and produces a factor $\mathbf{p}$!). For $\mathbf{M}_i > 0$ a proof is published (Mackrodt and Reeh 1997) A proof for massles particles (interpreted as limiting case) essentially is completed. Instead of going into details of the proof, I discuss two types of examples in the following.

3 Examples

The two transformations of the Schrödinger group besides of the Galilei group are

$$D : (t, \mathbf{x}) \to (e^{2\delta}t, e^{\delta}\mathbf{x}) , \qquad \text{“dilations”} , -\infty < \delta < \infty ,$$

$$A : (t, \mathbf{x}) \to (t' = t/(1+\alpha t) , \mathbf{x}' = \mathbf{x}/(1+\alpha t)) , \text{ “expansions”} , -\infty < \alpha < \infty .$$

According to the theorem, they may show up as summational invariants. They do show up, however, only in special cases, e.g., for systems with the action integral

$$W = \int \left\{ \sum_{i=1,\ldots,n} \frac{M_i}{2}(\dot{\mathbf{x}}_i)^2 - \sum_{i\neq j} \frac{\gamma_{ij}}{(\mathbf{x}_i - \mathbf{x}_j)^2} \right\} dt .$$

Invariance under D is obvious. Invariance under A follows after separation of a total time derivative under the integral,

$$dt' = (1+\alpha t)^{-2} dt \, ,$$

$$\left(\frac{dx}{dt}\right) 2dt' = \left(\dot{\mathbf{x}}^2 - \alpha \cdot \frac{d}{dt}\left[\mathbf{x}^2/(1+\alpha t)\right]\right) dt \, .$$

Constants of motion according to Noether's theorem for a two particle system are

$$A = \mathbf{s}_1^2/2M_1 + \mathbf{s}_2^2/2M_2 + t^2\gamma / \mid \mathbf{x}_1 - \mathbf{x}^2 \mid^2 ,$$

$$D = \mathbf{s}_1\mathbf{p}_1/M_1 + \mathbf{s}_2\mathbf{p}_2/M_2 - 2t\gamma / \mid \mathbf{x}_1 - \mathbf{x}_2 \mid^2 \, .$$

In D, the last non additive term vanishes for large t, i.e., on asymptotic trajectories. This is different for the non additive term in A. For $t \to -\infty$ we have

$$t^2 \cdot \gamma / \mid \mathbf{x}_1 - \mathbf{x}_2 \mid^2 \to \gamma / \mid \mathbf{v}_1 - \mathbf{v}_2 \mid^2 , \qquad (5)$$

where $\mathbf{v}_i$ denote the incoming asymptotic velocities of the particles. Energy conservation in the center of mass system implies that (5) is a separate collision invariant which, however, is not additive. We therefore may drop it and for the two particle system we are left indeed with what according to the theorem may show up as summational invariant.

For an n-particle system we have correspondingly the constants of motion

$$A = \sum_i \mathbf{s}_i^2/2M_i + t^2 \sum_{i \neq j} \gamma_{ij} / \mid \mathbf{x}_i - \mathbf{x}_j \mid^2 ,$$

$$D = \sum_i \mathbf{s}_i\mathbf{p}_i/M_i - 2t \sum_{i \neq j} \gamma_{ij} / \mid \mathbf{x}_i - \mathbf{x}_j \mid^2 ,$$

which for $t \to -\infty$ approach

$$A^{\mathrm{in}} = \sum (\mathbf{s}_i^{\mathrm{in}})^2/2M_i + \sum \gamma_{ij} / \mid \mathbf{v}_i^{\mathrm{in}} - \mathbf{v}_j^{\mathrm{in}} \mid^2 , \quad D^{\mathrm{in}} = \mathbf{s}_i^{\mathrm{in}}\mathbf{p}_1^{\mathrm{in}}/M_i \, .$$

In contrast to $n = 2$, the second, velocity dependent, term in A^{in} in general is not equal to the corresponding term in A^{out}. Hence here in general A is not asymptotically additive. For n $= 3$ and $\gamma_{12} = \gamma_{23} = \gamma_{31} = 1$ this may be seen either by numerical integration (Maison, private commun.) or as follows: Consider

$$F^{\mathrm{in}} : \frac{1}{\mid \mathbf{v}_1 - \mathbf{v}_2 \mid^2} + \frac{1}{\mid \mathbf{v}_2 - \mathbf{v}_3 \mid^2} + \frac{1}{\mid \mathbf{v}_3 - \mathbf{v}_1 \mid^2} \, .$$

Look at a situation in which particle 3 is far apart and does not change its velocity during collision. Go into a reference system in which $v_3 = v_3' = 0$, denoting the variables of the outgoing particles now with a prime. In this inertial system

$$F^{\text{in}} = \frac{1}{|\mathbf{v}_1-\mathbf{v}_2|^2} + \frac{1}{|\mathbf{v}_2|^2} + \frac{1}{|\mathbf{v}_1|^2}$$

$$= \frac{1}{2E-\mathbf{P}^2} + \frac{1}{|\mathbf{v}_1|^2} + \frac{1}{|\mathbf{v}_2|^2} = \frac{1}{2E-\mathbf{P}^2} + \frac{2E\cos^2\phi}{(\frac{1}{2}\mathbf{P}^2-E)^2}\,,$$

with $2E = \mathbf{v}_1^2 + \mathbf{v}_2^2$, $\mathbf{P} = \mathbf{v}_1 + \mathbf{v}_2$, $\mathbf{v}_1\mathbf{v}_2 = v_1 v_2 \cos\phi = 1/2\mathbf{P}^2 - E$. Now $\mathbf{P} = \mathbf{P}'$, $E = E'$, but in general $\phi \neq \phi'$ and $F^{\text{in}} \neq F^{\text{out}}$. (Look, e.g., at

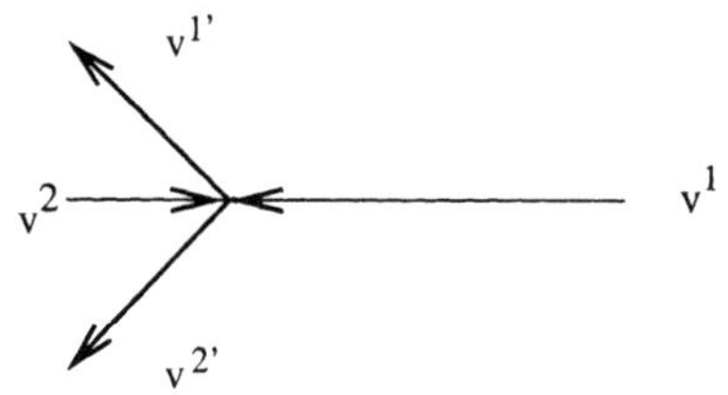

where $\phi = \pi$, $\phi' = \pi/2$.) The example shows that the situation is more complicated than the theorem and a superficial application of cluster properties might suggest.

It has been shown (Amigo and Reeh 1983) for two-particle systems with central symmetric interaction that conservation of the asymptotic D is related to the vanishing of the time delay during scattering and occurs only in case of potentials $\sim 1/r^2$. Examples for non central potential interaction are also known for which D is not accompanied by A (Baumann, K., private commun.). For the following example D and A are also conserved and asymptotically additive.

Let us now turn to an example where two-particle scattering is not restricted to a plane: Consider a particle (position $\mathbf{x}_1$) with electric charge e and another (at $\mathbf{x}_2$) with magnetic charge g, for simplicity both with mass 1: The non relativistic equations of motion are (putting $c = 1$)

$$\ddot{\mathbf{x}}_1 = eg \cdot (\dot{\mathbf{x}}_1 - \dot{\mathbf{x}}_2) \times (\mathbf{x}_1 - \mathbf{x}_2)/\mid \mathbf{x}_1 - \mathbf{x}_2 \mid^3,$$

$$\ddot{\mathbf{x}}_2 = -eg \cdot (\dot{\mathbf{x}}_2 - \dot{\mathbf{x}}_1) \times (\mathbf{x}_2 - \mathbf{x}_1)/\mid \mathbf{x}_2 - \mathbf{x}_1 \mid^3 .$$

(There is no canonical formulation without singularities corresponding to the "Dirac string" (Houard 1977)). The equations of motion are covariant under rotations and there is a cor- responding conservation law

$$\mathbf{J} = \mathbf{x}_1 \times \dot{\mathbf{x}}_1 + \mathbf{x}_2 \times \dot{\mathbf{x}}_2 + eg \cdot (\mathbf{x}_2 - \mathbf{x}_1)/\mid \mathbf{x}_2 - \mathbf{x}_1 \mid,$$

$$\frac{d}{dt}\mathbf{J} = 0\,,$$

by inspection. (The third term, of course, may be interpreted as the angular momentum of the electromagnetic field, $\int d^3x[\mathbf{x} \times (\mathbf{E} \times \mathbf{B})]$). Hence $\mathbf{J}$ is not asymptotically additive!

For an n-particle system of equal masses at $\mathbf{x_i}$, carrying electric and magnetic charges e_j, g_j we have correspondingly

$$\ddot{\mathbf{x}}_i = \sum_{k(\neq i)} (e_i g_k - e_k g_i)(\dot{\mathbf{x}}_i - \dot{\mathbf{x}}_k) \times (\mathbf{x}_i - \mathbf{x}_k) / \mid \mathbf{x}_i - \mathbf{x}_k \mid^3$$

$$+ \sum_{i(\neq k)} (g_i g_k + e_i e_k)(\mathbf{x}_i - \mathbf{x}_k)/|\mathbf{x}_i - \mathbf{x}_k|^3 ,$$

$$\mathbf{J} = \sum_i \mathbf{x}_i \times \dot{\mathbf{x}}_i + \sum_{i \neq k} g_i e_k \frac{\mathbf{x}_i - \mathbf{x}_k}{\mid \mathbf{x}_i - \mathbf{x}_k \mid} , \qquad \frac{d}{dt}\mathbf{J} = 0 . \tag{6}$$

$\mathbf{J}$ is a non asymptotically additive conservation law for the n-particle systems[2]

This example demonstrates that the assumption of a scattering plane for two point particles is essential for the theorem but can not, in general, be justified. The two particles seem to have sufficient clustering so that a scattering theory exists (Goldhaber 1965) because for large separation the interaction between a charge and a monopole vanishes faster than for Coulomb interaction due to the cross product.

The two types of examples at least show that there still is something to be done. I should like also to mention that for classical particles with intrinsic angular momentum like billiard balls there seem to be only partial results (Huber 1990). An investigation of collision invariants for systems of pointlike particles by a different method was given by M. Requardt (Requardt 1987).

References

Amigo, J.M. and Reeh, H., (1983): J. Math. Phys. **24**, 1544
Baumann, K., private communication
Goldhaber, Alfred S., (1965): Phys. Rev. **B140**, 1407
Hamel, G., (1905): Math. Ann. **60**, 459
Huber, W. (1990): J. Math. Phys. **31**, 1693
Houard, J.C., (1997): J. Math. Phys. **18**, 502
Łopuszański, Jan T., (1991): An Introduction to Symmetry and Supersymmetry in Quantum Field Theory, Singapore
Mackrodt, Ch. and Reeh, H., (1997): J. Math. Phys. **38**, 3012

[2] A classical model consisting of a magnetic monopole interacting with an electric dipole also be considered. It turns out, however, that there is no reasonable limit of vanishing length of the dipole with fixed dipole moment such that the intrinsic angular momentum could be neglected.

Maison, D., private communication
Niederer, U., (1972): Helv. Phys. Acta **45**, 802;
Helv. Phys. Acta **47**, (1974) 119 and 167
Requardt, M., (1987): J. Math. Phys. **28**, 1827

Relative Entropy Estimates in Statistical Mechanics and Field Theory

B. Zegarliński

Department of Mathematics,
Imperial College,
London, UK
b.zegarlinski@ic.ac.uk

*To Professor Jan Łopuszański
for his* 75th *birthday*

Abstract. We review numerous applications of relative entropy estimates in Statistical Mechanics and Field Theory.

The Particle Structure Implies Relative Entropy Bounds

It is well known that one can represent the physical Hilbert space $\mathcal{H}$ of the free scalar massive field theory as $\mathbb{L}_2(\mu_G)$ defined with a mean zero Gaussian measure of covariance $G = (-\Delta + m^2)^{-\frac{1}{2}}$. This Hilbert space has a natural Fock space structure

$$\mathcal{H} = \bigoplus_{n=0}^{\infty} \mathcal{H}_n$$

that is it can be represented as a direct sum of orthogonal n-particle subspaces $\mathcal{H}_n$ which are preserved by the semigroup $P_t \equiv e^{-tH}$, $t \geq 0$, where H denotes the physical Hamiltonian of the free field. At the end of sixties it has been discovered that this Particle Structure implies the following very special property of the semigroup P_t

$$||P_t f||_{\mathbb{L}_q} \leq ||f||_{\mathbb{L}_2} \tag{$\mathbb{H}$}$$

with $q \equiv q(t) = 1 + e^{2t/c}$ for some positive constant c. That means the semigroup is not only contractive in the physical Hilbert space, (which follows from the fact that physical Hamiltonian has non-negative spectrum), but maps this space into a strictly smaller subspaces consisting of more smooth vectors. Since then this property is called the Hypercontractivity. For the references to the related publications including those of J. Glimm, E. Nelson, B. Simon, R. Hoegh-Krohn and others, see e.g., (Simon 1974) and (Glimm and Jaffe 1987).

In (Federbush 1969), P. Federbush studied perturbation of free Hamiltonian by $\lambda : \varphi^4 :$ interaction. He has shown that the Hypercontractivity property implies the following infinitesimal condition, called Logarithmic Sobolev inequality,

$$\mu_G\left(f^2 \log f^2\right) \leq 2c\,\mu_G\left(fHf\right), \tag{$\mathbb{LS}$}$$

where f belongs to the quadratic form domain of the Hamiltonian and is normalised by $\mu_G f^2 = 1$. For later purposes we note that the quadratic form on the right hand side of ($\mathbb{LS}$) can be regarded as a Dirichlet form, that is it can be represented as an expectation of a square of (infinite dimensional) gradient, (Araki 1960), (Herbst 1976), (Albeverio and Hoegh-Krohn 1977). Later L. Gross showed, (Gross 1976), that actually this Relative Entropy bound is equivalent to the Hypercontractivity property.
We remark that, because of our normalisation condition, f^2 in the above inequality can be regarded as a probability density with respect to the measure μ_G and so the quantity on the left hand side of ($\mathbb{LS}$) can be interpreted as the relative entropy of the corresponding measures. Thus we see that in the Free Field Theory the Particle Structure implies the Relative Entropy bound.

1 The Relative Entropy Bounds for Gibbs Measures

One of the main properties of ($\mathbb{LS}$) is the fact that whenever it holds for any two measures, it is also true for their product. Thus such entropy bounds are naturally suitable for description of large or even infinite physical systems. At the time when ($\mathbb{LS}$) was introduced the only known examples of measures satisfying it were given by the Gaussian or some product measures. This situation persisted till the mid of eighties when Bakry and Emery introduced a very efficient criterion for the case when the underlying configuration space was given as $\Omega = \mathbb{M}^\Gamma$ with $\mathbb{M}$ being a Riemannian manifold with strictly positive Ricci curvature and Γ a countable set, (Bakry and Emery 1984). It has been applied in (Carlen and Stroock 1986) to show that ($\mathbb{LS}$) hold for infinite volume measures describing some continuous spin systems on a lattice at very high temperatures. A new idea which allowed to extend this result came at the end of eighties from the Statistical Mechanics (where some other relative entropy bounds proved to be a useful tool in the study of infinite systems). Studying a uniqueness problem for disordered spin systems, the author realised that one can use the Gibbs structure related to the spin systems to prove ($\mathbb{LS}$). It allowed him to show this relative entropy estimate not only for continuous spins when the single spin space do not satisfy $Ricc > 0$, (as for example in planar rotators), (Zegarliński 1990), but also for discrete spin systems, (Zegarliński 1990), (Zegarliński 1992). To describe the related idea and some results, we need to recall the basic notion of the Gibbsian description. We begin from introducing *the finite volume Gibbs measures*

$$\mu_\Lambda^\omega(f) \equiv \delta_\omega \left(\frac{\mu_0^\Lambda(e^{-U_\Lambda} f)}{\mu_0^\Lambda(e^{-U_\Lambda})} \right),$$

where δ_ω is the Dirac measure fixing the external configuration $\omega \in \Omega$ of the system outside a finite subset Λ of the lattice $\mathbb{Z}^d$, (denoted later on by $\Lambda \subset\subset \mathbb{Z}^d$); μ_0^Λ denotes a free product measure in Λ and the interaction energy is given by

$$U_\Lambda \equiv \sum_{X \cap \Lambda \neq \emptyset} \Phi_X(\sigma_X).$$

The potential $\Phi \equiv \{\Phi_X : X \subset\subset \mathbb{Z}^d, |X| < \infty\}$, for simplicity of the exposition, is assumed to be of finite range, that is $\Phi_X \equiv 0$ if $diam(X) > R$ for some fixed $R > 0$.
The infinite system is described by a Gibbs measure μ which by definition is a solution of the celebrated Dobrushin - Lanford - Ruelle equation

$$\mu\left(\mu_\Lambda^\cdot(f)\right) = \mu(f). \qquad (\mathbb{DLR})$$

Using this equation one can represent the relative entropy as follows

$$\mu\left(f^2 \log f^2/\mu f^2\right) = \mu\left(\mu_\Lambda^\cdot(f^2 \log f^2/\mu f^2)\right)$$

$$= \mu\left(\mu_\Lambda^\cdot(f^2)\log\frac{f^2}{\mu_\Lambda^\cdot(f^2)}\right) + \mu\left(\mu_\Lambda^\cdot(f^2)\log\left(\mu_\Lambda^\cdot(f^2)/\mu(\mu_\Lambda^\cdot(f^2))\right)\right).$$

In this way we split the estimate into two parts. The first one involves the local relative entropy estimate with the measure $\mu_\Lambda^\cdot$ and as a finite dimensional problem is usually easy. On the other hand the second term has a similar structure but involves a new density $\mu_\Lambda^\cdot(f^2)$ which is in some way smoother. Choosing another finite set, (given as a translation of Λ), we can apply the same idea to that second term. It is an interesting fact that under some mixing condition such procedure can be iterated and leads to convergent expansion which results with the desired relative entropy estimate.

2 Equivalence of Equilibrium and Non-Equilibrium Descriptions

An interesting outcome of the research on relative entropy estimates is contained in the following result.
Theorem: *The following conditions are equivalent*
(I) *Strong Mixing :* $\exists M > 0 \forall \Lambda \subset\subset \mathbb{Z}^d, \omega \in \Omega$

$$|\mu_\Lambda^\omega\left((f - \mu_\Lambda^\omega(f))(g - \mu_\Lambda^\omega(g))\right)| \leq C(f,g)e^{-M \cdot dist(\Lambda_f, \Lambda_g)}$$

for any local observable f and g localised in a bounded set Λ_f and Λ_g, respectively.

(II) *Spectral Gap :* $\exists m > 0 \forall \Lambda \subset\subset \mathbb{Z}^d, \omega \in \Omega$

$$m\,\mu_\Lambda^\omega(f-\mu_\Lambda^\omega(f))^2 \le \mu_\Lambda^\omega(\sum_i |\nabla_i f|^2)$$

for any f in the domain of the Dirichlet form.

(III) *Logarithmic Sobolev Inequality* : $\exists c>0\forall\Lambda\subset\subset\mathbb{Z}^d,\omega\in\Omega$

$$\mu_\Lambda^\omega\left(f^2\log f^2/\mu_\Lambda^\omega(f^2)\right) \le 2c\,\mu_\Lambda^\omega(\sum_i |\nabla_i f|^2)$$

for any f in the domain of the Dirichlet form.

(IV) *Asymptotic Sobolev Inequality* :
$\exists C>0\forall\Lambda\subset\subset\mathbb{Z}^d,\omega\in\Omega$, $\forall p\in[2,2+\frac{1}{|\Lambda|}]$

$$||f||^2_{\mathbb{L}_p(\mu_\Lambda^\omega)} \le ||f||^2_{\mathbb{L}_2(\mu_\Lambda^\omega)} + (p-2)C\mu_\Lambda^\omega(\sum_i |\nabla_i f|^2)$$

for any f in the domain of the Dirichlet form.

The equivalence of the (I)-(III) has been proven in (Stroock and Zegarliński 1992). The last point has been added only recently in (Zegarliński 1998).

The first condition is a statement from statistical mechanics which says that in the given systems one has a fast uniform decrease of correlations. It means, (Dobrushin and Shlosman 1985, 1987), that in such system one has no phase transitions in the strongest possible sense of analytic dependence of expectations on the potential.

The second has an interpretation within the spectral theory of the selfadjoint Markov generator described by the Dirichlet form on the right hand side of the inequality. It means that one has a gap at the bottom of the spectrum of this generator. Thus it carries an information about the ergodicity of the corresponding semigroup in the $\mathbb{L}_2$ sense.

The third statement is our relative entropy estimate. It is well known that ($\mathbb{LS}$) implies the spectral gap. On the other hand one can show examples when the spectral gap inequality is true, but ($\mathbb{LS}$) does not hold. The important point here is that the spectral gap is uniform with respect to the volume Λ and external conditions ω.

Finally the last property involves a classical Sobolev inequality which is one of the cornerstones of the twentieth century analysis. It tells us that given the $\mathbb{L}_2$ information about the gradient of a function we can estimate its $2+\delta$ moment with strictly positive delta δ. This improvement of square by a small root is much stronger than the logarithmic one in ($\mathbb{LS}$), but is relaxed in the thermodynamic limit. The equivalence of (III) and (IV) follows from the very special behaviour of the coefficient at the Dirichlet form.

The condition (I) is a condition of the equilibrium statistical mechanics. If we would think of (II) - (IV) as some features of dissipative dynamics with generator described by the corresponding Dirichlet form, we could say that the above Theorem establishes an equivalence between equilibrium and non-equilibrium description of a physical system.

3 Strong Decay to Equilibrium

One of interesting consequences of the hypercontractivity of the dissipative dynamics is the fact that the corresponding systems decays exponentially fast to the unique equilibrium.

Theorem: ((Stroock and Zegarliński 1995))
If a Gibbs measure μ *satisfies*

$$\mu\left(f^2 \log \frac{f^2}{\mu f^2}\right) \leq 2c\,\mu(fHf)$$

then

$$||e^{-tH}f - \mu f||_\infty \leq e^{-mt} C_V |||f|||$$

with a positive constant C_V *dependent only on a finite set* V *in which an observable* f *is localised, and with an arbitrary* $m \in (0, gap_2 H)$ *, provided some suitable seminorm* $|||f|||$ *of* f *is finite.*

It is interesting to remark that the region where ($\mathbb{LS}$) remains true, in many systems, extends to the critical point. (It includes for example the Ising ferromagnet with nearest neighbours interactions.)

Results about the decay to equilibrium, besides their theoretical and esthetical value, play an important role in numerical analysis. One should recall that the only practical way of making actual computations of equilibrium expectations of large system is via running a stochastic process on a computer. To illustrate the computational difficulty, consider the Ising model on ten by ten square of the integer lattice $\mathbb{Z}^2$. This certainly can not be regarded as a large system if we compared it with a small macroscopic piece of ferromagnet which is known to contain 10^{23} elements. Yet the corresponding configuration space contains $2^{100} \geq 10^{30}$ different configurations of the ± 1 spins. That is we would have to compute more than 10^{30} terms $\exp\{-U_\Lambda(\sigma)\}$ to get a value of expectation with a Gibbs measure. Computing 10^9 terms per second, (which is a very good speed !), one would need 10^{21} seconds. That is about 10^{14} years. Compare that with the age of the Universe which when estimated using the Big Bang theory is equal approximately 10^{10} years !

For related development see also (Martinelli and Olivieri 1994), (Lu and Yau 1993), (Zegarliński 1990)–(Zegarliński 1998) and references there in.

4 Strong Decay to Equilibrium in Disordered Systems

It is well known that the systems with random interactions can exhibit an interesting behaviour. A simplest example of such a system is given by the Edwards - Anderson model described by the following interaction energy

$$U = \sum_{|i-j|=1} J_{ij}\sigma_i\sigma_j \,,$$

where the couplings J_{ij} are random i.i.d. variable. If the couplings can take on arbitrary large values, the corresponding system exhibits a non-analytic behaviour even at the high temperature region. In view of the previous discussion it is natural to expect that one should have also different non-equilibrium behaviour. The first numerical evidence that at the high temperatures one should have a stretched exponential decay has been published in the mid of eighties by Ogielski, (Ogielski 1985). For a mathematical results one needed to wait for a long time. By adapting the strategy based on the hypercontractivity the following result has been proven for Glauber dynamics of two dimensional models.

Theorem ((Guionnet and Zegarliński 1996, 1997))
Almost surely

$$||e^{t\mathcal{L}_J} f - \mu_J f||_\infty \leq e^{-t^\alpha} C(J)|||f|||$$

with some $\alpha \in (0,1)$ *and a random variable* $C(J)$, *where* J *denotes the random configuration of the couplings.*

See also (Cesi et al. 1997) for further development on that subject.

5 The Relative Entropy Estimates in Quantum Systems

In the description of quantum spin systems we describe observables as elements of a C^* algebra $\mathcal{A} = \overline{\cup_{\Lambda\subset\subset\mathbb{Z}^d}\mathcal{A}_\Lambda}$, where $\mathcal{A}_\Lambda$ is isomorphic to $\mathbb{M}^\Lambda$ the algebra of all complex $n \times n$ matrices. The free state is given by a normalised trace Tr, satisfying the usual properties

$$Tr\mathbf{1} = 1, \qquad Tr(a^*a) > 0 \text{ for } a \neq 0 \text{ and } \qquad Tr(ab) = Tr(ba)\,.$$

With the trace Tr we can associate a family of partial traces Tr_X, $X \subset\subset \mathbb{Z}^d$ possessing all basic properties of conditional expectations. A Gibbs state ω on the algebra $\mathcal{A}$ is given by

$$\omega_{|\mathcal{A}_\Lambda}(f) = Tr(\rho_\Lambda f)\,,$$

where ρ_Λ is a density matrix. In this setting a dissipative dynamics is described as a Markov semigroup, that is a semigroup satisfying the following properties

$$P_t\mathbf{1} = \mathbf{1}, \qquad P_t f^* f \geq 0,$$

and possibly also the following *Feller Property*

$$P_t(\mathcal{A}) \subset \mathcal{A}\,,$$

Frequently we want to distinguish a'priori a family of invariant states that is the states satisfying

$$\omega(P_t f) = \omega(f) .$$

One convenient way of doing that is by assuming the following detailed balance condition with respect a scalar product associated to the state ω

$$< P_t f, g >_{\mathcal{H}_\omega} = < f, P_t g >_{\mathcal{H}_\omega} .$$

A construction of a dissipative dynamics preserving positivity in the algebra and simultaneously satisfying this symmetry condition constitutes one of the toughest problems of mathematical physics; for some progress in that direction see (Majewski et al. 1998) and references therein.
We recall that, unlike as in the classical case, in the non-commutative theory one can consider many scalar products associated to a given state $\omega = Tr(\rho \cdot)$. Some examples are given by

$$< f, g >_{\omega,s} \equiv Tr(\rho^{s/2} f \rho^{(1-s)/2})^* (\rho^{s/2} f \rho^{(1-s)/2}) .$$

In particular if we set $s = 0$ one gets the usual scalar product used in the GNS construction. An integral over $s \in [0,1]$ gives a scalar product relevant to the linear response theory.
Additionally one can associate to ω an interpolating family of $\mathbb{L}_p(\omega, s)$ spaces defined by the following norms

$$||f||^p_{\mathbb{L}_p(\omega,s)} \equiv Tr|\rho^{s/p} f \rho^{(1-s)/p}|^p .$$

Note that for $1 \leq p \leq q \leq \infty$ we have

$$\mathbb{L}_p \supset \mathbb{L}_q \supset \mathcal{A} .$$

We note that a symmetric in $\mathbb{L}_2(\omega, s)$ Feller - Markov semigroup can be extended to a contractive semigroup in all $\mathbb{L}_p(\omega, s)$

$$||P_t f||_{\mathbb{L}_p(\omega,s)} \equiv ||f||_{\mathbb{L}_p(\omega,s)}$$

in a full analogy to the classical theory.

6 Hypercontractivity in Noncommutative $\mathbb{L}_p$ Spaces

Given a family of noncommutative $\mathbb{L}_p(\omega, s)$ spaces, in a natural way we can define a Hypercontractive semigroup by the following condition

$$||P_t f||_{\mathbb{L}_p(\omega,s)} \leq ||f||_{\mathbb{L}_2(\omega,s)} ,$$

where $p = 1 + e^{\frac{2}{c}t}$, with some $c \in (0, \infty)$. Later on $s = \frac{1}{2}$.
Theorem: ((Olkiewicz and Zegarliński 1999))
Hypercontractivity in $\mathbb{L}_q(\omega, \frac{1}{2})$ spaces implies the following Quantum Relative Entropy bound

$$QE_p(f) \equiv Tr|\rho^{1/2p} f \rho^{1/2p}|^p \left(\log|\rho^{1/2p} f \rho^{1/2p}| - 1/p \log\rho\right)$$
$$-||f||^p_{\mathbb{L}_p(\omega,\frac{1}{2})} \log ||f||_{\mathbb{L}_p(\omega,\frac{1}{2})}$$
$$\leq c(p)\, \mathcal{E}_p(f)\,,$$

where $c(p) = \frac{cp}{2(p-1)}$ *and*

$$\mathcal{E}_p(f) = < I_{p,q}(f), \mathcal{L}_p f >_{\mathbb{L}_2(\omega,\frac{1}{2})}$$

with isometry $I_{p,q} : \mathbb{L}_p \to \mathbb{L}_q$.
Moreover if

$$\mathcal{E}_2(I_{2,p} f) \leq \frac{q^2}{4(q-1)} \mathcal{E}_q(f)$$

then

$$QE_2(f) \leq c\,\mathcal{E}_2(f)$$

implies hypercontractivity
The statement simply says that basically the quantum relative entropy estimate is equivalent to hypercontractivity in this more general noncommutative setting. Note that the theorem introduces a new kind of quantum relative entropy not considered before in the literature. In the particular case when the observable f is nonnegative and commutes with the density matrix ρ, the renormalization of the logarithm gives us the classical formula for the relative entropy.

7 Spectral Theory of Hypercontractive Semigroups

Suppose $P_t = e^{-t\mathcal{L}}$ is a symmetric Markov semigroup in $\mathbb{L}_2(\mu)$. If its generator would have a discrete spectrum, (as it happens for example in case of the Laplace - Beltrami operator on a compact Riemannian manifold), we would have the following representation

$$P_t f = \sum_n e^{-t\lambda_n} (\Psi_n, f) \Psi_n$$

with Ψ_n being a normalised eigenfunction corresponding to an eigenvalue λ_n. Using this representation an equivalent condition of the hypercontractivity property

$$\exists T \in (0,\infty) \forall t > T\,, \qquad ||P_t f||^4_{\mathbb{L}_4} \leq ||f||^4_{\mathbb{L}_2}$$

can be written as follows

$$\sum_{n_1,..,n_4} e^{-t(\lambda_{n_1}+..+\lambda_{n_4})} \mu(\Psi_{n_1}..\Psi_{n_4}) \prod_{l=1,..,4} (\Psi_{n_l}, f) \leq \sum_{n_1,n_2} (\Psi_{n_1}, f)^2 (\Psi_{n_2}, f)^2$$

for all $t > T$. This means that for hypercontractivity to be true we need very special properties of the spectrum and overlapping property of the eigenfuctions, (that is the random variables Ψ_n have to be in some sense weakly dependent and behave similarly to the random variables with Gaussian distribution). As we have mentioned at the beginning of this lecture, in case of free scalar massive field one can derive hypercontractivity from the particle structure of the theory. More precisely one uses the following properties.
(I) *Existence of Invariant Subspaces* $\quad \forall n \in \mathbb{Z}^+ \exists \mathcal{H}_n \subset \mathbb{L}_2(\mu)$

$$P_t\mathcal{H}_n \subset \mathcal{H}_n, \qquad \mathcal{H}_n \perp \mathcal{H}_{n'}, \qquad \text{and } \cup_n \mathcal{H}_n = \mathbb{L}_2(\mu)\,.$$

(II) *Particle Structure of the Spectrum*

$$\exists \varepsilon \in (0,\infty) \forall n \in \mathbb{N} \qquad \inf \sigma(H|\mathcal{H}_n) \geq n\varepsilon\,.$$

(III) *Gaussian Bounds*

$$\exists C \in (0,\infty) \forall n \in \mathbb{N} \quad \forall f \in \mathcal{H}_n \qquad ||f||_{\mathbb{L}_4(\mu)} \leq C^n ||f||_{\mathbb{L}_2(\mu)}\,.$$

We mention that recently the following further examples of such structure has been exhibited, (Bodineau and Zegarliński 1998).
Example A: The Glauber dynamics in $D = 1$ Ising model
– $\mathcal{H}_n = \overline{Span\{\sigma_X, |X| = n\}}$
– $\sigma(H|\mathcal{H}_n) \subset [\eta_- n, \eta_+ n]$, *with some constants* $0 < \eta_- < \eta_+ < \infty$, *(Minlos and Trishch 1994)*
– *Gaussian Bounds*

$$\exists C \in (0,\infty) \forall n \in \mathbb{N} \quad \forall f \in \overline{Span\{\sigma_X, |X| = n\}} \qquad ||f||_{\mathbb{L}_4(\mu)} \leq C^n ||f||_{\mathbb{L}_2(\mu)}\,,$$

where μ is the infinite volume Gibbs measure of the model.
This method is a bit simpler than the one used in (Zegarliński 1990), but also offers more precise estimates on the Logarithmic Sobolev coefficient c.
Example B: The Free Dynamics for Quantum Spin Systems
Let $\omega \equiv \otimes_i \omega_{\Lambda_i}$, where $\Lambda_i \cap \Lambda_j = \emptyset$ for $i \neq j$,ω_Λ is a state on $\mathbb{M}^\Lambda$. Then the Logarithmic Sobolev inequality holds with generator

$$Hf = \sum_i (f - \omega_{\Lambda_i} f)\,.$$

8 A Problem

The problem of the particle structure of a physical theory is one of the important problems which still remain weakly understood. Some partial results, (see references in (Glimm and Jaffe 1987)), show that in the two dimensional models of scalar fields with polynomial interactions one has the particle structure up to a level N provided the coupling constant $\lambda = \lambda(N) > 0$ is sufficiently small. Similar structure has been proven to exist for generators of Glauber dynamics in classical spin systems on the lattice in an interesting paper (Minlos 1996).
Taking into account the progress made in the last decade in understanding the relative entropy estimates and our discussion presented in this lecture, it would be very interesting to prove that under some reasonable general conditions present in the physical models, one has the following implication

$$\textbf{Relative Entropy Bound} \quad \Longrightarrow \quad \textbf{Particle Structure}$$

Acknowledgment

The author would like to thank the organisers for this interesting and memorable meeting and is pleased to acknowledge the support of British Council. The research described in this paper has been supported by EPSRC Grants GR/K95635, GR/L58873 GR/K76801 and EU Stochastic Analysis Network.

References

Araki, H., (1960): *Hamiltonian formalisms and the canonical commutation relations in quantum field theory*, J. Math. Phys. **1**, 492–504

Albeverio, S., Hoegh-Krohn, R., (1977): *Dirichlet forms and diffusion processes on rigged Hilbert spaces*, Z. Wahr. und verw. Geb. **40**, 1–57

Bakry, D., Emery, M., (1984): *Hypercontractivitè de semi - groupes des diffusion*, C.R. Acad. Sci. Paris Ser. **I 299** pp. 775–777; *Diffusions hypercontractives*, pp 177–206 in Sem. de Probabilites XIX, Azema J. and Yor M. (eds.), **LNM 1123**

Bodineau, T., Zegarliński, B., (1998): *Hypercontractivity via Spectral Theory*, IC Preprint

Carlen, E.A., Stroock, D.W., (1986): *An application of the Bakry - Emery criterion to infinite dimensional diffusions*, Sem.de Probabilites XX, Azema J. and Yor M. (eds.), LNM 1204, pp, 341–348

Cesi, F., Maes, C., Martinelli, F., (1997): *Relaxation to equilibrium for two dimensional disordered Ising systems in the Griffiths phase*, Commun. Math. Phys. **189**, 323–335

Dobrushin, R.L., Shlosman, S., (1985): *Completely analytical Gibbs fields*, pp. 371–403 in Statistical Physics and Dynamical Systems, Rigorous Results, Eds. Fritz, Jaffe and Szasz , Birkhäuser

Completely analytical interactions: constructive description, (1987): J. Stat. Phys. **46**, 983-1014

Federbush, P., (1969): *A partially alternative derivation of a result of Nelson*, J. Math. Phys **10** , 50–52

Glimm, J., Jaffe, A., (1987): *Quantum Physics: The functional integral point of view*, Springer-Verlag

Gross, L., (1976): *Logarithmic Sobolev inequalities*, Amer. J. Math. **97** , 1061–1083

Guionnet, A., Zegarliński, B., (1996): *Decay to Equilibrium in Random Spin Systems on a Lattice*, Commun. Math. Phys. **181**, 703–732; *Decay to Equilibrium in Random Spin Systems on a Lattice, II*, J. Stat. Phys. **86**, (1997) 899–904

Herbst, I., (1976): *On Canonical Quantum Field Theories*, J. Math. Phys. **17** , 1210–1221

Lu, S.L., Yau, H.T., (1993): *Spectral gap and logarithmic Sobolev inequality for Kawasaki and Glauber dynamics.* Commun. Math. Phys. **156**, 399–433

Majewski, A.W., Olkiewicz, R., Zegarliński, B., (1998): *Construction and Ergodicity of Dissipative Dynamics for Quantum Spin Systems on a Lattice*, J. Phys. **A31**, 2045-2056

Martinelli, F., Olivieri, E., (1994): *Approach to Equilibrium of Glauber Dynamics in the One Phase Region: I. The Attractive case/ II. The General Case.* Commun. Math. Phys. **161**, 447–486 / 487–514

Minlos, R.A., (1996): *Invariant subspaces of the stochastic Ising high temperature dynamics*, Markov Process and Rel. Fields **2**, no. 2, 263–284

Minlos, R.A., Trishch, A.G., (1994): *The complete spectral decomposition of a generator of Glauber dynamics for the one-dimensional Ising model*, Uspekhi Mat. Nauk **49**, no. 6(300), 210–211

Ogielski, A.T., (1985): *Dynamics of three dimensional Ising spin glasses in thermal equilibrium*, Phys. Rev. **B32**, No 11 7384

Olkiewicz, R., Zegarliński, B., (1999): *Hypercontractivity In Noncommutative $\mathbb{L}_p$ Spaces.* J. Funct. Anal. **161**, 246–285

Simon, B., (1974): *The $P(\varphi)_2$ Euclidean (Quantum) Field Theory*, Princeton Univ Press

Stroock, D.W., Zegarliński, B., (1992): *The Logarithmic Sobolev inequality for Continuous Spin Systems on a Lattice*, J. Funct. Anal. **104**, 299–326

Stroock, D.W., Zegarliński, B., (1992): *The Equivalence of the Logarithmic Sobolev Inequality and the Dobrushin - Shlosman Mixing Condition*, Commun. Math. Phys. **144**, 303–323

Stroock, D.W., Zegarliński, B., (1992): *The Logarithmic Sobolev inequality for Discrete Spin Systems on a Lattice*, Commun. Math. Phys. **149**, 5–193

Stroock, D.W., Zegarliński, B., (1995): *The Ergodic Properties of Glauber Dynamics*, J. Stat. Phys. **81**, 1007–1019

Zegarliński, B., (1990): *On log - Sobolev Inequalities for Infinite Lattice Systems*, Lett. Math. Phys. **20**, 173–182

Zegarliński, B., (1990): *Log - Sobolev Inequalities for Infinite One Dimensional Lattice Systems*, Commun. Math. Phys. **133**, 147–162

Zegarliński, B., (1992): *Dobrushin Uniqueness Theorem and Logarithmic Sobolev Inequalities*, J. Funct. Anal. **105**, 77–111
Zegarliński, B., (1995): *Ergodicity of Markov Semigroups*, pp. 312–337, in Proc. of the Conference: Stochastic Partial Differential Equations, Edinburgh 1994, Ed. A. Etheridge, LMS Lec. Notes 216, Cambridge University Press
Zegarliński, B., (1998): *All that is Sobolev Inequality*, IC Preprint

Partial *-Algebras: A Retrospective

J.-P. Antoine

Institut de Physique Théorique, Université Catholique de Louvain
B-1348 Louvain-la-Neuve, Belgium
E-mail: Antoine@fyma.ucl.ac.be

Abstract. We review the main points in the development of partial *-algebras during the last 15 years, at three different levels. (i) The algebraic structure stemming from the partial multiplication; (ii) The topological partial *-algebras; (iii) The partial *-algebras of closable operators in Hilbert spaces or partial O*-algebras, including the representation theory of the abstract partial *-algebras.

1 Prologue

Fifteen years ago, on the occasion of Prof. Łopuszanski's 60th birthday, Witold Karwowski suggested to look at the algebraic structure, if any, that would arise if one tried to multiply unbounded operators in a Hilbert space. Indeed there was a rich structure behind, and not only at the algebraic level (Antoine and Karwowski 1983, Antoine and Karwowski 1985; 1986). Since then, several researchers have joined this circle of ideas, and a full-fledged theory has emerged. This lecture presents a quick overview of this rather unforeseen development, following essentially (Antoine et al. 1996) and (Antoine et al. 1998b), where the original references may be found.

2 The Algebraic Structure

A *partial *-algebra* is a complex vector space $\mathfrak{A}$, endowed with an involution $x \mapsto x^*$ (that is, a bijection such that $x^{**} = x$) and a partial multiplication defined by a set $\Gamma \subset \mathfrak{A} \times \mathfrak{A}$ (a binary relation) such that:

(i) $(x, y) \in \Gamma$ implies $(y^*, x^*) \in \Gamma$;
(ii) $(x, y_1), (x, y_2) \in \Gamma$ implies $(x, \lambda y_1 + \mu y_2) \in \Gamma$, $\forall\, \lambda, \mu \in \mathbb{C}$;
(iii) for any $(x, y) \in \Gamma$, there is defined a product $xy \in \mathfrak{A}$, which is distributive w.r. to the addition and satisfies the relation $(xy)^* = y^* x^*$.

Notice that the partial multiplication is *not* required to be associative (and often it is not). We shall assume the partial *-algebra $\mathfrak{A}$ contains a unit e, *i.e.*, $e^* = e$, $(e, x) \in \Gamma$, $\forall\, x \in \mathfrak{A}$, and $ex = xe = x$, $\forall\, x \in \mathfrak{A}$. (If $\mathfrak{A}$ has no unit, it may always be embedded into a larger partial *-algebra with unit, in the standard fashion (Antoine and Mathot 1987).)

Given the defining set Γ, spaces of multipliers are defined in the obvious way:

$$\begin{aligned}(x,y) \in \Gamma &\Longleftrightarrow x \in L(y) \ \text{ or } x \text{ is a left multiplier of } y \\ &\Longleftrightarrow y \in R(x) \ \text{ or } y \text{ is a right multiplier of } x\,.\end{aligned}$$

For any subset $\mathfrak{N} \subset \mathfrak{A}$, we write

$$L\mathfrak{N} = \bigcap_{x \in \mathfrak{N}} L(x), \quad R\mathfrak{N} = \bigcap_{x \in \mathfrak{N}} R(x)\,,$$

and, of course, the involution exchanges the two:

$$(L\mathfrak{N})^* = R\mathfrak{N}^*, \quad (R\mathfrak{N})^* = L\mathfrak{N}^*\,.$$

Clearly all these multiplier spaces are vector subspaces of $\mathfrak{A}$, containing e.

The partial *-algebra is *abelian* if $L(x) = R(x)$, $\forall\, x \in \mathfrak{A}$, and then $xy = yx$, $\forall\, x \in L(y)$. In that case, we write simply for the multiplier spaces $L(x) = R(x) \equiv M(x)$, $L\mathfrak{N} = R\mathfrak{N} \equiv M\mathfrak{N}$ $(\mathfrak{N} \subset \mathfrak{A})$.

Now the crucial fact is that the couple of maps (L, R) defines a *Galois connection* (Antoine and Gustafson 1981) on the complete lattice of all vector subspaces of $\mathfrak{A}$ (ordered by inclusion), which means that (i) both L and R reverse order; and (ii) both LR and RL are closures, that is:

$$\mathfrak{N} \subset LR\mathfrak{N} \text{ and } LRL = L\,, \qquad \mathfrak{N} \subset RL\mathfrak{N} \text{ and } RLR = R\,.$$

Let us denote by $\mathcal{F}^L$, resp. $\mathcal{F}^R$, the set of all LR-closed, resp. RL-closed, subspaces of $\mathfrak{A}$:

$$\mathcal{F}^L = \{\mathfrak{N} \subset \mathfrak{A} : \mathfrak{N} = LR\mathfrak{N}\}\,, \quad \mathcal{F}^R = \{\mathfrak{N} \subset \mathfrak{A} : \mathfrak{N} = RL\mathfrak{N}\}\,.$$

both ordered by inclusion. Then, from standard results of universal algebra, one can deduce that the set $\mathcal{F}^R$, ordered by inclusion, is a complete lattice with lattice operations

$$\mathfrak{M} \wedge \mathfrak{N} = \mathfrak{M} \cap \mathfrak{N}\,, \quad \mathfrak{M} \vee \mathfrak{N} = RL(\mathfrak{M} + \mathfrak{N})\,.$$

The largest element is $\mathfrak{A}$, the smallest $R\mathfrak{A}$. A corresponding result holds for $\mathcal{F}^L$, exchanging L and R. Both $L : \mathcal{F}^R \to \mathcal{F}^L$ and $R : \mathcal{F}^L \to \mathcal{F}^R$ are lattice anti-isomorphisms: $L(\mathfrak{M} \wedge \mathfrak{N}) = L\mathfrak{M} \vee L\mathfrak{N}$, etc., and the involution $\mathfrak{N} \leftrightarrow \mathfrak{N}^*$ is a lattice isomorphism between $\mathcal{F}^L$ and $\mathcal{F}^R$.

As examples of partial *-algebras, some of which we will encounter below, we may cite partial *-algebras of polynomials, of functions or infinite matrices, topological quasi *-algebras, CQ*-algebras, and partial *-algebras of closable operators in a Hilbert space (partial O*-algebras).

The last case is the most important in practice. It will also be needed to set up a representation theory, because a representation of a partial *-algebra $\mathfrak{A}$ is a homomorphism of $\mathfrak{A}$ into some partial O*-algebra. Here a **-homomorphism* of a partial *-algebra $\mathfrak{A}$ into another one $\mathfrak{B}$ is a linear map $\rho : \mathfrak{A} \to \mathfrak{B}$ such that (i) $\rho(x^*) = \rho(x)^*$ for each $x \in \mathfrak{A}$, and (ii) whenever $x \in L(y)$ in $\mathfrak{A}$, then $\rho(x) \in L(\rho(y))$ in $\mathfrak{B}$ and $\rho(x)\rho(y) = \rho(xy)$. The map ρ is a **-isomorphism* if it is a bijection and $\rho^{-1} : \mathfrak{B} \to \mathfrak{A}$ is also a *-homomorphism.

3 Topological Partial *-Algebras

3.1 Basic Definitions

Let $\mathfrak{A}$ be a partial *-algebra with unit and assume it carries a locally convex, Hausdorff, topology τ, which makes it into a locally convex topological vector space $\mathfrak{A}[\tau]$ (that is, the vector space operations are τ-continuous).

The partial *-algebraic structure of $\mathfrak{A}$ is completely characterized by its spaces of left, resp. right, multipliers. Thus, quite naturally, we describe the topological structure of $\mathfrak{A}[\tau]$ by providing all spaces of multipliers with appropriate topologies. Our goal is to make the algebraic and the topological structure coincide as much as possible.

We start with the following observation. Let $\mathfrak{M} \in \mathcal{F}^R$. To every $x \in L\mathfrak{M}$, we associate a linear map T_x^L from $\mathfrak{M}$ into $\mathfrak{A}$:

$$T_x^L(a) = xa, \quad a \in \mathfrak{M}, \; x \in L\mathfrak{M}.$$

This allows to define the topology $\rho_{\mathfrak{M}}$ on $\mathfrak{M}$ as the weakest locally convex topology on $\mathfrak{M}$ such that all maps T_x^L, $x \in L\mathfrak{M}$, are continuous from $\mathfrak{M}$ into $\mathfrak{A}[\tau]$. This is of course a projective topology. In the same way, the topology $\lambda_{\mathfrak{N}}$ on $\mathfrak{N} \in \mathcal{F}^L$ is the weakest locally convex topology on $\mathfrak{N}$ such that all maps $T_y^R : a \mapsto ay$, $y \in R\mathfrak{N}$, are continuous from $\mathfrak{N}$ into $\mathfrak{A}[\tau]$.

It follows immediately from the definition that, whenever $\mathfrak{M}_1, \mathfrak{M}_2 \in \mathcal{F}^R$ are such that $\mathfrak{M}_1 \subset \mathfrak{M}_2$, then the topology $\rho_{\mathfrak{M}_1}$ is finer than the topology $(\rho_{\mathfrak{M}_2} \upharpoonright \mathfrak{M}_1)$ induced by $\mathfrak{M}_2$ on $\mathfrak{M}_1$. In other words, the embedding $\mathfrak{M}_1 \to \mathfrak{M}_2$ is a *continuous injection*.

Take now $\mathfrak{A}$ itself. It carries three topologies, τ, $\rho_{\mathfrak{A}}$ and $\lambda_{\mathfrak{A}}$, and it is easy to see that both $\rho_{\mathfrak{A}}$ and $\lambda_{\mathfrak{A}}$ are finer than τ. As a consequence, since τ was assumed to be Hausdorff, all topologies $\rho_{\mathfrak{M}}$, $\mathfrak{M} \in \mathcal{F}^R$, and $\lambda_{\mathfrak{N}}$, $\mathfrak{N} \in \mathcal{F}^L$, are Hausdorff.

Now, for reasons of coherence, it would be preferable that all three topologies on $\mathfrak{A}$, τ, $\rho_{\mathfrak{A}}$ and $\lambda_{\mathfrak{A}}$ be equivalent. Here is a handy criterion.

Lemma 3.1 – *Let $\mathfrak{A}[\tau]$ be a partial *-algebra with locally convex topology τ. Then the projective topology $\rho_{\mathfrak{A}}$ on $\mathfrak{A}$ is equivalent to τ iff, for each $x \in L\mathfrak{A}$, the map $T_x^L : a \mapsto xa$ is continuous from $\mathfrak{A}[\tau]$ into itself. Similarly, the projective topology $\lambda_{\mathfrak{A}}$ on $\mathfrak{A}$ is equivalent to τ iff, for each $y \in R\mathfrak{A}$, the map $T_y^R : a \mapsto ay$ is continuous from $\mathfrak{A}[\tau]$ into itself.*

Moreover, if the involution $x \mapsto x^*$ is τ-continuous, then it is continuous from $\mathfrak{M}[\rho_{\mathfrak{M}}]$ into $\mathfrak{M}^*[\lambda_{\mathfrak{M}^*}] \in \mathcal{F}^L$, for every $\mathfrak{M} \in \mathcal{F}^R$.

According to our goal, we will naturally require that all three topologies $\rho_{\mathfrak{A}}$, $\lambda_{\mathfrak{A}}$ and τ on a topological partial *-algebra coincide and that the involution be continuous. Let us now look at multiplier spaces $\mathfrak{M} \in \mathcal{F}^R$. If $\mathfrak{M}_1 \subset \mathfrak{M}_2$, we have seen that the embedding is continuous. In order to make the structure tighter, we should also require that $\mathfrak{M}_1$ be *dense* in $\mathfrak{M}_2[\rho_{\mathfrak{M}_2}]$.

This is true in many examples, typically the function spaces. Of course it is enough to require that $R\mathfrak{A}$ be dense in each $\mathfrak{M}[\rho_{\mathfrak{M}}] \in \mathcal{F}^R$. Indeed, if $R\mathfrak{A} \subset \mathfrak{M}_1 \subset \mathfrak{M}_2$, and $R\mathfrak{A}$ is dense in $\mathfrak{M}_2$ for $\tau_{\mathfrak{M}_2}$, so is a fortiori $\mathfrak{M}_1$. But this condition is still too strong (and hardly verifiable in practice, because $\mathcal{F}^R$ is too large). To go beyond, we introduce the notion of *generating family*, that is, a subset $\mathcal{I}^R$ of $\mathcal{F}^R$ such that (i) $R\mathfrak{A} \in \mathcal{I}^R$ and $\mathfrak{A} \in \mathcal{I}^R$, and (ii) $x \in L(y)$ iff $\exists \mathfrak{M} \in \mathcal{I}^R$ s.t. $y \in \mathfrak{M}, x \in L\mathfrak{M}$. A generating family for $\mathcal{F}^L$ is defined in a similar way. Clearly, if $\mathcal{I}^R$ is a generating family for $\mathcal{F}^R$, $\mathcal{I}^L = L\mathcal{I}^R = \{L\mathfrak{M} : \mathfrak{M} \in \mathcal{I}^R\}$ is generating for $\mathcal{F}^L$. The usefulness of this notion is twofold: (i) if $\mathcal{I}^R$ is generating for $\mathcal{F}^R$, so is the sublattice $\mathcal{J}^R$ of $\mathcal{F}^R$ generated from $\mathcal{I}^R$ by *finite* lattice operations $\vee$ and $\wedge$; (ii) if $\mathcal{I}^R$ is generating, the *complete* lattice generated by $\mathcal{I}^R$ is $\mathcal{F}^R$ itself. In other words, a generating family determines completely the partial multiplication.

We make immediate use of this last property for weakening the density condition.

Proposition 3.2 – *Let $\mathfrak{A}[\tau]$ be a partial *-algebra with topology τ. Assume there exists a generating family $\mathcal{I}^R$ for $\mathcal{F}^R$ such that $R\mathfrak{A}$ is dense in $\mathfrak{M}[\rho_{\mathfrak{M}}]$ for every $\mathfrak{M} \in \mathcal{I}^R$. Then, for any pair $\mathfrak{M}_1, \mathfrak{M}_2 \in \mathcal{F}^R$ such that $\mathfrak{M}_1 \subset \mathfrak{M}_2$, $\mathfrak{M}_1$ is dense in $\mathfrak{M}_2[\rho_{\mathfrak{M}_2}]$.*

Summarizing, we may now state our definition of topological partial *-algebra.

Definition 3.3 – Let $\mathfrak{A}[\tau]$ be a partial *-algebra, which is a TVS for the locally convex topology τ. Then $\mathfrak{A}[\tau]$ is called a *topological partial *-algebra* if the following two conditions are satisfied:

(i) the involution $a \mapsto a^*$ is τ-continuous;

(ii) the maps $a \mapsto xa$ and $a \mapsto ay$ are τ-continuous for all $x \in L\mathfrak{A}$ and $y \in R\mathfrak{A}$.

The topological partial *-algebra $\mathfrak{A}[\tau]$ is said to be *tight*, if, in addition,

(iii) there is a generating family $\mathcal{J}^R$ for $\mathcal{F}^R$ such that $R\mathfrak{A}$ is dense in $\mathfrak{M}[\rho_{\mathfrak{M}}]$, $\forall \mathfrak{M} \in \mathcal{J}^R$.

This definition seems natural, in the sense that it forces the topological structure determined by τ to be consistent with the multiplier structure of $\mathfrak{A}$. As an illustration, we consider two abstract examples.

(i) Topological Quasi *-Algebras. Let $(\mathfrak{A}, \mathfrak{A}_o)$ be a topological quasi-algebra, that is, $\mathfrak{A}_o$ is a topological *-algebra such that the multiplication is separately, but not jointly, continuous for the topology of $\mathfrak{A}_o$ and the latter is not complete, and $\mathfrak{A}$ is the completion of $\mathfrak{A}_o$. Thus $\mathfrak{A}$ is only a partial *-algebra: the product xy is defined only if either x or y belongs to $\mathfrak{A}_o$. Clearly, $(\mathfrak{A}, \mathfrak{A}_o)$ is a (trivial) partial *-algebra with $L\mathfrak{M} = R\mathfrak{M} = \mathfrak{A}_o$ and $\mathfrak{A}_o$ is dense in $\mathfrak{A}$. Thus every topological quasi *-algebra is a tight topological partial *-algebra.

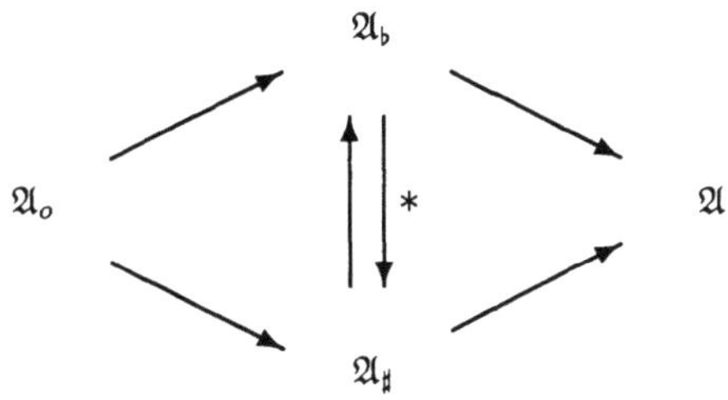

Fig. 1. Structure of a CQ*-algebra.

(ii) CQ*-Algebras. A CQ*-algebra is a Banach partial *-algebra with a very simple multiplication structure: the lattice of multipliers consists of four spaces only, as shown on Figure 1. In this diagram, $\mathfrak{A}_\flat$ is a C*-algebra, $\mathfrak{A}$ is a right Banach module over $\mathfrak{A}_\flat$, with isometric involution $*$, $\mathfrak{A}_\sharp = (\mathfrak{A}_\flat)^*$ (hence it is also a C*-algebra), $\mathfrak{A}_o = \mathfrak{A}_\flat \cap \mathfrak{A}_\sharp$ and each arrow denotes a continuous embedding with dense range. In addition, the C*-norm $\|\cdot\|_\flat$ on $\mathfrak{A}_\flat$ is related to the norm $\|\cdot\|$ of $\mathfrak{A}$ by the relation $\|B\|_\flat = \sup_{A\in\mathfrak{A}} \|AB\|$, $B \in \mathfrak{A}_\flat$. Thus $\mathfrak{A}$ is the completion of the C*-algebra $\mathfrak{A}_\flat$ with respect to the weaker norm $\|\cdot\|$. The product AB of two elements of $A, B \in \mathfrak{A}$ is defined only if either $A \in \mathfrak{A}_\sharp$ or $B \in \mathfrak{A}_\flat$. Hence $\mathfrak{A}_\flat = R\mathfrak{A}$, $\mathfrak{A}_\sharp = L\mathfrak{A}$ and $\mathfrak{A}_o = R\mathfrak{A} \cap L\mathfrak{A}$. Examples of such structures are given, for instance, by spaces of bounded operators in a (Gel'fand) triplet of Hilbert spaces $\mathcal{H}_\lambda \subset \mathcal{H} \subset \mathcal{H}_{\bar\lambda}$, where $\mathcal{H}_{\bar\lambda}$ is the antidual of $\mathcal{H}_\lambda$ with respect to the inner product of $\mathcal{H}$.

These CQ*-algebras appear as the natural extension of C*-algebras to the partial algebraic setting, and they may be viewed as a first step toward a more general study of *partial C*-algebras*, yet to be carried out. They are treated in detail in the lecture of C. Trapani.

3.2 Examples of Topological Partial *-Algebras

(i) L^p Spaces on a Finite Interval. Consider the chain $\mathcal{I} = \{L^p([0,1],dx), 1 \leqslant p \leqslant \infty\}$, with $L^p \subset L^q$, $p > q$. For $1 < p < \infty$, every space L^p is a reflexive Banach space with dual $L^{\bar p}$ $(1/p + 1/\bar p = 1)$. Notice that duality in the sense of Banach spaces coincides with duality for the inner product of L^2 thanks to Hölder's inequality.

Now, being a chain, $\mathcal{I}$ is of course a lattice, albeit not a complete one. The lattice completion of $\mathcal{I}$, denoted $\mathcal{F}$, may be characterized explicity. Define the two spaces :

$$L^{p-} = \bigcap_{1\leqslant q<p} L^q, \quad L^{p+} = \bigcup_{p<q\leqslant \infty} L^q .$$

Then for $1 < p \leqslant \infty$, L^{p-}, with the projective topology, is a non-normable reflexive Fréchet space, with dual $L^{\bar{p}+}$. And for $1 \leqslant p < \infty$, L^{p+}, with the inductive topology, is a nonmetrizable complete DF-space, with dual $L^{\bar{p}-}$. Finally the following inclusions are strict:

$$L^{p+} \subset L^p \subset L^{p-} \subset L^{q+} \quad (1 < q < p < \infty), \tag{3.1}$$

all embeddings in (3.1) are continuous and have dense range. Then the complete lattice $\mathcal{F}$ generated by $\mathcal{I}$ is also a chain, obtained by replacing each $L^p\,(1 < p < \infty)$ by the corresponding triplet as in (3.1) and adding the two spaces $L^{\infty-}$ and L^{1+}:

$$L^\infty \subset L^{\infty-} \subset \ldots \subset L^{p+} \subset L^p \subset L^{p-} \subset \ldots \subset L^{1+} \subset L^1 .$$

Now we turn to the partial *-algebra structure. The commutative partial multiplication on the space $L^1([0,1], dx)$ is defined as follows:

$$f \in M(g) \Leftrightarrow \exists q \in [1, \infty] \text{ such that } f \in L^q,\ g \in L^{\bar{q}},\ 1/q + 1/\bar{q} = 1, \tag{3.2}$$

i.e., $\mathcal{I}$ is a generating family. Then it is easy to see that

$$ML^p = L^{\bar{p}}, \quad ML^{p-} = L^{\bar{p}+}, \quad ML^{p+} = L^{\bar{p}-} .$$

As for the multiplier topologies, ρ_{L^p} is the L^p norm topology, $\rho_{L^{p-}}$ is the Fréchet projective topology on L^{p-} and $\rho_{L^{p+}}$ is the DF topology on L^{p+}. For both $\mathcal{I}$ and $\mathcal{F}$, the smallest space is $L^\infty = ML^1$, and it is dense in all the other ones. The involution $f \mapsto \bar{f}$ is of course L^1-continuous. The multiplication is continuous from $L^\infty \times L^1$ into L^1. In fact it is not only separately, but even jointly continuous, and similarly from $L^p \times L^{\bar{p}}$ and from $L^{p-} \times L^{\bar{p}+}$ into L^1, thanks to Hölder's inequality and the fact that all topologies are either Fréchet or DF.

In conclusion, the topological structure and the multiplier structure of $\mathcal{I}$ coincide, and we have an abelian tight topological partial *-algebra. In addition, the chain $\mathcal{I}$ is a partial inner product space (PIP-space) (Antoine and Grossmann 1976, Antoine 1980), and the latter structure coincides with the other two.

(ii) The Spaces $L^p(\mathbb{R}, dx)$. We turn now to the spaces $L^p(\mathbb{R}, dx)$ on the whole line. The difference with the previous case is that these no longer form a chain, no two of them being comparable. We have only

$$L^p \cap L^q \subset L^s, \forall s \text{ such that } p < s < q .$$

Hence we have to take the lattice generated by $\mathcal{I} = \{L^p(\mathbb{R}, dx), 1 \leqslant p \leqslant \infty\}$, that we call $\mathcal{J}$. The extreme spaces of the lattice are, respectively:

$$V_J^\# = \bigcap_{1 \leqslant q \leqslant \infty} L^q , \quad \text{and} \quad V_J = \bigcup_{1 \leqslant q \leqslant \infty} L^q = \sum_{1 \leqslant q \leqslant \infty} L^q .$$

Here too, the lattice structure allows to give to V_J a structure of topological partial *-algebra. The lattice operations on $\mathcal{J}$ are those familiar in interpolation theory (Bergh and Löfström 1976):

- $L^p \wedge L^q = L^p \cap L^q$ is a Banach space, with the projective norm $\|f\|_{p\wedge q} = \|f\|_p + \|f\|_q$.
- $L^p \vee L^q = L^p + L^q$ is a Banach space, with the inductive norm $\|f\|_{p\vee q} = \inf\left(\|g\|_p + \|h\|_q\right),\ f = g + h,\ g \in L^p,\ h \in L^q$.
- For $1 < p, q < \infty$, both spaces $L^p \wedge L^q$ and $L^p \vee L^q$ are reflexive and $(L^p \wedge L^q)' = L^{\bar{p}} \vee L^{\bar{q}}$.

At this stage, it is convenient to introduce a unified notation:

$$L^{(p,q)} = \begin{cases} L^p \wedge L^q, & \text{if } p \geqslant q\,, \\ L^p \vee L^q, & \text{if } p \leqslant q\,. \end{cases}$$

Thus, for $1 < p, q < \infty$, each space $L^{(p,q)}$ is a reflexive Banach space, with dual $L^{(\bar{p},\bar{q})}$. The modifications when p, q equal 1 or ∞ are obvious. In this notation, the set $\mathcal{J}$ of all spaces $L^{(p,q)}$ is partially ordered with the following rule:

$$L^{(p,q)} \subset L^{(p',q')} \iff (p,q) \leqslant (p',q') \iff p \geqslant p' \text{ and } q \leqslant q'. \quad (3.3)$$

Then it is easy to show that the family $\mathcal{J}$, generated by $\mathcal{I} = \{L^p\}$, is an involutive lattice with respect to the partial order (3.3), with operations:

$$\begin{aligned} (p,q) \wedge (p',q') &= (p \vee p', q \wedge q')\,, \\ (p,q) \vee (p',q') &= (p \wedge p', q \vee q')\,, \\ \overline{(p,q)} &= (\bar{p},\bar{q})\,, \end{aligned}$$

where, as usual, $p \wedge p' = \min\{p,p'\}$, $p \vee p' = \max\{p,p'\}$. Notice that the lattice $\mathcal{J}$ is already obtained at the first generation: for example, $L^{(r,s)} \wedge L^{(a,b)} = L^{(r\vee a, s\wedge b)}$. Furthermore, in the lattice $\mathcal{J}$, inclusion means continuous embedding with dense range. Thus, with the same partial multiplication (3.2), we obtain another tight topological partial *-algebra.

Two remarks are in order. First, here too, the lattice completion $\mathcal{F}$ of $\mathcal{J}$ and the multiplier spaces may be characterized explicitly. Second, another structure of topological partial *-algebra may be given to the family $\mathcal{J}$ of spaces, simply replacing multiplication by convolution, with similar results.

We note finally that the only difference between the two cases $\{L^p([0,1])\}$ and $\{L^p(\mathbb{R})\}$ lies in the type of order obtained: a chain $\mathcal{I}$ (total order) or a partially ordered lattice $\mathcal{J}$.

(iii) Amalgam Spaces. The lesson of the previous example is that an involutive lattice of (preferably reflexive) Banach spaces turns quite naturally into a (tight) topological partial *-algebra if it possesses a partial multiplication that satisfies a (generalized) Hölder inequality. A whole class of examples is given by the so-called *amalgam spaces* first introduced by N. Wiener (see (Fournier and Stewart 1985) for a review). The simplest ones are the spaces (L^p, ℓ^q), consisting of functions on $\mathbb{R}$ which are locally in L^p and have ℓ^q behavior at infinity, in the sense that the L^p norms over the intervals $(n, n+1)$ form an ℓ^q sequence. For $1 < p, q < \infty$, the corresponding norm

$$\|f\|_{p,q} = \left\{ \sum_{n=-\infty}^{\infty} \left[\int_n^{n+1} |f(x)|^p dx \right]^{q/p} \right\}^{1/q},$$

turns the space (L^p, ℓ^q) into a reflexive Banach space. The same is true for the obvious extensions to p and/or q equal to 1 or ∞. Notice that $(L^p, \ell^p) = L^p$. Once again, the set of all spaces (L^p, ℓ^q) may be partially ordered by inclusion, and it turns out to be a complete lattice. Thus one gets another topological partial *-algebra.

(iv) Topological Partial *-Algebras of Operators. A first example is the partial *-algebra of operators on a scale or a lattice of Hilbert spaces, with the usual operator multiplication (Antoine 1980). Familiar cases of such scales are the Hilbert scale built on the powers of a positive self-adjoint operator $H \geqslant 1$: $\mathcal{H}_n = D(H^n)$, with the graph norm $\|f\|_n = \|H^n f\|$, for $n \in \mathbb{N}$, and $\mathcal{H}_{-n} = \mathcal{H}_n^\times$, or the scale of Sobolev spaces $W_s^2(\mathbb{R})$, $s \in \mathbb{R}$, where $f \in W_s^2(\mathbb{R})$ if its Fourier transform $\widehat{f}$ satisfies the condition $(1 + |.|^2)^{s/2}\, \widehat{f} \in L^2(\mathbb{R})$.

A second example are partial *-algebras of closable operators in a Hilbert space. From now on, we will mostly concentrate on this class. We refer to (Antoine et al. 1996) for further details and the original references.

4 Partial *-Algebras of Closable Operators

Let $\mathcal{H}$ be a complex Hilbert space and $\mathcal{D}$ a dense subspace of $\mathcal{H}$. We denote by $\mathcal{L}^\dagger(\mathcal{D}, \mathcal{H})$ the set of all (closable) linear operators X such that $\mathcal{D}(X) = \mathcal{D}$, $\mathcal{D}(X^*) \supseteq \mathcal{D}$. The set $\mathcal{L}^\dagger(\mathcal{D}, \mathcal{H})$ is a partial *-algebra with respect to the following operations: the usual sum $X_1 + X_2$, the scalar multiplication λX, the involution $X \mapsto X^\dagger = X^* \lceil \mathcal{D}$ and the *(weak)* partial multiplication $X_1 \square X_2 = {X_1}^{\dagger *} X_2$, defined whenever X_2 is a weak right multiplier of X_1 (equivalently, X_1 is a weak left multiplier of X_2), that is, iff $X_2 \mathcal{D} \subset \mathcal{D}({X_1}^{\dagger *})$ and $X_1{}^* \mathcal{D} \subset \mathcal{D}(X_2{}^*)$ (we write $X_2 \in R^{\mathrm{w}}(X_1)$ or $X_1 \in L^{\mathrm{w}}(X_2)$). When we regard $\mathcal{L}^\dagger(\mathcal{D}, \mathcal{H})$ as a partial *-algebra with those operations, we denote it by $\mathcal{L}_{\mathrm{w}}^\dagger(\mathcal{D}, \mathcal{H})$.

A *partial O*-algebra* on $\mathcal{D}$ is a *-subalgebra $\mathfrak{M}$ of $\mathcal{L}_{\mathrm{w}}^\dagger(\mathcal{D}, \mathcal{H})$, that is, $\mathfrak{M}$ is a subspace of $\mathcal{L}_{\mathrm{w}}^\dagger(\mathcal{D}, \mathcal{H})$, containing the identity and such that $X^\dagger \in \mathfrak{M}$

whenever $X \in \mathfrak{M}$ and $X_1 \,\Box\, X_2 \in \mathfrak{M}$ for any $X_1, X_2 \in \mathfrak{M}$ such that $X_2 \in R^{\mathrm{w}}(X_1)$. Thus $\mathcal{L}^\dagger_{\mathrm{w}}(\mathcal{D},\mathcal{H})$ itself is the largest partial O*-algebra on the domain $\mathcal{D}$.

On the space $\mathcal{L}^\dagger(\mathcal{D},\mathcal{H})$ we will consider the *strong** topology τ_{s^*}, which is generated by the family of seminorms : $p_\xi{}^*(X) = \|X\xi\| + \|X^\dagger\xi\|, \xi \in \mathcal{D}$. The space $\mathcal{L}^\dagger(\mathcal{D},\mathcal{H})$ is complete for τ_{s^*}. For $\mathfrak{N} \subset \mathcal{L}^\dagger(\mathcal{D},\mathcal{H})$, we denote by $[\mathfrak{N}]^{s^*}$ the τ_{s^*}-closure of $\mathfrak{N}$.

We also need the *weak* topology τ_{w} on $\mathcal{L}^\dagger(\mathcal{D},\mathcal{H})$, which is generated by the family of seminorms $p_{f,g}(X) = |\langle f|Xg\rangle|$, $f, g \in \mathcal{D}$, and the *quasi-uniform* topology, τ_*, defined by the set of seminorms $p_{\mathcal{N}}(X) = \sup_{f\in\mathcal{N}}(\|Xf\| + \|X^\dagger f\|)$, where $\mathcal{N}$ is a bounded subset of $\mathcal{D}$, equipped with the projective topology determined by $\mathcal{L}^\dagger(\mathcal{D},\mathcal{H})$.

If we restrict ourselves to those operators in $\mathcal{L}^\dagger(\mathcal{D},\mathcal{H})$ that, together with their adjoint, leave the domain $\mathcal{D}$ invariant, we obtain a *-algebra, namely $\mathcal{L}^\dagger(\mathcal{D}) = \{A \in \mathcal{L}^\dagger(\mathcal{D},\mathcal{H}); A\mathcal{D} \subset \mathcal{D} \text{ and } A^*\mathcal{D} \subset \mathcal{D}\}$. Then an *O*-algebra* is defined as a *-subalgebra of $\mathcal{L}^\dagger(\mathcal{D})$; thus $\mathcal{L}^\dagger(\mathcal{D})$ is the maximal O*-algebra contained in $\mathcal{L}^\dagger(\mathcal{D},\mathcal{H})$ and it is τ_{s^*}-dense in $\mathcal{L}^\dagger(\mathcal{D},\mathcal{H})$, *i.e.*, $\mathcal{L}^\dagger(\mathcal{D},\mathcal{H}) = [\mathcal{L}^\dagger(\mathcal{D})]^{s^*}$. Clearly an O*-algebra is a particular case of a partial O*-algebra (see (Schmüdgen 1990) for a comprehensive study of partial O*-algebras).

Given a partial O*-algebra $\mathfrak{M}$, we define internal multipliers as $R(X) = R^{\mathrm{w}}(X) \cap \mathfrak{M}$ and $L(X) = L^{\mathrm{w}}(X) \cap \mathfrak{M}$. Then the universal right multipliers of $\mathfrak{M}$ are the elements of the set:

$$R\mathfrak{M} = R^{\mathrm{w}}(\mathfrak{M}) \cap \mathfrak{M} = \{Y \in \mathfrak{M};\ X \,\Box\, Y \text{ is well-defined, } \forall X \in \mathfrak{M}\}.$$

A †-invariant subset $\mathfrak{N}$ of $\mathcal{L}^\dagger(\mathcal{D},\mathcal{H})$ is called *fully closed* if $\mathcal{D} = \widehat{\mathcal{D}}(\mathfrak{N}) \equiv \bigcap_{X\in\mathfrak{N}} \mathcal{D}(\overline{X})$. If $\mathfrak{N}$ is not fully closed, its full closure is the smallest fully closed set that contains it, that is, $\widehat{\mathfrak{N}} = \{\hat\imath(X) \equiv \overline{X}\restriction\widehat{\mathcal{D}}(\mathfrak{N});\ X \in \mathfrak{N}\}$. Let $\mathfrak{M}$ be a partial O*-algebra. If it is not fully closed, it may be embedded into its full closure $\widehat{\mathfrak{M}} = \hat\imath(\mathfrak{M})$, which is a fully closed partial O*-algebra on the domain $\widehat{\mathcal{D}}(\mathfrak{M})$, isomorphic to $\mathfrak{M}$. Thus one may always restrict the analysis to fully closed partial O*-algebras without loss of generality. On the other hand, a partial O*-algebra $\mathfrak{M}$ is called *self-adjoint* if $\mathcal{D} = \mathcal{D}^*(\mathfrak{M}) \equiv \bigcap_{X\in\mathfrak{M}} \mathcal{D}(X^*)$, and this *is* a strong restriction.

Given a †-invariant subset $\mathfrak{N}$ of $\mathcal{L}^\dagger(\mathcal{D},\mathcal{H})$, we define, as usual, its weak unbounded commutant:

$$\mathfrak{N}'_\sigma = \{Y \in \mathcal{L}^\dagger(\mathcal{D},\mathcal{H}); (X\xi|Y\eta) = (Y^\dagger\xi|X^\dagger\eta) \quad \text{for each } \xi,\eta \in \mathcal{D} \text{ and } X \in \mathfrak{N}\} \tag{4.1}$$

and its weak bounded commutant:

$$\mathfrak{N}'_{\mathrm{w}} = \{C \in \mathcal{B}(\mathcal{H}); (CX\xi|\eta) = (C\xi|X^{\dagger}\eta)$$

$$\text{for each } \xi, \eta \in \mathcal{D} \text{ and } X \in \mathfrak{N}\}. \qquad (4.2)$$

The restriction to $\mathcal{D}$ of $\mathfrak{N}'_{\mathrm{w}}$ is the bounded part of $\mathfrak{N}'_{\sigma}$. Both $\mathfrak{N}'_{\sigma}$ and $\mathfrak{N}'_{\mathrm{w}}$ are weakly closed, †-invariant subspaces, but not necessarily algebras.

As for bicommutant, we consider the weak unbounded one, namely $\mathfrak{N}''_{\mathrm{w}\sigma} = (\mathfrak{N}'_{\mathrm{w}})'_{\sigma}$. Its bounded part is the (restriction to $\mathcal{D}$ of) $(\mathfrak{N}'_{\mathrm{w}})'$, where $\mathcal{B}'$ denotes the usual bounded commutant of a subset $\mathcal{B} \subset \mathcal{B}(\mathcal{H})$. We note the relation $(\mathfrak{N}''_{\mathrm{w}\sigma})''_{\mathrm{w}\sigma} = \mathfrak{N}''_{\mathrm{w}\sigma}$ and remark that $\mathfrak{N}''_{\mathrm{w}\sigma}$ is fully closed whenever $\mathfrak{N}$ is, because of the obvious inclusions $\mathcal{D} \subset \widehat{\mathcal{D}}(\mathfrak{N}''_{\mathrm{w}\sigma}) \subset \widehat{\mathcal{D}}(\mathfrak{N})$. The crucial fact is that, for any †-invariant subset $\mathfrak{N}$ of $\mathcal{L}^{\dagger}(\mathcal{D},\mathcal{H})$, $\mathfrak{N}'_{\mathrm{w}}$ is a von Neumann algebra if, and only if, $\mathfrak{N}''_{\mathrm{w}\sigma} = \left[(\mathfrak{N}'_{\mathrm{w}})' \upharpoonright \mathcal{D}\right]^{s^*}$.

A partial O*-algebra $\mathfrak{M}$ on $\mathcal{D}$ is said to be a *partial GW*-algebra* if it is fully closed and satisfies the two conditions $\mathfrak{M}'_{\mathrm{w}}\mathcal{D} = \mathcal{D}$ and $\mathfrak{M}''_{\mathrm{w}\sigma} = \mathfrak{M}$ (notice the analogy with the usual condition $\mathfrak{M}'' = \mathfrak{M}$ defining a von Neumann algebra). In that case, $\mathfrak{M}'_{\mathrm{w}}$ is a von Neumann algebra, the (closure of the) bounded part of $\mathfrak{M}$ is also a von Neumann algebra, namely $\mathfrak{M}_o \equiv (\mathfrak{M}'_{\mathrm{w}})'$, and $\mathfrak{M} = \left[(\mathfrak{M}'_{\mathrm{w}})' \upharpoonright \mathcal{D}\right]^{s^*}$. The good properties of partial GW*-algebras stem precisely from the fact that they contain a τ_{s^*}-dense subset of bounded operators.

The easiest way of constructing a partial GW*-algebra is to take a bicommutant. Indeed, if $\mathfrak{N}$ is a fully closed †-invariant subset of $\mathcal{L}^{\dagger}(\mathcal{D},\mathcal{H})$, then $\mathfrak{N}''_{\mathrm{w}\sigma}$ is a partial GW*-algebra on $\mathcal{D}$ iff $\mathfrak{N}'_{\mathrm{w}}\mathcal{D} = \mathcal{D}$. On the other hand, if $\mathfrak{M}$ is a partial O*-algebra on $\mathcal{D}$ (not necessarily fully closed), such that $\mathfrak{M}'_{\mathrm{w}}\mathcal{D} = \mathcal{D}$ and $\mathfrak{M}''_{\mathrm{w}\sigma} = \mathfrak{M}$, then $\widehat{\mathfrak{M}}$ is a partial GW*-algebra on $\widehat{\mathcal{D}}(\mathfrak{M})$.

As a last point, we may ask the question whether a partial O*-algebra is a (tight) topological partial *-algebra. The answer, of course, depends on which topology τ one chooses, and many different ones are available, the strong* τ_{s^*}, the quasi-uniform τ_*, the weak τ_{w}, etc. We will not enter into the technical details, for lack of space, but only indicate a few general results. First, if $\mathcal{L}^{\dagger}_{\mathrm{w}}(\mathcal{D},\mathcal{H})$ is self-adjoint, then it is a topological partial *-algebra for these three topologies, and it is complete for τ_{s^*} and τ_*. More generally, any self-adjoint partial O*-algebra $\mathfrak{M}$ is a topological partial *-algebra for the weak topology τ_{w}, and the same is true for τ_* if $R\mathfrak{M}$ contains only bounded operators. In all cases, tightness is open.

5 Representation Theory

5.1 Generalities

A *-representation of a partial *-algebra $\mathfrak{A}$ is a *-homomorphism of $\mathfrak{A}$ into $\mathcal{L}^{\dagger}_{\mathrm{w}}(\mathcal{D},\mathcal{H})$, for some pair $\mathcal{D} \subset \mathcal{H}$, that is, a linear map $\pi : \mathfrak{A} \to \mathcal{L}^{\dagger}_{\mathrm{w}}(\mathcal{D},\mathcal{H})$

such that : (i) $\pi(x^*) = \pi(x)^\dagger$ for every $x \in \mathfrak{A}$; (ii) $x \in L(y)$ in $\mathfrak{A}$ implies $\pi(x) \in L^{\mathrm{w}}(\pi(y))$ and $\pi(x) \,\Box\, \pi(y) = \pi(xy)$.

Let π be a *-representation of a partial *-algebra $\mathfrak{A}$. It is called *fully closed* if $\pi(\mathfrak{A})$ is fully closed. In any case, a *-representation π can always be extended to a fully closed *-representation $\widehat{\pi}(\mathfrak{A})$, namely $\widehat{\pi}(x) = \widehat{\pi(x)}, x \in \mathfrak{A}$, on the domain $\mathcal{D}(\widehat{\pi}) = \widehat{\mathcal{D}}(\pi(\mathfrak{A}))$. We also need a notion of cyclic vectors adapted to the representation context. A vector ξ is called *strongly $\hat{p}$-cyclic* for π if the set $\{a\xi : a \in R\mathfrak{A}, \pi(a)\xi \in \mathcal{D}(\widehat{\pi})\}$ is a core for every $\pi(x), x \in \mathfrak{A}$. (Notice that this is one possible definition among several, see (Antoine et al. 1996) for alternative ones).

Next we define the weak commutants of a *-representation of a partial *-algebra. Besides the usual weak bounded commutant $\pi(\mathfrak{A})'_{\mathrm{w}}$ of $\pi(\mathfrak{A})$, as defined in (4.2), we introduce a new one, called *quasi-weak*, which takes explicitly into account the possible lack of associativity :

$$\begin{aligned}\mathcal{C}_{\mathrm{qw}}(\pi) = \{C \in \pi(\mathfrak{A})'_{\mathrm{w}}; (C\pi(x_1{}^*)\xi|\pi(x_2)\eta) = (C\xi|\pi(x_1x_2)\eta)\,,\\ \text{for all } x_1, x_2 \in \mathfrak{A} \text{ such that } x_1 \in L(x_2) \text{ and all } \xi, \eta \in \mathcal{D}(\pi)\}\,. \quad (5.1)\end{aligned}$$

$\mathcal{C}_{\mathrm{qw}}(\pi)$ is a weakly closed *-invariant subspace of $\mathcal{B}(\mathcal{H})$ and, moreover, $\mathcal{C}_{\mathrm{qw}}(\widehat{\pi}) = \mathcal{C}_{\mathrm{qw}}(\pi)$.

Accordingly, a *-representation π of $\mathfrak{A}$ in $\mathcal{L}^\dagger_{\mathrm{w}}(\mathcal{D}, \mathcal{H})$ is said to be *irreducible* iff its bounded quasi-weak commutant $\mathcal{C}_{\mathrm{qw}}(\pi)$ is trivial, $\mathcal{C}_{\mathrm{qw}}(\pi) = \{\lambda I, \lambda \in \mathbb{C}\}$. This definition leads to the expected correspondence between pure states and irreducible GNS representations (see below). The same result does not hold, in general, if we replace the quasi-weak commutant $\mathcal{C}_{\mathrm{qw}}(\pi)$ by the weak bounded commutant $\pi(\mathfrak{A})'_{\mathrm{w}}$, and *a fortiori* by the weak unbounded one $\pi(\mathfrak{A})'_{\sigma}$.

5.2 The GNS Construction

As always, the crucial question is how to build concrete representations. For *-algebras, the Gel'fand-Naimark-Segal (GNS) construction is usually the answer (Bratteli and Robinson 1979). In order to extend it to partial *-algebras, we must first have an appropriate notion of state. In the case of a *-algebra $\mathfrak{A}$, a state is a normalized positive linear form on $\mathfrak{A}$. If $\mathfrak{A}$ is only a partial *-algebra, the positivity condition alone already requires the use of *sesquilinear* forms. Then, for a *-algebra $\mathfrak{A}$, the GNS construction works only if the starting sesquilinear form ϕ on $\mathfrak{A} \times \mathfrak{A}$ is *invariant*, in the sense that $\phi(x^*y, z) = \phi(y, xz)$, for all $x, y, z \in \mathfrak{A}$. Clearly this definition is inapplicable for a partial *-algebra, since the products x^*y and xz need not exist. An obvious solution is to impose this relation for $y, z \in R\mathfrak{A}$ only, and this gives us a good hint.

Let $\mathfrak{A}$ be a partial *-algebra. A sesquilinear form φ on $\mathfrak{A} \times \mathfrak{A}$ is called *positive* if $\varphi(x, x) \geq 0, \forall\, x \in \mathfrak{A}$. When $\mathfrak{A}$ has a unit e, a positive sesquilinear form φ on $\mathfrak{A} \times \mathfrak{A}$ is called a *state* if $\varphi(e, e) = 1$. For each positive sesquilinear form φ on $\mathfrak{A} \times \mathfrak{A}$, we have:

$$\varphi(x,y) = \overline{\varphi(y,x))}, \ \forall x,y \in \mathfrak{A} \,, \tag{5.2}$$

$$|\varphi(x,y)|^2 \leq \varphi(x,x)\,\varphi(y,y), \ \forall x,y \in \mathfrak{A} \,, \tag{5.3}$$

and hence

$$\mathfrak{N}_\varphi \equiv \{x \in \mathfrak{A}; \varphi(x,x) = 0\} = \{x \in \mathfrak{A}; \varphi(x,y) = 0 \text{ for all } y \in \mathfrak{A} \}, \tag{5.4}$$

and so $\mathfrak{N}_\varphi$ is a subspace of $\mathfrak{A}$. For each $x \in \mathfrak{A}$ we denote by $\lambda_\varphi(x)$ the coset of $\mathfrak{A}/\mathfrak{N}_\varphi$ which contains x, and define an inner product $(.|.)$ on $\lambda_\varphi(\mathfrak{A})$ by

$$(\lambda_\varphi(x)|\lambda_\varphi(y)) = \varphi(x,y), \ x,y \in \mathfrak{A}. \tag{5.5}$$

We denote by $\mathcal{H}_\varphi$ be the Hilbert space obtained by the completion of the pre-Hilbert space $\lambda_\varphi(\mathfrak{A})$. We are now ready to introduce our notion of invariance.

Definition 5.1 – A positive sesquilinear form φ on $\mathfrak{A} \times \mathfrak{A}$ is said to be $\mathfrak{B}$-*invariant* if there exists a subspace $\mathfrak{B}$ of $R\mathfrak{A}$ such that :

(1) $\lambda_\varphi(\mathfrak{B})$ is dense in $\mathcal{H}_\varphi$;
(2) $\varphi(xb_1, b_2) = \varphi(b_1, x^* b_2), \forall\, x \in \mathfrak{A}, \forall\, b_1, b_2 \in \mathfrak{B}$;
(3) $\varphi(x_1{}^* b_1, x_2 b_2) = \varphi(b_1, (x_1 x_2) b_2), \forall\, x_1 \in L(x_2), \forall\, b_1, b_2 \in \mathfrak{B}$;
(4) if $\mathfrak{A}$ has a unit e, then $e \in \mathfrak{B}$.

Condition (3) takes explicitly into account the possible lack of associativity of $\mathfrak{A}$. Note also that conditions (1)-(4) are not imposed to the whole set $R\mathfrak{A}$, but only to the subspace $\mathfrak{B}$ (with dense image $\lambda_\varphi(\mathfrak{B})$) in $\mathcal{H}_\varphi$. The reason is that $R\mathfrak{A}$ may be too large or difficult to characterize completely, whereas it is often easy to find a suitable subspace $\mathfrak{B}$ of $R\mathfrak{A}$.

Given φ, we denote by $\mathcal{F}_\varphi$ the family of subspaces $\mathfrak{B}$ satisfying the conditions (1)-(4). Then, for any $\mathfrak{B} \in \mathcal{F}_\varphi$, there exists a maximal subspace in $\mathcal{F}_\varphi$ containing $\mathfrak{B}$; we denote it by $[\mathfrak{B}]$.

The next theorem establishes the GNS construction, including the dependence on the choice of the subspace $\mathfrak{B} \in \mathcal{F}_\varphi$.

Theorem 5.2 – *Let φ be a $\mathfrak{B}$-invariant positive sesquilinear form on $\mathfrak{A} \times \mathfrak{A}$. Then:*

(1) Define

$$\pi_\varphi^{\mathfrak{B}}(x)\lambda_\varphi(b) = \lambda_\varphi(xb), \ x \in \mathfrak{A}, b \in \mathfrak{B}. \tag{5.6}$$

Then $\pi_\varphi^{\mathfrak{B}}$ is a *-representation of $\mathfrak{A}$ into $\mathcal{L}^\dagger_{\rm w}(\lambda_\varphi(\mathfrak{B}), \mathcal{H}_\varphi)$. If $\mathfrak{A}$ has a unit e, the vector $\Omega_\varphi = \lambda_\varphi(e)$ is strongly $\hat{\mathfrak{p}}$-cyclic for $\pi_\varphi^{\mathfrak{B}}$.

(2) $\widehat{\pi}_\varphi^{\mathfrak{B}} \subset \widehat{\pi}_\varphi^{[\mathfrak{B}]}$ for each $\mathfrak{B} \in \mathcal{F}_\varphi$ and $\widehat{\pi}_\varphi^{\mathfrak{B}} \neq \widehat{\pi}_\varphi^{[\mathfrak{B}]}$ in general.

(3) Let $\mathfrak{B}_1, \mathfrak{B}_2 \in \mathcal{F}_\varphi$. Then $[\mathfrak{B}_1] \neq [\mathfrak{B}_2]$ if and only if $\widehat{\pi}_\varphi^{[\mathfrak{B}_1]} \neq \widehat{\pi}_\varphi^{[\mathfrak{B}_2]}$.

We call the triple $(\pi_\varphi^{\mathfrak{B}}, \lambda_\varphi, \mathcal{H}_\varphi)$ the *GNS construction* for φ, based on $\mathfrak{B}$.

We assume now that $\mathfrak{A}$ has a unit e, and consider a $\mathfrak{B}$-invariant state φ on $\mathfrak{A} \times \mathfrak{A}$. As usual, we say that the state φ is *pure* if it cannot be written as a convex combination of two $\mathfrak{B}$-invariant states φ_1, φ_2:

$$\varphi \neq \lambda\varphi_1 + (1-\lambda)\varphi_2, \ 0 < \lambda < 1. \tag{5.7}$$

The interest of this concept is that the equivalence between the purity of a state φ and the irreducibility of its GNS representation $\pi_\varphi^{\mathfrak{B}}$ extends to partial *-algebras, essentially with the same proof:

Proposition 5.3 – *Let $\mathfrak{A}$ be a partial *-algebra with unit and φ a $\mathfrak{B}$-invariant state on $\mathfrak{A} \times \mathfrak{A}$. Then the GNS representation $\pi_\varphi^{\mathfrak{B}}$ is irreducible, in the sense that $C_{\mathrm{qw}}(\pi_\varphi^{\mathfrak{B}}) = \mathbb{C}I$, if and only if φ is pure.*

Notice that, if we define the irreducibility of $\pi_\varphi^{\mathfrak{B}}$ by the condition $\pi_\varphi^{\mathfrak{B}}(\mathfrak{A})'_{\mathrm{w}} = \mathbb{C}I$, then the proof breaks down, because the positive sesquilinear forms φ_1, φ_2 on $\mathfrak{A} \times \mathfrak{A}$ into which φ would decompose need not be $\mathfrak{B}$-invariant (Condition (3) of Definition 5.1 may fail).

In the usual case of *-algebras, weights (*i.e.*, unbounded functionals) are the most general objects that allow a GNS construction. It turns out that the notion of weight can be extended to partial *-algebras, in several ways (Antoine et al. 1995, Antoine et al. 1996).

Weights on a *-algebra $\mathfrak{A}$ are usually defined as functions from the positive cone of $\mathfrak{A}$, *i.e.*, $\mathfrak{A}_+ = \{x \in \mathfrak{A} \mid x = y^*y \text{ for some } y \in \mathfrak{A}\}$, into $[0, \infty]$, which preserve addition and multiplication by non-negative real numbers (Bratteli and Robinson 1979, Stratila and Zsidó 1979). Notice that weights are allowed to be infinite on some elements. In a partial *-algebra, however, we cannot consider positive elements, since x^*x need not be defined for arbitrary x! Their role will be played by the diagonal elements of $\mathfrak{A} \times \mathfrak{A}$, that is, $\{(x, y) \in \mathfrak{A} \times \mathfrak{A} \mid x = y\}$, and a weight will be nothing else than a non-negative function of them, which can then be extended to a suitable larger subset of $\mathfrak{A} \times \mathfrak{A}$ under some extra conditions. In addition, one has to introduce a notion of invariance with respect to a subspace $\mathfrak{B}$ of $R\mathfrak{A}$. This being done, it is possible to formulate a variant of the GNS theorem.

Another extension yet is obtained by replacing a weight on the partial *-algebra $\mathfrak{A}$ by a *quasi-weight*, meaning a weight (in the usual sense) which is defined and finite only on the positive cone generated by some left-ideal of $\mathfrak{A}$. In fact this concept of quasi-weight has been developed so far mostly for O*-algebras using the notion of standard generalized vector. In that case it is possible to build (by an appropriate Tomita-Takesaki construction (Bratteli and Robinson 1979)) quasi-weights satisfying a KMS-condition. If the O*-algebra represents the observable algebra of some physical system, then the KMS quasi-weights are good candidates for representing equilibrium states of the system. These results may in turn be generalized to partial O*-algebras, and in particular to partial GW*-algebras, with a suitable extension of standard generalized vectors (Antoine et al. 1997). On the other hand, the notion of invariant positive sesquilinear form developed above may also be extended further, leading to what seems the natural definition of sesquilinear weight on a partial *-algebra, appropriately called a *biweight*. In a nutshell, a biweight is a positive sesquilinear form which is defined only on a dense domain, and for such objects, the GNS construction goes through (Antoine et al. 1998a).

6 *-Automorphisms and Derivations of Partial O*-Algebras

6.1 *-Automorphisms

In the algebraic formulation of quantum theories, the observables of a physical system are represented by hermitian elements of a certain *-algebra $\mathfrak{A}$ and states by positive linear functionals on $\mathfrak{A}$. Then a symmetry of the system is realized by a *-automorphism σ of $\mathfrak{A}$, and a one parameter symmetry group by a *-automorphism group σ_t ($t \in \mathbb{R}$) of $\mathfrak{A}$. Given a state, the GNS construction yields a representation π of $\mathfrak{A}$ in a Hilbert space $\mathcal{H}_\pi$ and a *-automorphism σ^π (resp.*-automorphism group σ_t^π) of $\pi(\mathfrak{A})$. Then the question is whether σ^π is *spatial*, that is, whether there exists a unitary operator U in $\mathcal{H}_\pi$ such that $\sigma^\pi(A) = U^* A U$, for every $A \in \mathfrak{A}$. Even more interesting is the case where U itself can be taken in $\mathfrak{A}$, *i.e.*, the automorphism is *inner*. For a one parameter group σ_t, spatiality means that the automorphism group σ_t^π is unitarily implemented, *i.e.*, $\sigma_t^\pi(A) = e^{-iHt} A e^{iHt}$, where H is a self-adjoint operator. In particular, if the automorphism is inner, this means that $H \in \mathfrak{A}''$ (or H is affiliated to $\mathfrak{A}''$), in other words that the operator H is an observable. For instance, if σ_t represents the time evolution of the system, then 'σ_t^π is inner' means that the Hamiltonian exists as an observable in the (GNS) representation at hand (Borchers 1966, dell'Antonio 1966, Kadison and Ringrose 1967). Similar questions are commonly asked about other automorphisms, such as Bogoliubov transformations for the CCR (Bratteli and Robinson 1979).

At the infinitesimal level, the generator of a given *-automorphism group σ_t^π of symmetries is a *derivation*, that is, a map δ^π from $\mathfrak{A}$ into some class of operators on $\mathcal{H}_\pi$, which satisfies a Leibniz rule: $\delta^\pi(AB) = A\delta^\pi(B) + \delta^\pi(A)B,\ A, B \in \mathfrak{A}$. If σ_t^π is spatial, with generator H, then the corresponding derivation δ^π satisfies the relation $\delta^\pi(A) = i[H, A],\ \in \mathfrak{A}$, on appropriate domains, and naturally the derivation δ^π is then also said to be spatial.

Now, if one decides to describe the set of observables of a given physical system by some partial *-algebra, one must extend to that context the notions of *-automorphism and of derivation, and of spatiality as well.

Let $\mathfrak{M}$ be a partial O*-algebra on $\mathcal{D}$, obtained, for instance, from the partial *-algebra of observables by a GNS construction. According to the general definition, a **-automorphism* of $\mathfrak{M}$ is a linear bijection $\sigma : \mathfrak{M} \to \mathfrak{M}$ such that (i) $\sigma(X^\dagger) = \sigma(X)^\dagger, \forall X \in \mathfrak{M}$; (ii) $\sigma(Y) \in R^{\mathrm{w}}(\sigma(X))$ iff $Y \in R^{\mathrm{w}}(X)$ and then $\sigma(X \,\square\, Y) = \sigma(X) \,\square\, \sigma(Y)$ and (iii) the same relations hold for σ^{-1}.

It follows that $\sigma(\mathfrak{M}) = \mathfrak{M}$ and $\sigma(R\mathfrak{M}) = R\mathfrak{M}$. The *-automorphism σ is *spatial* if there exists a unitary operator $U \in \mathcal{H}$ such that $U_o \equiv U\restriction\mathcal{D} \in R^{\mathrm{w}}(\mathfrak{M})$ and $\sigma(X) = U_o{}^*(X \,\square\, U_o), \forall X \in \mathfrak{M}$. It is *inner* if in addition $U_o \in \mathfrak{M}$, *i.e.*, $U_o \in R\mathfrak{M}$. Notice that, if $\mathfrak{M}$ is self-adjoint and σ is spatial, then $U_o \in \mathcal{L}^\dagger(\mathcal{D})$ and $\sigma(X) = U_o{}^* X U_o, \forall X \in \mathfrak{M}$.

As expected, one gets stronger results if one assumes that $\mathfrak{M}$ is a partial GW*-algebra, because then many known results about the spatiality of *-automorphisms of von Neumann algebras will be lifted from the bounded part of $\mathfrak{M}$ to $\mathfrak{M}$ itself (Antoine et al. 1994). Thus we consider a partial GW*-algebra $\mathfrak{M}$ on the domain $\mathcal{D}$. Then its bounded part $\mathfrak{M}_b = \{X \in \mathfrak{M}; \overline{X} \in \mathcal{B}(\mathcal{H})\}$ is the restriction to $\mathcal{D}$ of a von Neumann algebra $\overline{\mathfrak{M}_b} = \{\overline{X}; X \in \mathfrak{M}_b\}$, with commutant $\overline{\mathfrak{M}_b}' = \mathfrak{M}'_{\mathrm{w}}$. The key observation is that every *-automorphism σ of $\mathfrak{M}$ induces a *-automorphism σ_b of the von Neumann algebra $\overline{\mathfrak{M}_b}$, by the simple relation $\sigma_b(\overline{A}) = \overline{\sigma(A)},\ A \in \mathfrak{M}_b$. Accordingly, one says that a *-automorphism σ of $\mathfrak{M}$ is *weakly spatial* (resp. *weakly inner*) if the corresponding *-automorphism σ_b of the von Neumann algebra $\overline{\mathfrak{M}_b}$ is spatial (resp. inner).

Now it is clear that a weakly spatial *-automorphism will be spatial as soon as there is enough continuity for lifting it from $\mathfrak{M}_b$ to $\mathfrak{M}$. This is indeed the case, and of course the relevant topology is the strong* one.

Theorem 6.1 — *Let $\mathfrak{M}$ be a partial GW*-algebra on $\mathcal{D}$ and σ a *-automorphism of $\mathfrak{M}$. If σ is weakly spatial (resp. weakly inner) and τ_{s^*}-continuous, then σ is spatial (resp. inner).*

If $\mathfrak{M}$ is self-adjoint, the two conditions are in fact equivalent:

Corollary 6.2 — *Let $\mathfrak{M}$ be a self-adjoint partial GW*-algebra on $\mathcal{D}$ and σ a *-automorphism of $\mathfrak{M}$. Then σ is spatial if and only if it is weakly spatial and τ_{s^*}-continuous.*

At this point, one may systematically list all the known results about spatial *-automorphisms of von Neumann algebras, as given in standard treatises such as (Stratila and Zsidó 1979 or (Dixmier 1957, 1969), and try to lift them to partial GW*-algebras. The following one is very simple.

Corollary 6.3 — *Let $\mathfrak{M}$ be a partial GW*-algebra with a cyclic and separating vector. Then every τ_{s^*}-continuous *-automorphism of $\mathfrak{M}$ is spatial.*

6.2 Automorphism Groups and *-Derivations

From the point of view of physical applications, a crucial role is played by automorphism groups: they describe either the time evolution of the system or physical symmetries. This is the reason why it is worth considering them in the context of partial O*-algebras.

Let $\mathfrak{M}$ be a partial O*-algebra. A one-parameter **-automorphism group* of $\mathfrak{M}$ is a map $\mathbb{R} \ni t \mapsto \alpha_t \in \mathrm{Aut}_*(\mathfrak{M})$ such that (i) $\alpha_0(X) = X, \quad \forall X \in \mathfrak{M}$;

and (ii) $\alpha_{s+t}(X) = \alpha_s(\alpha_t(X)), \quad \forall X \in \mathfrak{M}$. If τ is any topology on $\mathcal{L}^\dagger_{\mathrm{w}}(\mathcal{D},\mathcal{H})$, then the automorphism group α_t is called τ-continuous if $\mathbb{R} \ni t \mapsto \alpha_t(X)$ is continuous from $\mathbb{R}$ into $\mathcal{L}^\dagger_{\mathrm{w}}(\mathcal{D},\mathcal{H})[\tau]$, $\forall X \in \mathfrak{M}$.

If $t \mapsto \alpha_t(X)$ is τ-continuous, we define its *infinitesimal generator* as

$$\delta_\alpha(X) = \tau - \lim_{t\to 0} t^{-1}(\alpha_t(X) - X)\,.$$

on the domain $D_\tau(\delta_\alpha)$ which consists of all $X \in \mathfrak{M}$ for which the limit $\tau - \lim_{t\to 0} t^{-1}(\alpha_t(X) - X)$ exists in $\mathcal{L}^\dagger_{\mathrm{w}}(\mathcal{D},\mathcal{H})$. If the involution $X \mapsto X^\dagger$ is τ-continuous, then $X \in D_\tau(\delta_\alpha)$ implies $X^\dagger \in D_\tau(\delta_\alpha)$ and $\delta_\alpha(X^\dagger) = \delta_\alpha(X)^\dagger$.

In analogy with the C*-situation, one would expect that $D_\tau(\delta_\alpha)$ is a partial O*-algebra and that δ_α is a *-derivation of it, but for this we need a suitable form of the Leibniz rule. Motivated by the properties of infinitesimal generators, we define a *weak *-derivation* of $\mathfrak{M}$ as a linear map $\delta : \mathfrak{M} \to \mathcal{L}^\dagger_{\mathrm{w}}(\mathcal{D},\mathcal{H})$ satisfying the following conditions:

(i) $\delta(X)^\dagger = \delta(X^\dagger),\ \forall\, X \in \mathfrak{M}\,,$

(ii) $(\delta(X \,\square\, Y)\xi|\,\eta) = (Y\xi|\,\delta(X^\dagger)\,\eta) + (\delta(Y)\xi|\,X^\dagger\eta)\,,$

$\forall\, X, Y \in \mathfrak{M}$ such that $X \in L^{\mathrm{w}}(Y)$ and $\forall\, \xi, \eta \in \mathcal{D}\,.$

The weak *-derivation δ of $\mathfrak{M}$ is called *spatial*, resp. *inner*, if there exists an element $H_o = H_o^\dagger \in R^{\mathrm{w}}(\mathfrak{M})$, resp. $H \in R\mathfrak{M}$, such that

(i) H_o is the restriction to $\mathcal{D}$ of an operator $H \in \mathcal{L}(\mathfrak{M}\mathcal{D},\mathcal{H})$ that satisfies the relation $(HX^\dagger\xi|Y\eta) = (X^\dagger\xi|HY\eta)$, for all $X \in R^{\mathrm{w}}(\mathfrak{M})$ and $\xi, \eta \in \mathcal{D}$.

(ii) $\delta(X) = \delta_{H_o}(X) \equiv i(H \,\square\, X - X \,\square\, H),\ X \in \mathfrak{M}$.

Clearly, the properties of these derivations will depend both on the continuity properties of the automorphism group, and on the type of partial O*-algebra considered. As expected, partial GW*-algebras will again behave better. Indeed we have:

Proposition 6.4 — *Let $\mathfrak{M}$ be a partial GW^*-algebra. Let $t \mapsto \alpha_t$ be a strong*-continuous one parameter *-automorphism group of $\mathfrak{M}$ and δ_α the corresponding infinitesimal generator. Then $D(\delta_\alpha)$ is a partial O^*-algebra and δ_α is a weak *-derivation of $D(\delta_\alpha)$ satisfying $\delta_\alpha\,(D(\delta_\alpha)) \subset \mathfrak{M}$.*

We consider first derivations which are generators of the automorphism group α_t^H generated by a self-adjoint operator H in $\mathcal{H}$, that is,

$$\alpha_t^H(X) = e^{itH} X e^{-itH}, \quad X \in Ł,.$$

Proposition 6.5 - *Let $\mathfrak{M}$ be a partial GW^*-algebra and H a self-adjoint operator in $\mathcal{H}$. Assume that the corresponding unitary group $\{e^{itH}, t \in \mathbb{R}\}$ satisfies the following conditions:*

(i) $e^{itH}\mathcal{D} = \mathcal{D}, \quad \forall\, t \in \mathbb{R}$;

(ii) $e^{itH}\mathfrak{M}e^{-itH} = \mathfrak{M}, \quad \forall\, t \in \mathbb{R}$;

(iii) $t \mapsto e^{itH}\xi$ *is* $t_{\mathfrak{M}}$ *-continuous,* $\forall\, \xi \in \mathcal{D}$.
Then

(1) $\forall X \in \mathfrak{M}$, $\alpha_t^H(X)$ *is a strong*-continuous one parameter *-automorphism group of* $\mathfrak{M}$.

(2) $D_{\mathrm{w}}(\delta_\alpha)$ *is a partial O*-algebra and* δ_α *is a weak *-derivation of* $D_{\mathrm{w}}(\delta_\alpha)$.

(3) If either $D_{\mathrm{w}}(\delta_\alpha)\mathcal{D} \subset \mathcal{D}(H)$, *or* $\mathcal{D} \subset \mathcal{D}(H)$, *then* δ_α *is spatial, that is,* $\delta_\alpha(X) = \delta_{H_o}(X) \equiv i(H \,\square\, X - X \,\square\, H)$, $X \in \mathfrak{M}$, *where* $H_o \equiv H{\restriction}\mathcal{D}$.

Let now δ be an arbitrary weak *-derivation of $\mathfrak{M}$ such that $\delta(\mathfrak{M}) \subset \mathfrak{M}$. Then $\delta_b \equiv \delta{\restriction}\mathfrak{M}_b$ is a *-derivation of $\mathfrak{M}_b$ into $\mathfrak{M}$, which provides a simple tool for the study of δ.

Theorem 6.6 – *Let* $\mathfrak{M}$ *be a partial GW*-algebra and* δ *a weak *-derivation of* $\mathfrak{M}$ *such that* $\delta(\mathfrak{M}) \subset \mathfrak{M}$ *and* $\delta(\mathfrak{M}_b) \subset \mathfrak{M}_b$. *If* δ *is weakly continuous, then there exists* $H = H^\dagger \in \mathfrak{M}_b \cap R\mathfrak{M}$ *such that*

$$\delta(X)\xi = i(H \,\square\, X - X \,\square\, H)\xi,\ \forall X \in \mathfrak{M}, \forall\, \xi \in \mathcal{D}.$$

The converse is also true if $\mathfrak{M}$ *is self-adjoint.*

7 Epilogue

The conclusion of this rapid survey is that the theory of partial *-algebras has reached after fifteen years a reasonable stage of maturity. Many nontrivial examples have been studied, both abelian and nonabelian, although no classification has been made so far. The representation theory is well under control. In particular, many standard results extend to partial *-algebras, such as the GNS construction or various structure properties. Two offshoots, in particular, have undergone a rapid development, namely CQ*-algebras and partial O*-algebras. The latter, and among them partial GW*-algebras, are a far reaching generalization of *-algebras of operators, both bounded and unbounded. Their structure is quite complex, yet a substantial body of information is available. Besides the representation theory associated to various notions of generalized vectors and weights, progress has been achieved also in the study of *-automorphisms and *-derivations, in particular the spatial theory.

These last results point toward the most promising direction of research, namely the study of dynamical systems based on partial O*-algebras. In view of the results obtained so far, it is reasonable to expect progress for the case of partial GW*-algebras, since then the powerful theory of von Neumann algebras is available. In particular, the modular theory of Tomita–Takesaki extends, with suitable modifications, to partial GW*-algebras.

What about physical applications? So far spin systems with long range correlations are essentially the only systems where partial *-algebras have had a impact. However, the mathematical tool is there and may be developed for its own sake. Future research will decide which physical systems, if any, are complex enough to *require* the use of this approach.

References

Antoine, J-P., Grossmann, A., (1976): Partial inner product spaces. I. General properties. II. Operators, J. Funct. Anal. **23**, 369–378, 379–391

Antoine, J-P., (1980): Partial inner product spaces. III. Compatibility relations revisited. IV. Topological considerations, J. Math. Phys. **21**, 268–279, 2067–2079

Antoine, J-P., Gustafson, K., (1981): Partial inner product spaces and semi-inner product spaces, Adv. in Math. **41**, 281–300

Antoine, J-P., Karwowski, W., (1983): Partial *-algebras of closed operators, in Quantum Theory of Particles and Fields, pp. 13–30; B. Jancewicz and J. Lukierski (eds.), World Scientific, Singapore

Antoine, J-P., Karwowski, W., (1985): Partial *-algebras of closed linear operators in Hilbert space, Publ. RIMS, Kyoto Univ. **21**, 205–236; Add./Err. *ibid.* **22**, (1986) 507–511

Antoine, J-P., Mathot, F., (1987): Partial *-algebras of closed operators and their commutants. I. General structure, Ann. Inst. H. Poincaré **46**, 299–324

Antoine, J-P., Inoue, A., Trapani, C., (1994): Spatial theory of *-automorphisms on partial O*-algebras, J. Math. Phys. **35**, 3059–3073

Antoine, J-P., Soulet, Y., Trapani, C., (1995): Weights on partial *-algebras, J. Math. Anal. Appl. **192**, 920–941

Antoine, J-P., Inoue, A., Trapani, C., (1996): Partial *-algebras of closable operators: A review, Reviews Math. Phys. **8**, 1–42

Antoine, J-P., Inoue, A., Ogi, H., (1997): Standard generalized vectors for partial O*-algebras, Ann. Inst. H. Poincaré **67**, 223–258

Antoine, J-P., Inoue, A., Trapani, C., (1998): Biweights on partial *-algebras, preprint UCL-98-11, Louvain-la-Neuve (submitted)

Antoine, J-P., Bagarello, F., Trapani, C., (1998): Topological partial *-algebras: Basic properties and examples, Reviews Math. Phys. **10**, (to appear)

dell'Antonio, G.F., (1966): On some groups of automorphisms of physical observables, Commun. Math. Phys. **2**, 384–397

Bergh, J., Löfström, J., (1976): *Interpolation Spaces*, Springer, Berlin

Borchers, H.J., (1966): Energy and momentum as observables in quantum field theory, Commun. Math. Phys. **2**, 49–54

Bratteli, O., Robinson, D.W., (1979): *Operator Algebras and Quantum Statistical Mechanics I, II*, Springer-Verlag, Berlin

Dixmier, J., (1957), (1969): *Les algèbres d'opérateurs dans l'espace hilbertien (Algèbres de von Neumann)*, Gauthier-Villars, Paris

Fournier, J.J.F., Stewart, J., (1985): Amalgams of L^p and ℓ^q, Bull. Amer. Math. Soc. **13**, 1–21

Kadison, R.V., Ringrose, J.R., (1967): Derivations and automorphisms of operator algebras, Commun. Math. Phys. **4**, 32–63

Schmüdgen, K., (1990): *Unbounded Operator Algebras and Representation Theory*, Akademie-Verlag, Berlin

Stratila, S., Zsidó, L., (1979): *Lectures on von Neumann Algebras*, Abacus Press, Tunbridge Wells (England)

Banach Partial *-Algebras and Quantum Models

C. Trapani

Dipartimento di Scienze Fisiche ed Astronomiche
Università di Palermo
I-90123 Palermo (Italy)

1 Introduction

C^*-algebras are, as known, the basic mathematical ingredient of the Haag-Kastler (Haag and Kastler 1964) algebraic approach to quantum systems, with infinitely many degrees of freedom. The usual procedure starts, in fact, with associating to each bounded region V of the configuration space of the system the C^*-algebra $\mathcal{A}_V$ of local observables in V. The uniform completion $\mathcal{A}$ of the algebra generated by the $\mathcal{A}_V$'s is then considered as the C^*-algebra of observables of the system.

Several physical models, however, do not fit into the Haag-Kastler setup.

In many quantum statistical systems, in fact, the thermodynamical limit of some local observables, for instance the local Heisenberg dynamics, does not exist in the uniform topology and thus it is not an element of the observables algebra as defined before. This is the case, for instance, of the BCS model (Thirring and Wehrl 1967), and, in general, of any mean field model.

Also, in the Wightman formulation of quantum field theory, point-like fields are not, in general, elements of any C^*-algebra: the field $A(x)$ at a point $x \in \mathbb{R}^4$ is, in fact, an (unbounded) sesquilinear form on the domain $\mathcal{D}$ where all *smeared* fields $A(f)$, $f \in \mathcal{S}(\mathbb{R}^4)$ are defined. If $A(x)$ is, for each fixed $x \in \mathbb{R}^4$, a continuous map from $\mathcal{D}$ into its dual $\mathcal{D}'$ (Epifanio and Trapani 1987), then it is the limit of a sequence of observables localized in a shrinking sequence of space-time regions and, therefore, it belongs to a certain completion of the C^*-algebra $\mathcal{A}_0$ of local observables (Ascoli et al. 1970), (Fredenhagen and Hertel 1981).

In spite of these physical models, the elegant Haag-Kastler construction can be entirely preserved if the assumption that the observable algebra $\mathcal{A}_0$ is a C^*-algebra is conveniently weakened.

Two possibilities are then at hand. The first one occurs if there exists on $\mathcal{A}_0$ a weaker norm such that the completion of $\mathcal{A}_0$ with respect to this norm contains all 'objects' of physical interest. If this possibility fails, it could still happen that these 'objects' can be recovered by taking the completion of $\mathcal{A}_0$ with respect to the locally convex (*non-normable*) topology generated by a suitable family of seminorms.

Now, the completion of a locally convex algebra, where the multiplication is not jointly continuous, is the most typical instance of a structure with a partially defined multiplication: the multiplication in the completion is defined for pairs of elements one of which lies in the original algebra.

For this reason, in both of the cases discussed above one has to deal with *(topological) partial *-algebras*, introduced originally by Antoine and Karwowski (Antoine and Karwowski 1985) and studied by several authors (Mathot, Inoue, Ekhaguere, Bagarello and the author, see (Antoine et al. 1996, Antoine et al. to appear) for complete references). In view of applications, more than a general partial *-algebra it is sometimes useful to consider some relevant subclass of them such as *quasi *-algebras* (introduced by Lassner (Lassner 1981), (Lassner 1984)) or *CQ^*-algebras*, that will be discussed below.

The class of CQ^*-algebras and their possible applications in the study of some quantum model are the main subject of this paper. The study of CQ^*-algebras has been carried out in a series of papers by F. Bagarello and the author (Bagarello and Trapani 1994, Bagarello and Trapani 1996b, Bagarello and Trapani 1996c)

CQ^*-algebras are *Banach partial *-algebras* with a particularly simple multiplication structure: the lattice of multipliers consists of only four spaces. But they enjoy a lot of interesting topological properties that make of them a possible extension of the notion of C^*-algebra in the partial algebraic framework.

2 CQ*-Algebras

To begin with, let us give the basic aspects of the theory and discuss some examples.

Definition 2.1 *Let $\mathcal{A}$ be a right Banach module over the C^*-algebra $\mathcal{A}_\flat$, with isometric involution $*$ and such that $\mathcal{A}_\flat \subset \mathcal{A}$. Set $\mathcal{A}_\sharp = (\mathcal{A}_\flat)^*$. We say that $\{\mathcal{A}, *, \mathcal{A}_\flat, \flat\}$ is a CQ^*-algebra if*

(i) $\mathcal{A}_\flat$ is dense in $\mathcal{A}$ with respect to its norm $\| \ \|$
(ii) $\mathcal{A}_o := \mathcal{A}_\flat \cap \mathcal{A}_\sharp$ is dense in $\mathcal{A}_\flat$ with respect to its norm $\| \ \|_\flat$
(iii) $\|B\|_\flat = \sup_{A\in\mathcal{A}} \|AB\|, \quad B \in \mathcal{A}_\flat$

Since $*$ is isometric, the space $\mathcal{A}_\sharp$ is itself, as it is easily seen a C*-algebra with respect to the involution $X^\sharp := (X^*)^\flat$ and norm $\| X \|_\sharp := \| X^* \|_\flat$.
A CQ*-algebra is called *proper* if $\mathcal{A}_\sharp = \mathcal{A}_\flat$.

As is clear the algebraic structure of a CQ*-algebra is just that of a partial *-algebra: indeed the multiplication of two elements $A, B \in \mathcal{A}$ is defined if either $A \in \mathcal{A}_\sharp$ or $B \in \mathcal{A}_\flat$. So that the lattice of multipliers contains only $\mathcal{A}$, $\mathcal{A}_\flat = RA$, $\mathcal{A}_\sharp = LA$ and $\mathcal{A}_0 = RA \cap LA$.

From the topological point of view $\mathcal{A}$ is a Banach partial*-algebra with the additional assumption that $\mathcal{A}_\flat$ (and therefore also $\mathcal{A}_\sharp$) is a C*-algebra dense in it.

Examples of CQ*-algebras can be easily built up taking into account the following proposition proved in (Bagarello and Trapani 1994, Bagarello et al. unpublished):

Proposition 2.2 *Let $\mathcal{A}_\flat$ be a C^*-algebra, with norm $\| \, \|_\flat$ and involution $\flat$; let $\| \, \|$ be another norm on $\mathcal{A}$, weaker than $\| \, \|_\flat$ and such that*

(i) $\| AB \| \leq \| A \| \| B \|_\flat \qquad \forall A, B \in \mathcal{A}_\flat$
(ii) there exists a $\| \, \|_\flat$-dense subalgebra $\mathcal{A}_0$ of $\mathcal{A}_\flat$ where an involution * is defined with the property

$$\| A^* \| = \| A \| , \quad \forall A \in \mathcal{A}_0 ,$$

then if $\mathcal{A}$ denotes the $\| \, \|$-completion of $\mathcal{A}_\flat$, $(\mathcal{A}, *, \mathcal{A}_\flat, \flat)$ is a CQ^*-algebra.

If $\mathcal{A}_\flat = \mathcal{A}_0$ and $\flat = *$, then the CQ*-algebra we obtain is proper (Bagarello and Trapani 1994).

Thus, for instance, if I is a compact interval on the real line, the usual space $L^p(I, dx)$ (dx the usual Lebesgue measure on I) can be viewed as a proper CQ*-algebra over the C*-algebra $C(I)$ of all continuous function on I (Bagarello and Trapani 1996c).

Interesting examples of non-abelian CQ*-algebras can be obtained in several ways. One of them can be built up, as shown in (Bagarello at al. 1998), starting from an achieved left (or right) Hilbert algebra (this is, as known, the basic ingredient for the Tomita-Takesaki theory). Nevertheless, the most typical examples of non-commutative CQ*-algebras are provided by spaces of linear operators acting on a *scale* of Hilbert spaces. This will be explained in details in the next subsection.

2.1 CQ*-Algebras of Operators on Scales of Hilbert Spaces

Let $\mathcal{H}$ be a Hilbert space with scalar product $(.,.)$ and S a positive selfadjoint *unbounded* operator, with dense domain $D(S)$. The subspace $D(S)$ becomes a Hilbert space, denoted by $\mathcal{H}_{+1}$, with the scalar product

$$(f, g)_{+1} = (f, g) + (Sf, Sg), . \tag{1}$$

Let $\mathcal{H}_{-1}$ denote the conjugate dual of $\mathcal{H}_{+1}$. Then $\mathcal{H}_{-1}$ itself is a Hilbert space.

With this construction, we get, in canonical way a scale of Hilbert spaces

$$\mathcal{H}_{+1} \stackrel{i}{\to} \mathcal{H} \stackrel{j}{\to} \mathcal{H}_{-1} , \tag{2}$$

where i is the identity map i of $\mathcal{H}_{+1}$ into $\mathcal{H}$ and j is the canonical embedding of $\mathcal{H}$ into $\mathcal{H}_{-1}$ (both these maps are continuous and with dense range).

With obvious identifications, we can read (2) as a chain of topological inclusions

$$\mathcal{H}_{+1} \subset \mathcal{H} \subset \mathcal{H}_{-1}\,.$$

Let $\mathcal{B}(\mathcal{H}_{+1}, \mathcal{H}_{-1})$ be the Banach space of bounded operators from $\mathcal{H}_{+1}$ into $\mathcal{H}_{-1}$ with its natural norm $\| \cdot \|_{+1,-1}$.

In $\mathcal{B}(\mathcal{H}_{+1}, \mathcal{H}_{-1})$ define an involution $A \mapsto A^*$ by:

$$(A^* f, g) = \overline{(Ag, f)} \quad \forall f, g \in \mathcal{H}_{+1}\,.$$

Then $A^* \in \mathcal{B}(\mathcal{H}_{+1}, \mathcal{H}_{-1})$ and $\| A^* \|_{+1,-1} = \| A \|_{+1,-1} \quad \forall A \in \mathcal{B}(\mathcal{H}_{+1}, \mathcal{H}_{-1})$.

Let $\mathcal{B}(\mathcal{H}_{+1})$ be the C^*-algebra of all bounded operators on $\mathcal{H}_{+1}$. Its natural involution is denoted here as $\flat$ and its C*-norm as $\| \cdot \|_\flat$.
Furthermore, let $\mathcal{B}(\mathcal{H}_{-1})$ be the C^*-algebra of all bounded operators on $\mathcal{H}_{-1}$ with involution $\sharp$ and C*-norm $\| \cdot \|_\sharp$.

Then $\mathcal{B}(\mathcal{H}_{+1})$ and $\mathcal{B}(\mathcal{H}_{-1})$ are (isomorphic to) subspaces of $\mathcal{B}(\mathcal{H}_{+1}, \mathcal{H}_{-1})$, and $A \in \mathcal{B}(\mathcal{H}_{+1})$ if, and only if, $A^* \in \mathcal{B}(\mathcal{H}_{-1})$.

There is a *distinguished* *-algebra of $\mathcal{B}(\mathcal{H}_{+1}, \mathcal{H}_{-1})$ is

$$\mathcal{B}^+(\mathcal{H}_{+1}) = \{A \in \mathcal{B}(\mathcal{H}_{+1}, \mathcal{H}_{-1}) \;\; : A, A^* \in \mathcal{B}(\mathcal{H}_{+1})\,\}\,.$$

Clearly, if $A \in \mathcal{B}(\mathcal{H}_{+1}, \mathcal{H}_{-1})$ and $B \in \mathcal{B}(\mathcal{H}_{+1})$, then AB is well-defined and $AB \in \mathcal{B}(\mathcal{H}_{+1}, \mathcal{H}_{-1})$. Analogously, if $C \in \mathcal{B}(\mathcal{H}_{-1})$, CA is well-defined and $CA \in \mathcal{B}(\mathcal{H}_{+1}, \mathcal{H}_{-1})$.

Therefore, $\mathcal{B}(\mathcal{H}_{+1}, \mathcal{H}_{-1})$ is a *right Banach module* over the C*-algebra $\mathcal{B}(\mathcal{H}_{+1})$.
Then, $(\mathcal{B}(\mathcal{H}_{+1}, \mathcal{H}_{-1}), *, \mathcal{B}(\mathcal{H}_{+1}), \flat)$ is a CQ*-algebra, whenever the density conditions (i),(ii) of Definition 2.1 are satisfied. This is, however, not true in general (in contrast with the claim in (Bagarello and Trapani 1994)): additional assumptions on the operator S generating the scale of Hilbert spaces are needed. Nevertheless, Proposition 2.2 always allows the construction of a CQ*-algebra of operators acting in the given scale of Hilbert spaces, that can be, however, *smaller* than $(\mathcal{B}(\mathcal{H}_{+1}, \mathcal{H}_{-1}), *, \mathcal{B}(\mathcal{H}_{+1}), \flat)$ (Trapani 1998). We denote this CQ*-algebra with $(\mathcal{B}_c(\mathcal{H}_{+1}, \mathcal{H}_{-1}), *, \mathcal{B}_c(\mathcal{H}_{+1}), \flat)$ and refer to it as *the CQ*-algebra of operators acting in the scale of Hilbert spaces.*

2.2 Mathematical Properties of CQ*-Algebras

In this subsection we will shortly overview the structure properties of CQ*-algebras. This study has been carried out in (Bagarello and Trapani 1996b) and shows that, under certain respects, CQ*-algebras are a good generalization of the notion of C*-algebras to the unbounded case.

The GNS-Construction. At this point, in analogy with the C*-case, the question arises as to whether a given CQ*-algebra $(\mathcal{A}, *, \mathcal{A}_\flat, \flat)$ can be represented by means of bounded operators acting on a scale of Hilbert spaces. In other words: is a GNS-like construction possible also in this case?
The answer is positive, provided that we start from a suitably chosen *state.*

Let $(\mathcal{A}, *, \mathcal{A}_\flat, \flat)$ be a CQ*-algebra. A linear functional and ω on $\mathcal{A}$ is called *admissible* (for GNS) if

(i) $\omega(X^*X) \geq 0, \quad \forall X \in \mathcal{A}_\flat$
(ii) $\omega(\mathbb{I}) = 1$
(iii) $|\omega(X^*AY)|^2 \leq \|A\| \omega(X^\flat X)\omega(Y^\flat Y), \quad \forall A \in \mathcal{A}, \forall X, Y \in \mathcal{A}_\flat$
(iv) If $\{X_n\} \subset \mathcal{A}_\flat$ is a sequence such that $\lim_{n\to\infty} \omega(X_n^* X_n) = 0$ and $\omega((X_n - X_m)^\flat (X_n - X_m)) \to 0$, then $\lim_{n\to\infty} \omega(X_n^\flat X_n) = 0$ results

If ω is admissible, then it is also $\flat$-positive, i.e., $\omega(X^\flat X) \geq 0, \quad \forall X \in \mathcal{A}_\flat$; thus the usual GNS representation of $\mathcal{A}_\flat$ exists. The condition *(iii)* is then the key for the definition of the representation on the whole $\mathcal{A}$. The details are too technical to be reported here. We only mention the fact that a GNS-construction was also given in (Bagarello and Trapani 1994) under stronger assumptions.

***-Semisemplicity.** In contrast with the case of C*-algebras, *-semisemplicity is not automatic for CQ*-algebras. For C*-algebras this notion can be given in several equivalent ways: in few words it means that the intersection of the kernels of all states is the null subspace.

Definition 2.3 *Let $(\mathcal{A}, *, \mathcal{A}_\flat, \flat)$ be a CQ*-algebra. We call $\mathcal{S}(\mathcal{A})$ the set of sesquilinear forms on $\mathcal{A} \times \mathcal{A}$ with the following properties:*

(i) $\Omega(A, A) \geq 0, \quad \forall A \in \mathcal{A};$
(ii) $\Omega(AB, C) = \Omega(B, A^*C), \quad \forall A \in \mathcal{A}, \forall B, C \in \mathcal{A}_\flat;$
(iii) $|\Omega(A, B)| \leq \|A\| \, \|B\|, \quad \forall A, B \in \mathcal{A}.$

The family $\mathcal{S}(\mathcal{A})$ of sesquilinear forms on $(\mathcal{A}, *, \mathcal{A}_\flat, \flat)$ plays here the same role as the family of *states* on C*-algebras; for this reason it is used to generalize the notion of *-semisemplicity.

Let $(\mathcal{A}, *, \mathcal{A}_\flat, \flat)$ be a CQ*-algebra. The set

$$\mathfrak{R}^{(*)} = \{X \in \mathcal{A} \, : \, \Omega(X, X) = 0 \quad \forall \Omega \in \mathcal{S}(\mathcal{A})\} \tag{3}$$

is called the **-radical* of $\mathcal{A}$.

Definition 2.4 *We call *-semisimple any CQ*-algebra $(\mathcal{A}, *, \mathcal{A}_\flat, \flat)$ such that $\mathfrak{R}^{(*)} = \{0\}$.*

Examples of abelian and non-abelian *-semisimple CQ*-algebras do really exist. For instance, as shown in (Bagarello and Trapani 1996c), the CQ*-algebra $(L^p(I), C(I))$ is *-semisimple if, and only if, $p \geq 2$. *-Semisimple CQ*-algebras exhibit a number of interesting properties, that we summarize here.

- Possibility of refining the multiplication in $\mathcal{A}$.
  We can define for certain pairs X, Y of elements of $\mathcal{A}$ a *weak* product $X \circ Y$ (this extends the natural multiplication in the Banach module). We get in this way a *richer* partial *-algebra, in the sense that the new lattice of multipliers contains much more than four elements.

- One can introduce auxiliary norms such as

$$\|X\|_\alpha^2 \equiv \sup_{\Omega \in \mathcal{S}(\mathcal{A})} \Omega(X, X)\,, \tag{4}$$

  or

$$\|X\|_\beta \equiv \sup\{|\Omega(XB, B)|;\ \Omega \in \mathcal{S}(\mathcal{A}), B \in \mathcal{A}_\flat,\ \|B\|_\flat \leq 1\} \tag{5}$$

  and shows, for instance a C*-like property:
  if $X^* \circ X$ is defined for a certain $X \in \mathcal{A}$ then

$$\|\ X\ \|_\alpha^2 = \|X^* \circ X\|_\beta \leq \|X^* \circ X\|_\alpha\,. \tag{6}$$

  It is worth remarking that if these two norms on $\mathcal{A}$ coincide with the original norm of $\mathcal{A}$ then $\mathcal{A}$ is a C*-algebra.

- Some results of the functional calculus for C*-algebras extends to *-semisimple CQ*-algebras.

3 CQ*-Algebras and Spin Lattice Systems

In this Section, following (Bagarello and Trapani 1996a), we will show how the mathematical structures discussed above can be used in the study of spin systems. Let V be a finite region of a d-dimensional lattice and $|V|$ the number of points in V. The local C^*-algebra $\mathcal{A}_V$ is generated by the Pauli operators $\sigma_p = (\sigma_p^1, \sigma_p^2, \sigma_p^3)$ and by the unit 2×2 matrix I_p at every point $p \in V$. The σ_p's are copies of the Pauli matrices

$$\sigma^1 = \begin{pmatrix} 0 & 1 \\ 1 & 0 \end{pmatrix}, \quad \sigma^2 = \begin{pmatrix} 0 & -i \\ i & 0 \end{pmatrix}, \quad \sigma^3 = \begin{pmatrix} 1 & 0 \\ 0 & -1 \end{pmatrix}.$$

$\mathcal{A}_V$ is isomorphic to the C^*-algebra of all $2^{|V|} \times 2^{|V|}$-matrices on the $2^{|V|}$-dimensional complex Hilbert space $\mathcal{H}_V = \bigotimes_{p \in V} C_p^2$, where C_p^2 is the 2-dimensional complex Hilbert space at p. If $V \subset V'$ and $A_V \in \mathcal{A}_V$, then

$A_V \to A_{V'} = A_V \otimes (\underset{p \in V' \setminus V}{\otimes} I_p)$ defines the natural embedding of $\mathcal{A}_V$ into $\mathcal{A}_{V'}$.

Let $n = (n_1, n_2, n_3)$ be a unit vector in $\mathbb{R}^3$, and put

$$(\sigma \cdot n) = n_1\sigma^1 + n_2\sigma^2 + n_3\sigma^3.$$

Then, the spectrum $Sp(\sigma \cdot n)$ of $\sigma \cdot n$ is the set $\{1, -1\}$. Let $|n\rangle$ be a unit eigenvector associated with 1. Let $\{n\} = \{n_1, n_2, \cdots\}$ be an infinite sequence of unit vectors in $\mathbb{R}^3$ and $|\{n\}\rangle = \underset{p}{\otimes} |n_p\rangle$ the corresponding unit vector in the infinite tensor product $\mathcal{H}_\infty = \underset{p}{\otimes} C_p^2$. We put

$$\mathcal{A}_0 = \bigcup_V \mathcal{A}_V$$

and

$$\mathcal{D}^0_{\{n\}} = \mathcal{A}_0 |\{n\}\rangle$$

and we denote the closure of $\mathcal{D}^0_{\{n\}}$ in $\mathcal{H}_\infty$ by $\mathcal{H}_{\{n\}}$. Then, to any sequence $\{n\}$ of three-vectors it corresponds a state $|\{n\}\rangle$ of the system. Each of these states defines a faithful representation $\pi_{\{n\}}(\mathcal{A}_0)$ in the Hilbert space $\mathcal{H}_{\{n\}}$ in the following way. First, one starts with constructing a special basis for $\mathcal{H}_{\{n\}}$ by *flipping* a finite number of spins in the *ground* state $|\{n\}\rangle$.

Let (n, n^1, n^2) be an orthonormal basis of $\mathbb{R}^3$. We put

$$n^\pm = \frac{1}{2}(n^1 \pm in^2)$$

and

$$|m, n\rangle = (\sigma \cdot n^-)^m |n\rangle \qquad (m = 0, 1).$$

Then we have

$$(\sigma \cdot n)|m, n\rangle = (-1)^m |m, n\rangle \qquad (m = 0, 1)\,.$$

Thus the set $\{|\{m\}, \{n\}\rangle = \underset{p}{\otimes} |m_p, n_p\rangle;\ m_p = 0, 1,\ \sum_p m_p < \infty\}$ forms an orthonormal basis in $\mathcal{H}_{\{n\}}$, (Thirring and Wehrl 1967). In this space we define the unbounded self-adjoint operator M by

$$M|\{m\}, \{n\}\rangle = \left(\sum_p m_p\right) |\{m\}, \{n\}\rangle\,.$$

M counts the number of the flipped spins in $|\{m\}, \{n\}\rangle$ with respect to the ground state $|\{n\}\rangle$. Of course M depends on $\{n\}$, but we will not explicitly indicate this fact whenever no confusion arises.

The representation $\pi_{\{n\}}$ is then defined on the basis vectors $\{|\{m\}, \{n\}\rangle\}$ by

$$\pi_{\{n\}}(\sigma_p^i) \mid \{m\},\{n\}\rangle = \sigma_p^i \mid m_p, n_p\rangle \otimes (\prod_{p' \neq p} \otimes \mid m_{p'}, n_{p'}\rangle) \quad (i = 1,2,3).$$

and extended in obvious way to the whole space $\mathcal{H}_{\{n\}}$.

Even if $\pi_{\{n\}}$ is a *bounded* representation of $\mathcal{A}_0$ into $\mathcal{H}_{\{n\}}$, it is more convenient, in view of the construction of a topology where thermodynamical limits can be handled, to consider more appropriate domains for the representation $\pi_{\{n\}}$. Since the operator M is a number operator, the operator e^M is a densely defined self-adjoint operator. Let $\mathcal{D}$ denote its domain. Then $\mathcal{D}$ can be made into a Hilbert space, denoted as $\mathcal{H}_M$, in canonical way. The norm in $\mathcal{H}_M$ is given by

$$\| f \|_M = \| e^M f \|, \quad f \in \mathcal{H}_M .$$

Taking the conjugate dual $\mathcal{H}_{\overline{M}}$ of $\mathcal{H}_M$, with respect to the scalar product of $\mathcal{H}_{\{n\}}$, we get the triplet of Hilbert spaces

$$\mathcal{H}_M \subset \mathcal{H}_{\{n\}} \subset \mathcal{H}_{\overline{M}} .$$

Now we can consider the CQ*-algebra $(\mathcal{B}_c(\mathcal{H}_M, \mathcal{H}_{\overline{M}}), *, \mathcal{B}_c(\mathcal{H}_M), \flat)$ of bounded operators acting in the triplet.

The norm of $\mathcal{B}_c(\mathcal{H}_M, \mathcal{H}_{\overline{M}})$ can be written in terms of the norm in $\mathcal{B}(\mathcal{H}_{\{n\}})$ by

$$\| X \|_{M,\overline{M}} = \| e^{-M} X e^{-M} \|, \quad X \in \mathcal{B}_c(\mathcal{H}_M, \mathcal{H}_{\overline{M}}) .$$

Similarly the norm in $\mathcal{B}_c(\mathcal{H}_M)$ becomes

$$\| X \|_M = \| e^M X e^{-M} \|, \quad X \in \mathcal{B}_c(\mathcal{H}_M) .$$

Also in this case it is not difficult to prove that $\pi_{\{n\}}(A)$ maps $\mathcal{H}_M$ into itself, for each $A \in \mathcal{A}_0$. Then if we look at $\pi_{\{n\}}(A)$ as a bounded operator in $\mathcal{H}_M$, we get, taking into account the fact that $\pi_{\{n\}}$ can be viewed as the restriction to $\mathcal{A}_0$ of a *-representation of the simple C*-algebra $\mathcal{A}_S$,

$$\| \pi_{\{n\}}(A) \|_M = \| A \|, \quad \forall A \in \mathcal{A}_0 ,$$

where the norm on the right hand side is the C*-norm of $\mathcal{A}_0$.

On the other hand, the CQ*-norm

$$\| \pi_{\{n\}}(A) \|_{M,\overline{M}} = \| e^{-M} \pi_{\{n\}}(A) e^{-M} \|$$

is, in general different from $\| A \|$.

Of course, to each $\{n\}$ it corresponds a CQ*-algebra of the kind discussed above. This algebraic set-up could also be taken as a reasonable framework where discussing problems like the existence of the thermodynamical limit of the local Heisenberg dynamics. The results would, however, depend on $\{n\}$ and hence on the representation $\pi_{\{n\}}$. For some models, like for instance -the so-called *almost-mean field* (Bagarello and Trapani 1991, Bagarello and Trapani 1993), the dependence on the representation is not crucial at all, provided that the states $\{n\}$ are chosen in a suitable family of *relevant* states (Bagarello and Trapani 1996a).

There is, however, no simple way of eliminating this dependence on the representation for more general models. Therefore it may be convenient to define a topology which takes into account the whole class $\mathcal{F}$ of states under consideration. This topology, which we call *weak physical topology* and denote by $\tau_{\mathcal{F}}$ is defined by the family of seminorms (we will explicitly write the dependence of M on $\{n\}$):

$$A \in \mathcal{A}_0 \longrightarrow \| e^{-M_{\{n\}}} \pi_{\{n\}}(A) e^{-M_{\{n\}}} \| , \quad \{n\} \in \mathcal{F} .$$

Of course $\tau_{\mathcal{F}}$ is the weakest locally convex topology such that each $\pi_{\{n\}}$ is continuous from $\mathcal{A}_0$ into $(\mathcal{B}_c(\mathcal{H}_M, \mathcal{H}_{\overline{M}}), *, \mathcal{B}_c(\mathcal{H}_M), \natural)$. The completion of $\mathcal{A}_0[\tau_{\mathcal{F}}]$ will be denoted with $\mathcal{A}$. The following proposition is now obvious.

Proposition 3.1 *Let $\mathcal{A}$ denote the completion of $\mathcal{A}_0[\tau_{\mathcal{F}}]$. Then $(\mathcal{A}[\tau_{\mathcal{F}}], \mathcal{A}_0)$ is a topological quasi *-algebra.*

3.1 Thermodynamical Limits

When we deal with a specific model, the key of the construction of a topology $\tau_{\mathcal{F}}$ that allows a successful discussion of the thermodynamical limits, is just the family $\mathcal{F}$ of *relevant* states. A clever choice of $\mathcal{F}$ should be made in close connection with the specific form of the local hamiltonian. Of course, also *extreme* situations are allowed: for instance, if the hamiltonian is good enough, they may be freely chosen (for instance, for short range interactions). The other extreme (i.e., a very poor family $\mathcal{F}$) is also, in principle, possible.

If H_V is the finite volume hamiltonian of the system, we put, as usual,

$$\delta_V(A) = i[H_V, A] , \quad A \in \mathcal{A}_0$$

and

$$\delta_V^k(A) = i[H_V, A]_k , \quad A \in \mathcal{A}_0 ,$$

where $[H_V, A]_1 = [H_V, A]$ and $[H_V, A]_k = [H_V, [H_V, A]_{k-1}]$.

Proposition 3.2 *Let the finite volume hamiltonian H_V of the system be a polynomial $p_V(S_V^i, \sigma_k^j)$ in the variables S_V^i with $S_V^i \in \mathcal{A}_0$, $i = 1, 2, \ldots, N$ and a continuous function of the σ_k^j's. Assume that*

(i) $\forall A \in \mathcal{A}_0$, $[H_V, A]$ depends on V only through the S_V^i's;
(ii) $\lim_{|V| \to \infty} (S_V^i)^n$ exists in $\mathcal{A}[\tau_{\mathcal{F}}]$, $\forall n \in \mathbb{N}$, $i = 1, 2, \ldots, N$.

Then for each $A \in \mathcal{A}_0$ and for any $k \in \mathbb{N}$ the limits

$$\lim_{|V| \to \infty} \delta_V^k(A)$$

exist in $\mathcal{A}[\tau_{\mathcal{F}}]$. Setting $\delta(A) = \tau_{\mathcal{F}} - \lim_{|V| \to \infty} \delta_V(A)$, $A \in \mathcal{A}_0$, then δ has the properties:

(a) $\delta(A^*) = \delta(A)^*, \quad \forall A \in \mathcal{A}_0;$
(b) $\delta(AB) = \delta(A)B + A\delta(B), \quad \forall A, B \in \mathcal{A}_0.$

The derivation δ describes, at infinitesimal level, the infinite volume dynamics of the system. What we expect is that it is the generator of a one-parameter group of automorphisms of the completion of $\mathcal{A}_0$ in a suitably chosen topology. Since δ_V is bounded in $\mathcal{A}_0$, we also get

$$\alpha_V^t(A) = \sum_{k=0}^{\infty} \frac{(it)^k}{k!} \delta_V^k(A), \quad A \in \mathcal{A}_0$$

and we know that $\delta_V^k(A)$ is $\tau_{\mathcal{F}}$-convergent to $\delta^{(k)}(A)$. The problem of finding the limit for $|V| \to \infty$ of $\alpha_V^t(A)$ is then solved if the behavior of $\|\pi_{\{n\}}(\delta_V^k(A))\|_{M,\overline{M}}$, for $|V| \to \infty$ can be conveniently controlled. For this, additional conditions on H_V are needed.

Proposition 3.3 *Assume that the local hamiltonian H_V of a spin model can be written in the form*

$$H_V = |V| p(S_V^1, S_V^2, S_V^3),$$

where p is a polynomial and

$$S_V^i = \frac{1}{|V|} \sum_{p \in V} \sigma_p^i, \quad i = 1, 2, 3.$$

Then for each $A \in \mathcal{A}_0$ there exists a positive number r_A such that the local dynamics

$$\alpha_V^t(A) = e^{iH_V t} A e^{-iH_V t}$$

converges in the topology $\tau_{\mathcal{F}}$, when $|V| \to \infty$, to a limit which we denote by $\alpha^t(A)$, for any t with $|t| < r_A$.

It is worth mentioning that, in spite of their appearance, the assumptions made on H_V are fulfilled by several interesting models such as the mean field spin models. For the Heisenberg model, for instance, the local hamiltonian has the form

$$H_V = \frac{J}{|V|} \sum_{p,q \in V} \sum_{i=1}^{3} \sigma_p^i \sigma_q^i, \tag{7}$$

and the operators S_V^i are defined as

$$S_V^i = \frac{1}{|V|} \sum_{p \in V} \sigma_p^i, \quad i = 1, 2, 3.$$

Another possible approach, that may be helpful if the assumptions on H_V fail, makes use of the possibility of defining an *effective hamiltonian*, by which we mean that for each $\{n\} \in \mathcal{F}$ there exists a self-adjoint operator $H_{\mathit{eff}}^{\{n\}}$ in $\mathcal{H}_{\{n\}}$ such that the derivation $\pi_{\{n\}} \circ \delta$ is weakly-spatial, i.e.,

$$< \pi_{\{n\}}(\delta(A))\phi, \psi > = i\left\{< \pi_{\{n\}}(A)\phi, H_{\mathit{eff}}^{\{n\}}\psi > - < H_{\mathit{eff}}^{\{n\}}\phi, \pi_{\{n\}}(A^*)\psi >\right\},$$
$$\forall \phi, \psi \in D(H_{\mathit{eff}}^{\{n\}}),\ \forall A \in \mathcal{A}_0 .$$

In this case, we can consider the subset $\mathcal{D}$ of $D(\delta) = \{A \in \mathcal{A}_0 : \delta(A) \in \mathcal{A}_0\}$ consisting of all the *generalized analytic elements* of δ, i.e., $A \in \mathcal{D}$ iff

$$e^{iH_{\mathit{eff}}^{\{n\}}t}\pi_{\{n\}}(A)e^{-iH_{\mathit{eff}}^{\{n\}}t} = \sum_{k=0}^{\infty} \frac{(it)^k}{k!}\pi_{\{n\}}(\delta^{(k)}(A)),$$

where the series on the right hand side is understood to converge with respect to $\| \, \|_{M_{\{n\}},\overline{M}_{\{n\}}}$.

Proposition 3.4 *Let $\mathcal{F}_0 \subset \mathcal{F}$ be a set of states such that for each $\{n\} \in \mathcal{F}_0$, $M_{\{n\}}$ and $H_{\mathit{eff}}^{\{n\}}$ commute strongly (i.e., their spectral families commute). Then, for each $A \in \mathcal{D}$, the series $\sum_{k=0}^{\infty} \frac{(it)^k}{k!}(\delta^{(k)}(A))$ converges, with respect to the topology $\tau_{\mathcal{F}_0}$, to an element of $\mathcal{A}$ which we call $\alpha_t(A)$. Moreover, α_t can be extended to the closure $\overline{\mathcal{D}}$ of $\mathcal{D}$ in $\mathcal{A}[\tau_{\mathcal{F}_0}]$.*
If $\overline{\mathcal{D}} = \mathcal{A}$, then α_t is a one-parameter group of automorphisms of $\mathcal{A}$.

Effective Hamiltonians do really exists, for instance, in the case of the mean field models (Thirring and Wehrl 1967) and also for the *almost* mean field models considered in (Bagarello and Trapani 1991, Bagarello and Trapani 1993).

4 Conclusions

For several quantum models, many alternative possible ways of overcoming the problems that cannot be directly solved in the C*-approach, have been proposed in the literature.
The approach sketched in this paper represents only an additional contribute to this research area.
However, in our opinion, it has the advantage that it can be used regardless to the specific model: indeed, it provides a common language for classes of models enjoying sufficiently many properties to allow the construction of families of CQ*-algebras.
From a merely mathematical point of view, it puts on the stage the possibility of extending the structure properties of C*-algebras to certain classes of topological partial *-algebras.
A lot of work is still to be done first from the mathematical side, where a more detailed analysis of the structure properties of topological partial *-algebras

is still in order.
From the point of view of physical applications, there is, first the need of investigating several aspects of the theory (automorphisms groups, KMS-states ...) that would make of CQ*-algebras a more useful tool for the study of concrete models. On the other hand, the problem of finding quantum models for which this approach appears to be strictly needed is still open.

References

Antoine, J.-P. Karwowski, W., (1985): *Partial *-Algebras of Closed Linear Operators in Hilbert Space*, Publ. RIMS, Kyoto Univ. **21**, 205-236

Antoine, J.-P., Inoue, A., Trapani, C., (1996): *Partial *-Algebras of Closable Operators: a review*, Rev. Math. Phys., **8**, 1–42

Antoine, J.-P., Bagarello, F., Trapani, C., *Topological partial *-algebras: Basic properties and examples*, Rev. Math. Phys., to appear

Ascoli, R., Epifanio, G., Restivo, A., (1970): *On the Mathematical Description of Quantized Fields*, Commun. Math. Phys. **18**, 291–300

Bagarello, F., Trapani, C., (1991): *'Almost' Mean Field Ising Model: an Algebraic Approach*, J. Statistical Phys. **65**, 469–482

Bagarello, F., Trapani, C., (1993): *A Note on the Algebraic Approach to the Almost Mean Field Heisenberg Model*, Nuovo Cimento, **108B**, 779–784

Bagarello, F., Trapani, C., (1994): *States and representations of CQ* -algebras*, Ann. Inst. H. Poincaré **61**, 103–133

Bagarello, F., Trapani, C., (1996a): *The Heisenberg Dynamics of Spin Systems: A Quasi *-Algebras approach*, J. Math. Phys. **37**, 4219–4234

Bagarello, F., Trapani, C., (1996b): *CQ*-algebras: structure properties* Publ. RIMS, Kyoto Univ. **32**, (1996) 85–116

Bagarello, F., Trapani, C., (1996c): *L^p-spaces as quasi *-algebras*, J. Math. Anal. Appl. **197**, 810–824

Bagarello, F., Inoue, A., Trapani, C., (1998): *Standard CQ*-algebras*, Pre-print n. 63, Dipartimento di Matematica e Applicazioni, Palermo

Bagarello, F., Inoue, A., Trapani, C., *Isomorphisms of CQ*-algebras*, unpublished note

Epifanio, G., Trapani, C., (1987): *Quasi *-Algebras valued Quantized Fields*, Ann. Inst. H. Poincar **46**, 175–185

Fredenhagen, K., Hertel, J., (1981): *Local Algebras of Observables and Pointlike Localized Fields*, Commun. Math. Phys. **80**, 555–561

Haag, R., Kastler, D., (1964): *An Algebraic Approach to Quantum Field Theory* J. Math. Phys. **5**, 848–861

Lassner, G., (1981): *Topological algebras and their applications in quantum statistics*, Wiss, Z. KMU, Leipzig, Math.-Nat. R. **6**, 572–595

Lassner, G., (1984): *Algebras of Unbounded Operators and Quantum Dynamics*, Physica, **124A**, 471–480

Thirring, W., Wehrl, A., (1967): *On the Mathematical Structure of the B.C.S.-Model*, Commun. Math. Phys. **4**, 303–314

Trapani, C., (1998): *Some results on CQ*-algebras of operators*, Proceedings of the International Workshop on Operator Theory, Cefalù, Italy 1997, Suppl. Rand. Circ. Mat. Palermo, **56** 199–205

Part III

Supersymmetry and Quantum Deformations

New Superparticle Models Outside the HLS Supersymmetry Scheme

I. Bandos[1] and J. Lukierski[2] *

[1] Institute for Theoretical Physics,
NSC Kharkov Institute of Physics and Technology,
310108 Kharkov, Ukraine
e-mail: bandos@tph32.tuwien.ac.at
[2] Institute for Theoretical Physics,
University of Wrocław,
50-204 Wrocław, Poland

Abstract. We consider the superparticle models invariant under the supersymmetries with tensorial central charges, which were not included in $D = 4$ Haag-Lopuszanski-Sohnius (HLS) supersymmetry scheme.

We present firstly a generalization of $D = 4$ Ferber-Shirafuji (FS) model with fundamental bosonic spinors and tensorial central charge coordinates. The model contains four fermionic coordinates and possesses three κ-symmetries thus providing the BPS configuration preserving 3/4 of the target space supersymmetries. We show that the physical degrees of freedom (8 real bosonic and 1 real Grassmann variable) of our model can be described by $OSp(8|1)$ supertwistor. Then we propose a higher dimensional generalization of our model with one real fundamental bosonic spinor. $D = 10$ model describes massless superparticle with composite tensorial central charges and in $D = 11$ we obtain 0-superbrane model with nonvanishing mass which is generated dynamically. The introduction of $D = 11$ Lorentz harmonics provides the possibility to construct massless $D = 11$ superparticle model which can be formulated in a way preserving 1/2, 17/32, 18/32, ..., 31/32 supersymmetries. In a special case we obtain the twistor-like formulation of the usual massless $D = 11$ superparticle proposed recently by Bergshoeff and Townsend.

1 Introduction

It is our great pleasure to contribute this article to the volume dedicated to Professor Jan Lopuszanski on his 75-th birthday. He is one of the founders of algebraic background for present supersymmetric theories. In seventies, when in 1975 he published fundamental paper with Haag and Sohnius (see (Haag et al. 1995)) it was however assumed that the relativistic superalgebra should contain in its bosonic sector a direct summ of space-time symmetry generators (Poincaré, de-Sitter, conformal) and internal symmetry generators, i.e., the space-time bosonic generators and internal bosonic generators should commute. As a consequence the internal Abelian generators, called

* Supported in part by KBN grant 2P03B13012

also central charges, had to be scalar. Recently however this conclusion has been relaxed, and in present algebraic framework of SUSY appear generalized central charges - tensorial (van Holten and van Proyen 1982–(Hewson and Perry 1997; Hewson 1997) or even spinorial (d'Auria and Fré 1982, Sezgin 1997) ones. The best example can be provided by D=11 supersymmetry algebra, containing topological contributions from M2 and M5 superbranes:

$$\{Q_\alpha, Q_\beta\} = P_m \Gamma^m_{\alpha\beta} + Z_{m_1 m_2} \Gamma^{m_1 m_2}_{\beta\alpha} + Z_{m_1 \ldots m_5} \Gamma^{m_1 \ldots m_5}_{\beta\alpha} . \quad (1.1)$$

In this lecture we shall consider the new superparticle models, invariant under SUSY with tensor charge generators. We shall formulate such a model following the ideas of supertwistor formulation by Ferber and Shirafuji (Ferber 1978, Shirafuji 1983). In Sect 2 we shall consider the D=4 model which is invariant under the following D=4 SUSY algebra

$$\{Q_A, Q_B\} = Z_{AB}, \qquad \{\bar{Q}_{\dot{A}}, \bar{Q}_{\dot{B}}\} = \bar{Z}_{\dot{A}\dot{B}} , \quad (1.2)$$
$$\{Q_A, \bar{Q}_{\dot{B}}\} = P_{A\dot{B}} ,$$

where $(Q_A)^* = \bar{Q}_{\dot{A}}$, $(P_{A\dot{B}})^* = P_{B\dot{A}}$, $(Z_{AB})^* = \bar{Z}_{\dot{A}\dot{B}}$ and six real commuting central charges $Z_{\mu\nu} = -Z_{\nu\mu}$ are related to the symmetric complex spin-tensor Z_{AB} by[1]

$$Z_{\mu\nu} = \frac{i}{2} \left(\bar{Z}_{\dot{A}\dot{B}} \tilde{\sigma}^{\dot{A}\dot{B}}_{\mu\nu} - Z_{AB} \sigma^{AB}_{\mu\nu} \right) . \quad (1.3)$$

Thus the spin-tensors Z_{AB} and $Z_{\dot{A}\dot{B}}$

$$Z_{AB} = \frac{i}{4} Z_{\mu\nu} \sigma^{\mu\nu}_{AB} , \qquad \bar{Z}_{\dot{A}\dot{B}} = -\frac{i}{4} Z_{\mu\nu} \tilde{\sigma}^{\mu\nu}_{\dot{A}\dot{B}} ,$$

represent the self-dual and anti-self-dual parts of the central charge matrices. It should be stressed that the superalgebra (1.2-3) goes outside of the HLS scheme.

The $D = 4$ model considered in Section 2 can be reformulated in terms of two Weyl spinors λ_A, μ_A and one real Grassmann variable ζ expressed by the generalization of supersymmetric Penrose–Ferber relations (Ferber 1978, Shirafuji 1983, Bengtsson et al. 1987; Lukierski and Nowicki 1988) between supertwistor and superspace coordinates. Such reformulation is described by $OSp(8|1)$ invariant free supertwistor model with the action

$$S = -\frac{1}{2} \int d\tau Y_{\mathcal{A}} G^{\mathcal{A}\mathcal{B}} \dot{Y}_{\mathcal{B}} , \quad (1.4)$$

where $Y_{\mathcal{A}} = (y_1, \ldots, y_8; \zeta) \equiv (\lambda_\alpha, \mu^\alpha, \zeta)$ is the real $SO(8|1)$ supertwistor (see e.g., (Lukierski 1979; Heidenreich and Lukierski 1990)) and

[1] For two-component $D = 4$ Weyl spinor formalism see e.g., (Corson 1953). We have
$(\sigma_{mn})_A{}^B = \frac{1}{2i} \left((\sigma_\mu)_{A\dot{B}} \tilde{\sigma}_\nu^{\dot{B}B} - (\sigma_\nu)_{A\dot{B}} \tilde{\sigma}_\mu^{\dot{B}B} \right) = -\frac{i}{2} e_{\mu\nu\rho l} (\sigma^{\rho l})_A{}^B = [(\tilde{\sigma}_{\mu\nu})^{\dot{B}}{}_{\dot{A}}]^*$.

$$G^{AB} = \begin{pmatrix} \omega^{(8)} & 0 \\ 0 & 2i \end{pmatrix} = \left(\begin{array}{c|c} \begin{pmatrix} 0_2 & I_2 & 0_2 & 0_2 \\ -I_2 & 0_2 & 0_2 & 0_2 \\ 0_2 & 0_2 & 0_2 & I_2 \\ 0_2 & 0_2 & -I_2 & 0_2 \end{pmatrix} & 0 \\ \hline 0 & i \end{array} \right) \tag{1.5}$$

is the $OSp(8|1)$ supersymplectic structure with bosonic $Sp(8)$ symplectic metric
$\omega^{(8)} = -(\omega^{(8)})^T$. It should be mentioned therefore that due to the presence of tensorial central charges the standard $SU(2,2|1)$ supertwistor description (Ferber 1978, Shirafuji 1983, Bengtsson et al. 1987; Lukierski and Nowicki 1988, Sorokin et al. 1989; Volkov and Zheltukhin 1988, 1990; Sorokin et al. 1989, Gumenchuk and Sorokin 1990, Townsend 1991) of the Brink–Schwarz (BS) massless superparticle (Brink and Schwarz 1981) with one complex Grassmann coordinate is replaced by a model with $OSp(8|1)$ invariance and one real Grassmann degree of freedom.

It should be stressed that by the use of spinor coordinates in the presence of tensorial central charges

- we do not increase the initial number of spinor degrees of freedom (four complex or eight real components) in comparison with the model without tensorial central charges;
- we keep the manifest Lorentz invariance despite the presence of tensorial central charges.

In fact, when we use our formulae (see Section **3**)

$$P_{A\dot{B}} = \lambda_A \bar{\lambda}_{\dot{B}}, \qquad Z_{AB} = \lambda_A \lambda_B, \qquad \bar{Z}_{\dot{A}\dot{B}} = \bar{\lambda}_{\dot{A}} \bar{\lambda}_{\dot{B}}\,, \tag{1.6}$$

we find that, in comparison with standard FS model ($P_{A\dot{B}} = \lambda_A \bar{l}_{\dot{B}}$, $Z_{AB} = \bar{Z}_{\dot{A}\dot{B}} = 0$), only the phase of spinor λ_A becomes an additional physical bosonic degree of freedom.

In Section 3 we shall consider the D=10 and D=11 models described by multidimensional extensions of FS model with one fundamental spinor coordinates. The D=11 model is invariant under the superalgebra (1.1). It appears that D=10 model is massless (due to the famous Fierz identities for D=10 gamma matrices) and D=11 is generally a massive one with a mass generated dynamically. In Section 4 we shall consider the large family of D=11 massless models with particular fundamental spinor coordinates described by Lorentz haımonics.

We would like to add that the results presented in Sections 2 and 3 can also be found in our recent article (Bandos and Lukierski 1998), but all the results from Section 4 are new.

2 Generalization of Ferber–Shirafuji Superparticle Model: Spinor Fundamental Variables and Central Charges

We generalize the model presented in (Shirafuji 1983) as follows

$$S = \int d\tau \left(\lambda_A \bar{\lambda}_{\dot{B}} \Pi_\tau^{A\dot{B}} + \lambda_A \lambda_B \Pi_\tau^{AB} + \bar{\lambda}_{\dot{A}} \bar{\lambda}_{\dot{B}}\, \Pi_\tau^{\dot{A}\dot{B}} \right) , \tag{2.1}$$

where

$$\begin{aligned} \Pi^{A\dot{B}} &\equiv d\tau \Pi_\tau^{A\dot{B}} = dX^{A\dot{B}} + i \left(d\Theta^A \bar{\Theta}^{\dot{B}} - \Theta^A d\bar{\Theta}^{\dot{B}} \right) , \\ \Pi^{AB} &\equiv d\tau \Pi_\tau^{AB} = dz^{AB} - \, i\, \Theta^{(A}\, d\Theta^{B)} , \\ \bar{\Pi}^{\dot{A}\dot{B}} &\equiv d\tau \bar{\Pi}_\tau^{\dot{A}\dot{B}} = d\bar{z}^{\dot{A}\dot{B}} - \, i\, \bar{\Theta}^{(\dot{A}}\, d\bar{\Theta}^{\,\dot{B})} , \end{aligned} \tag{2.2}$$

are the supercovariant one–forms in $D = 4$, $N = 1$ generalized flat superspace

$$M^{(4+6|4)} = \{Y^M\} \equiv \{(X^{A\dot{A}}, z^{AB}, \bar{z}^{\dot{A}\dot{B}}; \Theta^A, \bar{\Theta}^{\dot{A}})\} , \tag{2.3}$$

with tensorial central charge coordinates $z^{mn} = (z^{AB}, \bar{z}^{\dot{A}\dot{B}})$ (see (1.3)). The complete configuration space of the model (2.1) contains additionally the complex-conjugate pair $(\lambda_A, \bar{l}_{\dot{A}})$ of Weyl spinors

$$\begin{aligned} \mathcal{M}^{(4+6+4|4)} &= \{q^{\mathcal{M}}\} \equiv \{(Y^M; \lambda^A, \bar{l}^{\dot{A}})\} \\ &= \{(X^{A\dot{A}}, z^{AB}, \bar{z}^{\dot{A}\dot{B}}; \lambda^A, \bar{l}^{\dot{A}}; \Theta^A, \bar{\Theta}^{\dot{A}})\} , \end{aligned} \tag{2.4}$$

Calculating the canonical momenta

$$\mathcal{P}_{\mathcal{M}} = \frac{\partial L}{\partial \dot{q}^{\mathcal{M}}} = (P_{A\dot{A}}, Z_{AB}, \bar{Z}_{\dot{A}\dot{B}}; P^A, \bar{P}^{\dot{A}}; \pi^A, \bar{\pi}^{\dot{A}}) , \tag{2.5}$$

we obtain the following set of the primary constraints

$$\Phi_{A\dot{B}} \equiv P_{A\dot{B}} - \lambda_A \bar{\lambda}_{\dot{B}} = 0 , \tag{2.6}$$
$$\Phi_{AB} \equiv Z_{AB} - \lambda_A \lambda_B = 0 , \tag{2.7}$$
$$\Phi_{\dot{A}\dot{B}} \equiv \bar{Z}_{\dot{A}\dot{B}} - \bar{\lambda}_{\dot{A}} \bar{\lambda}_{\dot{B}} = 0 , \tag{2.8}$$
$$P_A = 0, \qquad \bar{P}_{\dot{A}} = 0 , \tag{2.9}$$
$$D_A \equiv -\pi_A + i P_{A\dot{B}} \bar{\Theta}^{\dot{B}} + i Z_{AB} \Theta^B = 0 , \tag{2.10}$$
$$\bar{D}_{\dot{A}} \equiv \bar{\pi}_{\dot{A}} - i \Theta^B P_{B\dot{A}} - i \bar{Z}_{\dot{A}\dot{B}} \bar{\Theta}^{\dot{B}} = 0 . \tag{2.11}$$

Because the action (2.1) is invariant under the world line reparametrization, the canonical Hamiltonian vanishes

$$H \equiv \dot{q}^{\mathcal{M}} \mathcal{P}_{\mathcal{M}} - L(q^{\mathcal{M}}, \dot{q}^{\mathcal{M}}) = 0 . \tag{2.12}$$

It can be deduced that the set (2.6)–(2.11) of 14 bosonic and 4 fermionic constraints contains 6 bosonic and 3 fermionic first class constraints

$$B_1 = l^A \bar{l}^{\dot{B}} P_{A\dot{B}} = 0\,, \tag{2.13}$$

$$B_2 = l^A \hat{\bar{\mu}}^{\dot{B}} P_{A\dot{B}} - \lambda^A \hat{\mu}^B Z_{AB} = 0\,, \tag{2.14}$$

$$B_3 \equiv (B_2)^* = \hat{\mu}^A \bar{l}^{\dot{B}} P_{A\dot{B}} - \bar{l}^{\dot{A}} \hat{\bar{\mu}}^{\dot{B}} \bar{Z}_{\dot{A}\dot{B}} = 0\,, \tag{2.15}$$

$$B_4 = 2\hat{\mu}^A \hat{\bar{\mu}}^{\dot{B}} P_{A\dot{B}} - \hat{\mu}^A \hat{\mu}^B Z_{AB} - \hat{\bar{\mu}}^{\dot{A}} \hat{\bar{\mu}}^{\dot{B}} \bar{Z}_{\dot{A}\dot{B}} = 0\,, \tag{2.16}$$

$$B_5 = l^A \bar{l}^B Z_{AB} = 0\,, \tag{2.17}$$

$$B_6 \equiv (B_5)^* = \bar{l}^{\dot{A}} \bar{l}^{\dot{B}} \bar{Z}_{\dot{A}\dot{B}} = 0\,, \tag{2.18}$$

$$F_1 = l^A D_A = 0\,, \tag{2.19}$$

$$F_2 \equiv (F_1)^* = \bar{l}^{\dot{A}} \bar{D}_{\dot{A}} = 0\,, \tag{2.20}$$

$$F_3 = \hat{\mu}^A D_A + \hat{\bar{\mu}}^{\dot{A}} \bar{D}_{\dot{A}} = 0\,, \tag{2.21}$$

where we assume that $\lambda^A \mu_A \neq 0$ and

$$\hat{\mu}^A = \frac{\mu^A}{\lambda^B \mu_B}\,, \qquad \hat{\bar{\mu}}^{\dot{A}} = \frac{\bar{\mu}^{\dot{A}}}{\lambda^{\dot{B}} \mu_{\dot{B}}}\,, \tag{2.22}$$

i.e., $\lambda^A \hat{\mu}_A = \bar{l}^{\dot{A}} \hat{\bar{\mu}} = 1$. One can show[2] that our first class constraints (2.13)–(2.21) can be chosen for any particular form of the second spinor μ^A as a function of canonical variables $(q^{\mathcal{M}}, \mathcal{P}_{\mathcal{M}})$. Further we shall propose and motivate the choice for μ^A, $\bar{\mu}^{\dot{A}}$.

The remaining 8 bosonic and 1 fermionic constraints are the second class ones. They are

$$l^A \hat{\bar{\mu}}^{\dot{B}} P_{A\dot{B}} + \lambda^A \hat{\mu}^B Z_{AB} = 0\,, \quad \hat{\mu}^A \bar{l}^{\dot{B}} P_{A\dot{B}} + \bar{l}^{\dot{A}} \hat{\bar{\mu}}^{\dot{B}} \bar{Z}_{\dot{A}\dot{B}} = 0\,, \tag{2.23}$$

$$\hat{\mu}^A \hat{\mu}^B Z_{AB} - 1 = 0\,, \quad \hat{\bar{\mu}}^{\dot{A}} \hat{\bar{\mu}}^{\dot{B}} \bar{Z}_{\dot{A}\dot{B}} - 1 = 0\,, \tag{2.24}$$

$$P_A = 0, \quad \bar{P}_{\dot{A}} = 0\,, \tag{2.25}$$

$$S_F \equiv \hat{\mu}^A D_A - \hat{\bar{\mu}}^{\dot{A}} \bar{D}_{\dot{A}} = 0\,. \tag{2.26}$$

We see that the number # of on-shell phase space degrees of freedom in our model is

$$\# = (28_B + 8_F) - 2 \times (6_B + 3_F) - (8_B + 1_F) = 8_B + 1_F\,, \tag{2.27}$$

in distinction with the standard massless superparticle model of Brink–Schwarz (Brink and Schwarz 1981) or Ferber-Shirafuji (Ferber 1978, Shirafuji 1983) containing $6_B + 2_F$ physical degrees of freedom.

[2] We recall (Dirac 1967) that the first class constraints are defined as those whose Poisson brackets with all constraints weakly vanish. Then one can show (Dirac 1967) that the first class constraints form the closed algebra.

In order to explain the difference in the number of fermionic constraints, let us write down the matrices of Poisson brackets for the fermionic constraints (2.10), (2.11). In our case it has the form

$$C_{\alpha\beta} = \begin{pmatrix} \{D_A, D_B\}_P & \{D_A, \bar{D}_{\dot{B}}\}_P \\ \{\bar{D}_{\dot{A}}, D_B\}_P & \{\bar{D}_{\dot{A}}, \bar{D}_{\dot{B}}\}_P \end{pmatrix} = \begin{pmatrix} \lambda_A \lambda_B & \lambda_A \bar{l}_{\dot{B}} \\ \bar{l}_{\dot{A}} \lambda_B & \bar{l}_{\dot{A}} \bar{l}_{\dot{B}} \end{pmatrix}, \tag{2.28}$$

while for the standard FS model (Ferber 1978, Shirafuji 1983) we obtain

$$C^{FS}_{\alpha\beta} = \begin{pmatrix} 0 & \lambda_A \bar{l}_{\dot{B}} \\ \bar{l}_{\dot{A}} \lambda_B & 0 \end{pmatrix}. \tag{2.29}$$

Now it is evident that in our case the rank of the matrix C is one, while for FS model it is equal to two

$$rank(C) = 1, \qquad rank(C^{FS}) = 2\,.$$

Consequently, in our model there are three fermionic first class constraints generating three κ–symmetries (Azcarraga and Lukierski 1982), one more than in the FS model.

In order to clarify the meaning of the superparticle model (2.1) and present an explicit representation for its physical degrees of freedom, we shall demonstrate that it admits the supertwistor representation in terms of independent bosonic spinor λ^A, bosonic spinor μ^A being composed of λ^A and superspace variables

$$\mu^A = \left(X^{A\dot{B}} + i\Theta^A \bar{\Theta}^{\dot{B}}\right) \bar{\lambda}_{\dot{B}} + 2z^{AB}\lambda^B + i\Theta_A(\Theta^B \lambda_B)\,, \tag{2.30}$$

$$\bar{\mu}^{\dot{A}} = \left(X^{B\dot{A}} - i\Theta^B \bar{\Theta}^{\dot{A}}\right) \lambda_B + 2\bar{z}^{\dot{A}\dot{B}}\bar{\lambda}_{\dot{B}} - i\bar{\Theta}^{\dot{A}}\bar{\Theta}^{\dot{B}}\bar{\lambda}_{\dot{B}}\,, \tag{2.31}$$

and one real fermionic composite Grassmann variable ζ

$$\zeta = \Theta^A \lambda_A + \bar{\Theta}^{\dot{A}} \bar{\lambda}_{\dot{A}}\,. \tag{2.32}$$

Equations (2.30)–(2.32) describe $OSp(8|1)$–supersymmetric generalization of the Penrose correspondence which is alternative to the previously known $SU(2,2|1)$ correspondence, firstly proposed by Ferber (Ferber 1978). Performing integration by parts and neglecting boundary terms we can express our action (2.1) in terms of $OSp(8|1)$ supertwistor variables as follows:

$$S = -\int \left(\mu^A d\lambda_A + \bar{\mu}^{\dot{A}}\, d\bar{\lambda}_{\dot{A}} + id\zeta\, \zeta\right). \tag{2.33}$$

Equation (2.33) presents the free $OSp(8|1)$ supertwistor action. It can be rewritten in the form (1.4) with real coordinates $Y^A = (\mu^\alpha, \lambda^\alpha, \zeta)$ where real Majorana spinors $\mu^\alpha, \lambda^\alpha$ are obtained from the Weyl spinors $(\mu^A, \bar{\mu}^{\dot{A}})$, $(\lambda^A, \bar{l}^{\dot{A}})$ by a linear transformation changing for the $D = 4$ Dirac matrices the complex Weyl to real Majorana representation.

The action (2.33) produces only the second class constraints

$$P^{(\lambda)}_A - \mu_A = 0, \qquad P^{(\mu)}_A = 0\,, \tag{2.34}$$

$$\bar{P}^{(\lambda)}_{\dot{A}} - \mu_{\dot{A}} = 0\,, \qquad \bar{P}^{(\mu)}_{\dot{A}} = 0\,, \tag{2.35}$$

$$\pi^{(\zeta)} = i\zeta\,. \tag{2.36}$$

The Dirac brackets for the $OSp(8|1)$ supertwistor coordinates are

$$[\mu_A, \lambda^B]_D = \delta_A^{\ B}, \qquad [\bar{\mu}_{\dot{A}}, \bar{l}^{\dot{B}}]_D = \delta_{\dot{A}}^{\ \dot{B}}\,, \tag{2.37}$$

$$\{\zeta, \zeta\}_D = -\frac{i}{2}\,. \tag{2.38}$$

They can be also obtained after the analysis of the Hamiltonian system described by the original action (2.1). For this result one should firstly perform gauge fixing for all the gauge symmetries, arriving at the dynamical system which contains only second class constraints, and then pass to the Dirac brackets in a proper way (see (Gumenchuk and Sorokin 1990) for corresponding analysis of the BS superparticle model). This means that the generalization of the Penrose correspondence (2.30), (2.31), (2.32) should be regarded as coming from the second class constraints (primary and obtained from the gauge fixing) of the original system and, thus, should be considered as a relations hold in the strong sense (i.e., as operator identities after quantization) (Dirac 1967). Hence, after the quantization performed in the frame of supertwistor approach, the generalized Penrose relations (2.30), (2.31), (2.32) can be substituted into the wave function in order to obtain the $D = 4$ superspace description of our quantum system.

We shall discuss now the relation of (2.30), (2.31), (2.32), (2.33) with the known FS $SU(2,2|1)$ supertwistor description of the BS superparticle (Ferber 1978, Shirafuji 1983, Bengtsson et al. 1987; Lukierski and Nowicki 1988, Sorokin et al. 1989; Volkov and Zheltukhin 1988, 1990; Sorokin et al. 1989, Gumenchuk and Sorokin 1990, Townsend 1991). The standard FS description is given by the action

$$S = -\int \left(\mu^A d\lambda_A + \bar{\mu}^{\dot{A}}\, d\bar{\lambda}_{\dot{A}} + id\xi\ \bar{\xi}\right)\,, \tag{2.39}$$

supplemented by the first class constraint

$$\mu^A \lambda_A - \bar{\mu}^{\dot{A}}\, \bar{\lambda}_{\dot{A}} + 2i\xi\bar{\xi} = 0\,. \tag{2.40}$$

The $SU(2,2|1)$ supertwistor $(\lambda^A, \bar{\mu}_{\dot{A}}, \bar{\xi})$, contains complex Grassmann variable ξ and the supersymmetric Penrose–Ferber correspondence is given by

$$\bar{\mu}^{\dot{A}} = \left(X^{B\dot{A}} - i\Theta^B \bar{\Theta}^{\dot{A}}\right) \lambda_B\,, \tag{2.41}$$

$$\xi = \Theta^A \lambda_A, \qquad \bar{\xi} = \bar{\Theta}^{\dot{A}} \bar{\lambda}_{\dot{A}}\,. \tag{2.42}$$

Comparing equations (2.39)–(2.42) with our $OSp(8|1)$ supertwistor description (2.30)–(2.33) of the superparticle (2.1) with additional central charge coordinates, we note that

- Besides additional terms proportional to tensorial central charge coordinates z^{AB}, $\bar{z}^{\dot{A}\dot{B}}$, there is present in (2.31) the second term quadratic in Grassmann variables. This second term, however, does not contribute to the invariant $\mu^A \lambda_A$.
- In our model we get

$$\mu^A \lambda_A - \bar{\mu}^{\dot{A}} \bar{\lambda}_{\dot{A}} = 2\lambda_A \lambda_B z^{AB} - 2\bar{l}_{\dot{A}} \bar{l}_{\dot{B}} \bar{z}^{\dot{A}\dot{B}} + 2i\Theta^A \lambda_A \bar{\Theta}^{\dot{A}} \bar{l}_{\dot{A}} , \tag{2.43}$$

  i.e., we do not have additional first class constraint generating $U(1)$ symmetry (compare to (2.40) of the standard supertwistor formulation). Thus our action (2.33) is not singular in distinction to (2.39), where the first class constraint (2.40) should be taken into account, e.g., by introducing it into the action with Lagrange multiplier (Townsend 1991).
- The complex Grassmann variable ξ (2.42) of FS formalism is replaced in our case by the real one ζ (2.32). This difference implies that in our supertwistor formalism the limit $z^{AB} \to 0$, $\bar{z}^{\dot{A}\dot{B}} \to 0$ does not reproduce the standard $SU(2,2|1)$ supertwistor formalism. Indeed, this is not surprising if we take into account that, from algebraic point of view, $SU(2,2|1)$ is not a subsupergroup of $OSp(8|1)$.

The model (2.1) can be slightly generalized as follows

$$S = \int d\tau \left(\lambda_A \bar{l}_{\dot{B}} \Pi_\tau^{A\dot{B}} + Z \lambda_A l_B \Pi_\tau^{AB} + \bar{Z} \bar{l}_{\dot{A}} \bar{l}_{\dot{B}} \bar{\Pi}_\tau^{\dot{B}\dot{B}} \right) , \tag{2.44}$$

where Z, $\bar{Z}$ are complex numerical constants. It appears that for all values of $Z \neq 1$ the model (2.44) will have only two κ-symmetries, and only for particular value $Z = 1$ we obtain three κ-symmetries. The quantization of the model (2.44) is now under consideration (Azcarraga et al. in preparation).

3 $D = 10$ and $D = 11$ Models with One Fundamental Spinor

Recently the most general superparticle model associated with space–time superalgebra (1.1) was proposed by Rudychev and Sezgin (Sezgin and Rudychev 1997). Introducing generalized real superspace $(X^{\alpha\beta}, \Theta^\alpha)$ they consider the following action

$$S = \int d\tau L = \int d\tau \left(P_{\alpha\beta} \, \Pi_\tau^{\alpha\beta} + \frac{1}{2} e_{\alpha\beta} \, P^{\alpha\gamma} \, C_{\gamma\delta} \, P^{\delta\beta} \right) , \tag{3.1}$$

where $\Pi_\tau^{\alpha\beta} = \dot{X}^{\alpha\beta} - \dot{\theta}^{(\alpha}\theta^{\beta)}$ $(\dot{a} \equiv \frac{da}{d\tau})$, C is the charge conjugation matrix and $e_{\alpha\beta}$ is the set of Lagrange multipliers, generalizing einbein in the action for standard Brink-Schwarz massless superparticle (Brink and Schwarz 1981).

Generalized mass shell condition, obtained by varying $e_{\alpha\beta}$ in (3.1), takes the form

$$P^{\alpha\gamma} C_{\gamma\delta} P^{\delta\beta} = 0 . \tag{3.2}$$

We shall look for $P^{\alpha\beta}$ expressing it as spinor belinears and satisfying the generalized mass shell condition (3.2). Particular solution is provided by the following extension of our representation (1.6) to any dimension $D > 4$ with the use of one real D-dimensional Majorana spinor λ_α $(\alpha = 1, ..., 2^k,\ k = 4$ for $D = 10$, $k = 5$ for $D = 11)$:

$$P_{\alpha\beta} = \lambda_\alpha \lambda_\beta , \qquad (\lambda_\alpha)^* = \lambda_\alpha , \tag{3.3}$$

where (1.6) is obtained if $k = 2$. The expression (3.3) solves the BPS condition $\det P_{\alpha\beta} = 0$ as well as more strong Rudychev-Sezgin generalized mass shell constraint (3.2) valid in the model (3.1) with antisymmetric charge conjugation matrix C $(C_{\alpha\beta} = -C_{\beta\alpha})$.

Using (3.3) we get the multidimensional generalization of our action (2.1) which reads

$$S = \int_{\mathcal{M}^1} \lambda_\alpha \lambda_\beta \Pi^{\alpha\beta} , \tag{3.4}$$

$$\Pi^{\alpha\beta} = dX^{\alpha\beta} - i d\Theta^{(\alpha}\Theta^{\beta)} ,$$

$$\alpha = 1, ..., 2^k ,$$

and for $k = 2$ we get the action (2.1).

The case $k = 4$ can be treated as describing spinorial $D = 10$ massless superparticle model with 126 composite tensorial central charges $Z_{m_1 \ldots m_5}$ (cf. with (van Holten and van Proyen 1982, Eisenberg and Solomon 1989)). Indeed, using the basis of antisymmetric products of $D = 10$ sigma matrices we obtain

$$\lambda_\alpha \lambda_\beta \equiv P_{\alpha\beta} = P_m \sigma^m_{\alpha\beta} + Z_{m_1 \ldots m_5} \sigma_{\beta\alpha}^{m_1 \ldots m_5} . \tag{3.5}$$

Contraction of this equation with $\tilde{\sigma}^{m\alpha\beta}$ produces the expression for momenta in terms of bosonic spinors

$$P_m = \frac{1}{16} \lambda_\alpha \sigma_m^{\alpha\beta} \lambda_\beta \qquad \Rightarrow \qquad P_m P^m = 0 . \tag{3.6}$$

The mass shell condition $P_m P^m = 0$ appears then as a result of the $D = 10$ identity $(\sigma_m)_{(\alpha\beta}(\sigma^m)_{\gamma)\delta} = 0$.

The action (3.4) for $k = 8$ can be treated as describing a 0–superbrane model in $D = 11$ superspace with 517 composite tensorial central charge described by 32 components of one real Majorana $D = 11$ bosonic spinor.

In distinction to the above case such model does not produce a massless superparticle[3]). Indeed, decomposing (3.3) in the basis of products of $D = 11$ gamma matrices, one gets

$$\lambda_\alpha \lambda_\beta = P_m \Gamma^m_{\alpha\beta} + Z_{m_1 m_2} \Gamma_{\beta\alpha}^{m_1 m_2} + Z_{m_1 \dots m_5} \Gamma_{\beta\alpha}^{m_1 \dots m_5} . \tag{3.7}$$

The $D = 11$ energy-momentum vector is then given by

$$P_m = \frac{1}{32} \lambda_\alpha \Gamma^{m\alpha\beta} \lambda_\beta , \tag{3.8}$$

and the $D = 11$ mass-shell condition reads

$$M^2 = P_m P^m = \frac{1}{1024} (\lambda \Gamma^m \lambda) \, (\lambda \Gamma^m \lambda) . \tag{3.9}$$

Using the $D = 11$ Fierz identities one can prove that the mass shell condition acquires the form

$$M^2 = P_m P^m = 2 \, Z^{mn} Z_{mn} - 32{\cdot}5! \; Z^{m_1 \dots m_5} Z_{m_1 \dots m_5} , \tag{3.10}$$

with $Z_{mn} = -\frac{1}{64} \lambda \Gamma_{mn} l, \quad Z_{m_1 \dots m_5} = \frac{1}{32 \cdot 5!} \lambda \Gamma_{m_1 \dots m_5} l.$

If we take into consideration that the equations of motion for our model (3.4) imply that the bosonic spinor λ^α is constant ($d\lambda^\alpha = 0$), we have to conclude that (3.4) with $k = 8$ provides the $D = 11$ superparticle model with mass generated dynamically in a way similar to the tension generating mechanism, studied in superstring and higher branes in (Townsend 1992; Bergshoeff et al. 1992; Townsend 1997; Cederwall and Townsend 1997; Cederwall and Westenberg 1998; Bergshoeff and Townsend 1998).

Performing the integration by parts we can rewrite the action (3.4) in the $OSp(1|2^k)$ (i.e., $OSp(1|16)$ for $D = 10$ and $OSp(1|32)$ for $D = 11$) supertwistor $Y^{\mathcal{A}} = (\mu^\alpha, \zeta)$ components:

$$S = - \int (\mu^\alpha d\lambda_\alpha + i d\zeta \, \zeta), \qquad \alpha = 1, \dots, 2^k . \tag{3.11}$$

The generalized Penrose–Ferber correspondence between real supertwistors and real generalized superspace looks as follows

$$\mu^\alpha = X^{\alpha\beta} \lambda_\beta - i \Theta^\alpha (\Theta^\beta \lambda_{\underline{)}}, \qquad \zeta = \Theta^\alpha \lambda_\alpha . \tag{3.12}$$

4 A Set of $D = 11$ Massless Superparticle Models with Conservation of More than 1/2 Target Space Supersymmetries

In order to formulate the model we need to describe $SO(1,10)/(SO(1,1) \otimes SO(9) \otimes K_9)$ Lorentz harmonic formalism.

[3] Note, that the $D = 11$ Green–Schwarz superparticle model does exist and was presented in (Bergshoeff and Townsend 1997

4.1 $\frac{SO(1,10)}{SO(1,1)\otimes SO(9) \otimes K_9}$ Spinor Moving Frame

The $SO(1,10)$ valued moving frame matrix $u_{\underline{m}}^{\underline{a}}$ splits into two light–like and 9 space–like vectors (Sokatchev 1987)

$$u_{\underline{m}}^{\underline{a}} = (u_{\underline{m}}^{++}, u_{\underline{m}}^{--}, u_{\underline{m}}^{I}) \qquad \in \qquad SO(1,10)\,, \tag{4.13}$$

$$\Leftrightarrow \qquad u^{\underline{a}\underline{m}} u_{\underline{m}}^{\underline{b}} = \eta^{\underline{a}\underline{b}} \qquad \Leftrightarrow \qquad \begin{cases} u^{++\underline{m}} u_{\underline{m}}^{++} = 0\,, \\ u^{--\underline{m}} u_{\underline{m}}^{--} = 0\,, \\ u^{\pm\pm\underline{m}} u_{\underline{m}}^{I} = 0\,, \\ u^{I\underline{m}} u_{\underline{m}}^{J} = -\delta^{IJ}\,, \end{cases}$$

where $I = 1,...,9$ is $SO(9)$ vector index.

The $Spin(1,10)$ valued spinor moving frame matrix $v_{\underline{\mu}}^{\underline{\alpha}}$ representing the same Lorentz rotation

$$u_{\underline{m}}^{\underline{a}} \Gamma_{\underline{\mu}\underline{\nu}}^{\underline{m}} = v_{\underline{\mu}}^{\ \underline{\alpha}} \Gamma_{\underline{\alpha}\underline{\beta}}^{\underline{a}} v_{\underline{\nu}}^{\ \underline{\beta}}\,, \tag{4.14}$$

$$u_{\underline{m}}^{\underline{a}} \Gamma_{\underline{a}}^{\underline{\alpha}\underline{\beta}} = v_{\underline{\mu}}^{\ \underline{\alpha}} \Gamma_{\underline{a}}^{\underline{\mu}\underline{\nu}} v_{\underline{\nu}}^{\ \underline{\beta}}\,, \tag{4.15}$$

splits into two rectangular blocks

$$v_{\underline{\mu}}^{\underline{\alpha}} = (v_{\underline{\mu}A}^{\ +}, v_{\underline{\mu}A}^{\ -}) \qquad \in \qquad Spin(1,10)\,, \tag{4.16}$$

where $A = 1,...,16$ is $SO(9)$ spinor index and the sign superscripts denote the $SO(1,1)$ weight of the vector and spinor harmonics.

As the $Spin(1,10)$ transformations keep invariant not only the gamma matrices (4.14), but the $D = 11$ charge conjugation matrix as well

$$v_{\underline{\mu}}^{\ \underline{\alpha}} C^{\underline{\mu}\underline{\nu}} v_{\underline{\nu}}^{\ \underline{\beta}} = C^{\underline{\alpha}\underline{\beta}}\,, \tag{4.17}$$

the spinor harmonics (4.16) are normalized by

$$v_A^{+\underline{\mu}} v_{\underline{\mu}B}^{\ -} = -v_A^{-\underline{\mu}} v_{\underline{\mu}B}^{\ +} = -i\delta_{AB}\,, \qquad v_A^{-\underline{\mu}} v_{\underline{\mu}B}^{\ -} = 0\,, \qquad v_A^{+\underline{\mu}} v_{\underline{\mu}B}^{\ +} = 0\,. \tag{4.18}$$

Equations (4.18) is equivalent to the following decomposition of 32×32 unity matrix[4]

$$\delta_{\underline{\nu}}^{\underline{\mu}} = i v_{\underline{\nu}A}^{\ -} v_A^{+\underline{\mu}} - i v_{\underline{\nu}A}^{\ -} v_A^{+\underline{\mu}}\,. \tag{4.19}$$

In a suitable $SO(1,1) \otimes SO(9) \otimes K_9$ invariant representation for $D = 11$ gamma matrices the (4.14) acquire the form

---

[4] The appearance of multiplier i in (4.18), (4.19) is due to the fact that $D = 11$ charge conjugation matrix is imaginary for our choice of notations and signature $\eta^{\underline{a}\underline{b}} = diag(+1, -1, \ldots, -1)$

$$u_{\underline{m}}^{++}\Gamma^{\underline{m}}_{\underline{\mu\nu}} = 2v^{+}_{\underline{\mu}A}v^{+}_{\underline{\nu}A}, \qquad u_{\underline{m}}^{--}\Gamma^{\underline{m}}_{\underline{\mu\nu}} = 2v^{-}_{\underline{\mu}A}v^{-}_{\underline{\nu}A}, \qquad u_{\underline{m}}^{I}\Gamma^{\underline{m}}_{\underline{\mu\nu}} = 2v^{+}_{\{\underline{\mu}|A}\Gamma^{I}_{AB}v^{-}_{\underline{\nu}|B]}, \tag{4.20}$$

(compare e.g., with $D = 10$ cases from Refs. (Galperin et al. 1992b, Galperin et al. 1992a, Bandos and Zheltukhin 1991, 1992, 1993, 1994, Bandos et al. 1995)). The decomposition of the relations (4.15) includes, in particular

$$v^{-}_{\underline{\mu}A}\Gamma^{\underline{\mu\nu}}_{\underline{m}}v^{-}_{\underline{\nu}B} = 2\delta_{AB}u^{--}_{\underline{m}}, \tag{4.21}$$

4.2 Action for $D = 11$ Massless Superparticle with Tensorial Central Charge Coordinates

The twistor-like action for $D = 11$ massless superparticle with tensorial central charge coordinates has the form

$$S = \int_{\mathcal{M}^1} P^{++}_{AB}v^{-}_{A\mu}v^{-}_{B\nu}\Pi^{\mu\nu}, \tag{4.22}$$

with

$$\Pi^{\mu\nu} = dX^{\mu\nu} - id\Theta^{(\mu}\Theta^{\nu)},$$

and symmetric $SO(9)$ spin-tensor Lagrange multiplier P^{++}_{AB}.

The canonical momenta

$$P_{\mu\nu} = \frac{\partial L}{\partial \dot{X}^{\mu\nu}} = P^{++}_{AB}v^{-}_{A\mu}v^{-}_{B\nu}, \tag{4.23}$$

evidently satisfy the BPS condition

$$\det(P_{\mu\nu}) = 0,$$

as well as the more strong Rudychev-Sezgin generalized mass shell constraint

$$P_{\mu\rho}C^{\rho\sigma}P_{\sigma\nu} = 0.$$

The rank of the matrix $P_{\mu\nu}$ is less or equal to 16, equal in fact to the rank of the matrix P^{++}_{AB}. As we will demonstrate just this rank defines the number of preserved target space supersymmetries.

The variation of the action (4.22) with respect to the coordinate fields

$$\delta S = \int_{\mathcal{M}^1} P^{++}_{AB}v^{-}_{A\mu}v^{-}_{B\nu}(di_\delta\Pi^{\mu\nu} - 2id\Theta^{(\mu}\delta\Theta^{\nu)}), \tag{4.24}$$

$$i_\delta\Pi^{\mu\nu} = \delta X^{\mu\nu} - i\delta\Theta^{(\mu}\Theta^{\nu)},$$

includes effectively the $\delta\Theta^\mu$ variation only in the combination

$$d\Theta^\nu v^{-}_{A\nu}P^{++}_{AB}\delta\Theta^\mu v^{-}_{A\mu}.$$

Thus the half of Θ variations $\delta\Theta^\mu v^{+}_{A\mu}$ are not involved in the variation of action and, therefore, parametrize the 16 kappa symmetries.

When $\det(P^{++}_{AB}) \neq 0$, the rest 16 of the 32 Grassmann variations $\delta\Theta^{\mu} v^{-}_{A\mu}$ acts effectively and produce nontrivial equations of motion

$$d\Theta^{\nu} v^{-}_{A\nu} P^{++}_{AB} = 0, \qquad \Rightarrow \qquad d\Theta^{\nu} v^{-}_{A\nu} = 0 .$$

We see that there are only 16 kappa symmetries in such dynamical system and so it describes the BPS state preserving 1/2 of the $D = 11$ target space supersymmetry.

We obtain an important particular case of the model (4.22) with $\det(P^{++}_{AB}) \neq 0$ when the Lagrange multiplier P^{++}_{AB} is proportional to the unity matrix $P^{++}_{AB} = P^{++}\delta_{AB}$. Due to the properties (4.20) of the Lorentz harmonic, the product of spinor harmonics $v^{-}_{A\mu} v^{-}_{A\nu}$ is proportional to the gamma matrix $\Gamma_{m\mu\nu}$, hence it does not contain components proportional to $\Gamma^{mn}_{\mu\nu}$, $\Gamma^{mnklp}_{\mu\nu}$. Thus the central charge coordinates disappear from the action which in this case can be equivalently rewritten as

$$S = \frac{1}{32} \int_{\mathcal{M}^1} P_{--} v^{-}_{A\mu} v^{-}_{A\nu} \Gamma^{\mu\nu}_{m} \Pi^{m} , \tag{4.25}$$

$$\Pi^{m} = dX^{m} - i d\Theta^{\mu} \Gamma^{m}_{\mu\nu} \Theta^{\nu} .$$

The formula (4.25) provides the twistor-like formulation of the action for the 'standard' $D = 11$ massless superparticle (without tensorial central charge coordinates), whose 'standard' (Brink–Schwarz type) action was proposed recently in Ref. (Bergshoeff and Townsend 1997).

The generic case of nondegenerate P^{++}_{AB} matrix corresponds the model with central charge coordinates and half of 32 space time supersymmetries conserved.

The case with the matrix P^{++}_{AB} having the rank 1 can be described by

$$P^{++}_{AB} = \lambda^{+}_{A} \lambda^{+}_{B} ,$$

with one bosonic $SO(16)$ spinor λ^{+}_{A}. The action (4.22) in this case reduces to

$$S = \int_{\mathcal{M}^1} (\lambda^{+}_{A} v^{-}_{A\mu})(\lambda^{+}_{A} v^{-}_{B\nu}) \Pi^{\mu\nu} . \tag{4.26}$$

If one denotes $\lambda^{+}_{A} v^{-}_{A\mu} = \lambda_{\mu}$, one arrives to the expression $S = \int_{\mathcal{M}^1} \lambda_{\mu} \lambda_{\nu} \Pi^{\mu\nu}$ which formally coincides with the action proposed in (Bandos and Lukierski 1998). But the composite nature of the bosonic spinor λ_{μ} in the action (4.26) results in the relation

$$32 P_{m} \equiv \lambda_{\mu} \Gamma^{m}_{\mu\nu} \lambda_{\nu} = (\lambda^{+}_{A} v^{-}_{A\mu})(\lambda^{+}_{A} v^{-}_{B\nu}) \Gamma^{\mu\nu}_{m} = \lambda^{+}_{A} \lambda^{+}_{A} u^{--}_{m} , \tag{4.27}$$

where u^{--}_{m} is a light-like harmonic vector $u^{--m} u^{--}_{m} = 0$. Thus $P_{m} P^{m} = 0$ and we conclude that (4.26) describes a *massless* $D = 11$ superparticle with central charge coordinate in distinction with the $D = 11$ model described by (3.4) (Bandos and Lukierski 1998), where, in general, the particle is massive with mass generated dynamically (Townsend 1992; Bergshoeff et al. 1992; Townsend 1997; Cederwall and Townsend 1997; Cederwall and Westenberg 1998; Bergshoeff and Townsend 1998).

Nevertheless both the models (3.4) and (4.26) describe BPS configurations with preservation of 31/32 part of the $D = 11$ target space supersymmetries.

Indeed the variation of the action (4.26) includes effectively only one Grassmann variation $\delta\Theta^{\mu}\lambda_{\mu}$ (with λ_{μ} composed from harmonic and $SO(16)$ spinor as in (4.26)), which remains the same for the action (3.4), where the λ_{μ} spinor is fundamental (see (Bandos and Lukierski 1998)).

The matrix P^{++}_{AB} of the rank r, $1 < r < 8$ can be represented as

$$P^{++}_{AB} = \lambda^{+s}_{A}\lambda^{+s}_{B}, \qquad s = 1, ..., r, \qquad 1 < r < 8\,. \tag{4.28}$$

It is easy to see that such a model describes the BPS states preserving $\frac{(32-r)}{32}$ supersymmetries.

5 Final Remarks

We would like to recall that in the 'M-theoretic' approach (see e.g., (Azcarraga et al. 1989, Townsend 1995; 1996; 1997, Sorokin and Townsend 1997)) the tensorial central charges $Z_{m_1 \ldots m_p}$ are considered as carried by p-branes. Following such treatment, one should interpret e.g., in $D = 4$ central charges $Z_{\mu\nu}$ as an indication of presence of $D = 4$ supermembrane ($p = 2$). The relation of our superparticle model with such $D = 4$ membrane states is not clear now and can be regarded as an interesting subject for further study. Here we should only guess that there should be some singular point–like limit of supermembrane, which should keep the nontrivial topological charge and increase the number of preserved (realized linearly) $D = 4$ target space supersymmetries. Similar limiting prescription should be possible e.g., for 5–branes in $D = 10, 11$ leading to the $D = 10$ and $D = 11$ superparticle actions (3.4) with the relation (3.3) describing composite tensor charges.

At the end of the paper we proposed a generalized FS model for $D > 4$. The straightforward generalization provides us with $D = 10$ massless superparticle model preserving 15/16 supersymmetries and $D = 11$ superparticle model with arbitrary, in general nonvanishing, mass generated dynamically (Townsend 1992; Bergshoeff et al. 1992; Townsend 1997; Cederwall and Townsend 1997; Cederwall and Westenberg 1998; Bergshoeff and Townsend 1998). The latter concerves 31/32 of the target space supersymmetries. Then we introduce spinor harmonics and formulate *massless* $D = 11$ superparticle model preserving 1/2, 17/32, 18/32, ..., 31/32 supersymmetries dependent on the rank of the Lagrange multiplier matrix P^{++}_{AB}. The case with 1/2 corresponds to nondegenerate matrix P^{++}_{AB}: $\det(P^{++}_{AB}) \neq 0$. For the choice $P^{++}_{AB} \propto \delta_{AB}$, the dependence on central charge coordinates disappears and we arrive at the twistor-like formulation of the usual massless $D = 11$ superparticle proposed recently by Bergshoeff and Townsend.

It should be also mentioned that the superparticle model invariant under superPoincare symmetries with central charges can be obtained as a contraction limit of superparticle model defined on the orthosymplectic supergroup manifolds. The $D = 4$ case ($OSp(4|1)$ model) is now under consideration (Bandos et al. in preparation).

Acknowledgments

The authors would like to thank D. Sorokin for useful discussions. The authors are grateful for the hospitality at the Departamento de Fisica Teorica and financial support of the Universidad de Valencia (J.L.) and at the Dipartimento di Fisica "Galileo Galilei", Universita Degli Study di Padova ed INFN, Sezione di Padova (I.B.), which permited to complete the present paper.

The work was supported by the Austrian Science Foundation in the form of the Lise Meitner Fellowship under the Project M472–TPH, by the INTAS grants INTAS-96-308, INTAS-93-127-EXT and KBN grant 2P03B13012.

References

d'Auria, P., Fré, P., (1982): Nucl. Phys. **B201**, 101

de Azcarraga, J.A., Lukierski, J., (1982): Phys. Lett. **B113**, 170; Siegel, W., (1883): Phys. Lett. **B128**, 397

de Azcarraga, J.A., Gauntlett, J.P., Izquierdo, J.M., Townsend, P.K., (1989): Phys. Rev. Lett. **63**, 2443

de Azcarraga, J.A., Bandos, I., Lukierski, J., Sorokin, D., (in preparation)

Bandos, I.A. (1990): Sov. J. Nucl. Phys. **51**, 906; JETP. Lett. **52**, (1990) 205; Bandos, I.A., Zheltukhin, A.A., (1992): Phys. Lett. **B77**, 77; Class. Quantum Grav. **12**, (1995) 609;
Bandos, I.A., Pasti, P., Sorokin, D., Tonin, M., Volkov, D.V., (1995): Nucl. Phys. **B446**, 79

Bandos, I.A., Zheltukhin, A.A., (1991): JETP. Lett. **54**, 421; Phys. Lett. **B288**, (1992) 77; Sov. J. Nucl. Phys. **56**, (1993) 198; J. Elem. Part. Atom. Nucl. **25**, (1994) 453

Bandos, I.A., Zheltukhin, A.A., (1995): Class. Quantum Grav. **12**, 609

Bandos, I., Pasti, P., Sorokin, D., Tonin, M., Volkov, D., (1995): Nucl. Phys. . **B446**, 79

Bandos, I., Lukierski, J., (1998): *Tensorial Central Charges and New Superparticle Models with Fundamental Spinor Coordinates*, hep–th/9811022

Bandos, I., Lukierski, J., Sorokin, D., (in preparation)

Bars, I., (1996): Phys. Lett. **B373**, 68; Phys. Rev. **D54**, (1996) 5203

Bengtsson, A., Bengtsson, I., Cederwall, M., Linden, N., (1987): Phys. Rev. **36D**, 1766;
Lukierski, J., Nowicki, A., (1988): Phys. Lett. **B211**, 276

Bergshoeff, E., Townsend, P.K., (1997): Nucl. Phys. **B490**, 145

Bergshoeff, E., Townsend, P.K., (1997): *Super–D–branes*, Nucl. Phys. **B490**, 145–162, hep–th/9611173

Brink, L. Schwarz, J., (1981): Phys. Lett. **100B**, 310

Corson, E.M., (1953): *Introduction to Tensors, Spinors and Relativistic Wave Equations*, Blodue and Son Limited, London and Glasgow

Dirac, P.A.M., (1967): *Lectures on Quantum Mechanics*, Academic Press, NY

Eisenberg, Y., Solomon, S., (1989): Phys. Lett. **B220**, 562

Ferber, A., (1978): Nucl. Phys. **B132**, 55

Ferrara, S., Savoy, C.A., Zumino, B., (1981): Phys. Lett. **B100**, 393

Galperin, A., Ivanov, E., Kalitzin, S., Ogievetsky, V. Sokatchev, E., (1984): Class. Quantum Grav. **1**, 498; *Class. Quantum Grav.* **2**, (1985) 155

Galperin, A.S., Howe, P., Stelle, K.S., (1992): Nucl. Phys. **B368**, 248–280

Galperin, A.S., Delduc, F., Sokatchev, E., (1992): Nucl. Phys. **B368**, 280

Gumenchuk, A., Sorokin, D., (1990): Sov. J. Nucl. Phys. **51**, 350

Haag, R., Lopuszański, J., Sohnius, M., (1995): Nucl. Phys. **B88**, 257

Hewson, S.F., Perry, M.J., (1997): Nucl. Phys. **B492**, 249;
Hewson, S.F. (1997): Nucl. Phys. **B501**, 445; *An approach to F–theory*, hep–th/9712017

van Holten, J., van Proyen, A., (1982): J. Phys. **A15**, 3763

Lukierski, J., (1979): Lett. Nuov. Cim. **24**, 309;
Heidenreich, W., Lukierski, J., (1990): Mod. Phys. Lett. **A5**, 439

Sezgin, E., (1997): Phys. Lett. **B392**, 323

Sezgin, E., Rudychev, I., (1997): *Superparticles, p-form Coordinates and the BPS Condition*, hep–th/9711128

Shirafuji, T., (1983): Progr. Theor. Phys. **70**, 18

Sokatchev, E., (1987): Phys. Lett. **B169**, (1987) 209; Class. Quantum Grav. **4**, 237 *and refs. in Bandos et al. 1995*

Sorokin, D., Tkach, V., Volkov, D., (1989): Mod. Phys. Lett. **A4**, 901;
Volkov, D., Zheltukhin, A., (1988): Lett. Math. Phys. **48**, 61; Nucl. Phys. **B335**, (1990) 723;
Sorokin, D., Tkach, V., Volkov, D., Zheltukhin, A., (1989): Phys. Lett. **216B**, 302

Sorokin, D., Townsend, P.K., (1997): Phys. Lett. **B412**, 265

Townsend, P.K., (1991): Phys. Lett. **B261**, 65

Townsend, P.K., (1992): Phys. Lett. **B277**, 285;
Bergshoeff, E., London, J.A.J., Townsend, P.K., (1992): Class. Quantum Grav. **9**, 2545;
Townsend, P.K. (1997): Phys. Lett. **B409**, 131–135, hep–th/9705160;
Cederwall, M., Townsend, P.K., (1997): J. High Energy Phys. **09**, 003, hep–th/9709002;
Cederwall, M., Westenberg, A., (1998): J. High Energy Phys. **02**, 004, hep–th/9710160;
Bergshoeff, E., Townsend, P.K., (1998): *Super-Dp-branes revisited*, hep–th/9804011

Townsend, P.K., (1995), (1996), (1997): *P–brane democracy*, hep–th/9507048; *Four lectures on M–theory*, hep–th/9612121; *M–theory from its superalgebra*, hep–th/9712004

Zumino, B., (1962): J. Math. Phys. **3**, 1055–1057

On the Implementation of Supersymmetry

D. Buchholz

Institut für Theoretische Physik, Universität Göttingen,
Bunsenstraße 9, D-37073 Göttingen, Germany

Dedicated to Jan Łopuszański

Abstract. The implementation of supersymmetry transformations by Hilbert space operators is discussed in the framework of supersymmetric C*–dynamical systems. It is shown that the only states admitting such an implementation are pure supersymmetric ground states or mixtures and elementary excitations thereof. Faithful states, such as KMS–states, are never supersymmetric.

1 Introduction

Supersymmetry is an intriguing mathematical concept which has become a basic ingredient in many branches of modern theoretical physics. In spite of its still lacking physical evidence, its far–reaching theoretical implications uphold the belief that supersymmetry plays a prominent role in the fundamental laws of nature.

As for the theory of elementary particles, the possible manifestations of unbroken supersymmetry have been fully clarified by Haag, Łopuszanski and Sohnius (Haag et al., 1975, Łopuszański 1991). On the other hand it is known that supersymmetry is inevitably broken in thermal states. As a matter of fact, this breakdown is much stronger than that of internal bosonic symmetries: one may speak of a spontaneous collapse of supersymmetry (Buchholz and Ojima 1997).

These facts seem to indicate that supersymmetry is only implementable in states describing elementary systems. It is the aim of the present article to clarify this point for general C*–dynamical systems. Apart from supersymmetry, the only ingredient in our analysis is the assumption that the dynamics is asymptotically abelian (see below for precise definitions). So our framework covers also non–local theories.

We shall show in the subsequent section that supersymmetric states are always ground states. If these states are mixed (not pure), they can be decomposed into pure states which are also supersymmetric. At the other extreme, faithful states (such as KMS–states) are never supersymmetric. States which are not supersymmetric but still admit an implementation of the supersymmetry transformations by Hilbert space operators coincide asymptotically with supersymmetric ground states and may thus be regarded as excitations thereof. The physical significance of these results is discussed in the conclusions.

2 Implementations of Odd Derivations

We discuss in this section the consequences of unbroken supersymmetry. As our results do not rely on a specific physical interpretation we present them in the general mathematical setting of C*–dynamical systems (Sakai 1991).

Definition: Let $\mathcal{F} = \mathcal{F}_+ \oplus \mathcal{F}_-$ be a graded C^*–algebra, let α_t, $t \in \mathbb{R}$, be a group of automorphisms of $\mathcal{F}$ which respects the grading and let $\mathcal{A} \subset \mathcal{F}$ be the dense subalgebra of analytic elements with respect to the action of α. The dynamical system $(\mathcal{F}, \alpha)$ is said to be supersymmetric if the (skew symmetric) generator of α

$$\delta_0 \doteq -i\tfrac{d}{dt}\alpha_t \mid_{t=0} \tag{2.1}$$

can be represented in the form

$$\delta_0 = \tfrac{1}{2}(\delta \cdot \overline{\delta} + \overline{\delta} \cdot \delta) \ , \tag{2.2}$$

where δ is a closable odd derivation which is defined on $\mathcal{A}$ and commutes with α, and the associated linear map $\overline{\delta}$ on $\mathcal{A}$ is fixed by

$$\overline{\delta}(F_\pm) \doteq \mp\, \delta(F_\pm{}^*)^* \tag{2.3}$$

for $F_\pm \in \mathcal{A}_\pm \doteq \mathcal{A} \cap \mathcal{F}_\pm$.

The even and odd parts $\mathcal{F}_\pm$ of $\mathcal{F}$ may be interpreted as the Bose and Fermi parts of some field algebra. There holds in particular $\mathcal{F}_+ \cdot \mathcal{F}_- = \mathcal{F}_- \cdot \mathcal{F}_+ \subset \mathcal{F}_-$ and $\mathcal{F}_\pm \cdot \mathcal{F}_\pm \subset \mathcal{F}_+$. We recall that an odd derivation is a densely defined linear mapping which maps even operators into odd ones and *vice versa*, and which satisfies the graded Leibniz rule

$$\delta(F_\pm\, G) = \delta(F_\pm)\, G \pm\ F_\pm\, \delta(G) \tag{2.4}$$

for $F_\pm \in \mathcal{A}_\pm$ and $G \in \mathcal{A}$. It is easily checked that $\overline{\delta}$ is also an odd derivation.

Note that the right hand side of relation (2.2) always defines an even derivation. Hence, given δ, one can determine a corresponding δ_0 and if the latter derivation is sufficiently well behaved it is the generator of a group of automorphisms α satisfying relation (2.1) ((Sakai 1991 Ch. 3.4)). In this sense the whole structure is fixed by δ. We turn now to the analysis of supersymmetric states.

Definition: A state ω on $\mathcal{F}$ is said to be supersymmetric if $\omega \cdot \delta = 0$.

The following result on the implementability of derivations in representations induced by symmetric states is well known in the even case (Sakai 1991). Its straightforward generalization to odd derivations is given here for completeness.

Lemma 2.1 *Let ω be a supersymmetric state on $\mathcal{F}$ and let $(\pi, \mathcal{H}, \Omega)$ be its induced GNS–representation. The operator Q given by*

$$Q\,\pi(F)\,\Omega \doteq \pi(\delta(F))\,\Omega \quad \textit{for} \quad F \in \mathcal{A} \tag{2.5}$$

is well defined and closable. Moreover, there holds on its domain $\pi(\mathcal{A})\,\Omega$

$$\pi(\delta(F_\pm)) = Q\pi(F_\pm) \mp \pi(F_\pm)Q \quad \textit{for} \quad F_\pm \in \mathcal{A}_\pm \ . \tag{2.6}$$

Proof: By relation (2.4) and the supersymmetry of ω there holds for any $F \in \mathcal{A}$ and $G_\pm \in \mathcal{A}_\pm$

$$\begin{aligned}(\pi(G_\pm)\,\Omega,\, \pi(\delta(F))\,\Omega) &= \omega({G_\pm}^*\,\delta(F)) = \omega(\mp\,\delta({G_\pm}^*)\,F)\\ &= (\pi(\mp\,\delta({G_\pm}^*)^*)\,\Omega,\, \pi(F)\,\Omega)\\ &= (\pi(\overline{\delta}(G_\pm))\,\Omega,\, \pi(F)\,\Omega)\ .\end{aligned} \tag{2.7}$$

As in the case of even derivations one concludes from this equality that Q is a well defined linear operator which is closable. In fact, its adjoint Q^* is also defined on $\pi(\mathcal{A})\,\Omega$ and $Q^*\,\pi(F)\,\Omega = \pi(\overline{\delta}(F))\,\Omega$, $F \in \mathcal{A}$. The second part of the statement follows from relations (2.4) and (2.5) after a routine computation. □

Next we show, by making use of arguments in (Buchholz and Ojima 1997), that supersymmetric states are ground states with respect to the group α. If they are mixed, all states appearing in their decomposition are also supersymmetric.

Proposition 2.2 *Let ω be a supersymmetric state on $\mathcal{F}$. Then the group of automorphisms α is implemented in the corresponding GNS–representation $(\pi, \mathcal{H}, \Omega)$ by a continuous unitary group U with positive generator and Ω is invariant under the action of U. If ω is a mixed state, any component (sub–ensemble) $\omega_<$ appearing in its decomposition is also supersymmetric.*

Proof: As $\omega \cdot \delta = 0$ there holds $\omega \cdot \overline{\delta} = 0$, hence $\omega \cdot \delta_0 = 0$. It therefore follows from standard arguments that α_t, $t \in \mathbb{R}$, is implemented by a continuous unitary group $U(t)$, $t \in \mathbb{R}$, which leaves Ω invariant. Now for any $\sigma, \tau \in \{\pm\}$ and $F_\sigma \in \mathcal{A}_\sigma$, $G_\tau \in \mathcal{A}_\tau$ we have

$$\begin{aligned}\delta({F_\sigma}^*\,\overline{\delta}(G_\tau)) &= \delta\,({F_\sigma}^*)\,\overline{\delta}(G_\tau) + \sigma\,{F_\sigma}^*\,(\delta\cdot\overline{\delta}\,(G_\tau))\\ &= -\sigma\,\overline{\delta}\,(F_\sigma)^*\,\overline{\delta}(G_\tau) + \sigma\,{F_\sigma}^*\,(\delta\cdot\overline{\delta}\,(G_\tau))\ ,\end{aligned} \tag{2.8}$$

and consequently $\omega({F_\sigma}^*\,(\delta\cdot\overline{\delta}\,(G_\tau))) = \omega(\overline{\delta}(F_\sigma)^*\,\overline{\delta}(G_\tau))$. By interchanging the role of δ and $\overline{\delta}$ we also get $\omega({F_\sigma}^*\,(\overline{\delta}\cdot\delta\,(G_\tau))) = \omega(\delta(F_\sigma)^*\,\delta(G_\tau))$. Since $\delta, \overline{\delta}$ are linear we conclude that for $F \in \mathcal{A}$

$$\begin{aligned}\omega(F^*\delta_0(F)) &= \tfrac{1}{2}\,\omega(F^*(\delta\cdot\overline{\delta}\,(F))) + \tfrac{1}{2}\,\omega(F^*(\overline{\delta}\cdot\delta\,(F)))\\ &= \tfrac{1}{2}\,\omega(\overline{\delta}(F)^*\,\overline{\delta}(F)) + \tfrac{1}{2}\,\omega(\delta(F)^*\,\delta(F)) \geq 0\ ,\end{aligned} \tag{2.9}$$

proving that the generator of U is a positive selfadjoint operator ((Sakai 1991 Ch. 4.2)).

Next, if $t \mapsto f(t)$ is any absolutely integrable function whose Fourier transform has support in $\mathbb{R}_-$ and if $F \in \mathcal{F}$ we put $\alpha_f(F) \doteq \int dt\, f(t)\, \alpha_t(F)$. Since $t \mapsto \alpha_t(F)$ is strongly continuous there holds $\alpha_f(F) \in \mathcal{F}$. It follows from the preceding result that $\omega(\alpha_f(F)^* \, \alpha_f(F)) = 0$. Hence if $\omega_< \leq c \cdot \omega$ for some positive constant c, there holds $\omega_<(\alpha_f(F)^* \, \alpha_f(F)) = 0$ and consequently $\omega_<$ is also invariant under the action of α. Setting $F_T \doteq T^{-1} \int\limits_0^T dt\, \alpha_t(F)$, $F \in \mathcal{A}$, we get $\omega_<(\delta(F)) = \omega_<(\delta(F)_T) = \omega_<(\delta(F_T))$ and making use of relation (2.9) we obtain the inequality

$$
\begin{aligned}
|\omega_<(\delta(F))|^2 &= |\omega_<(\delta(F_T))|^2 \leq \omega_<(\delta(F_T)^* \, \delta(F_T)) \\
&\leq c \cdot \omega(\delta(F_T)^* \, \delta(F_T)) + c \cdot \omega(\overline{\delta}(F_T)^* \, \overline{\delta}(F_T)) \\
&= 2c \cdot \omega(F_T^* \, \delta_0(F_T)) \ . \qquad (2.10)
\end{aligned}
$$

As $\delta_0(F_T) = -iT^{-1}(\alpha_T(F) - F)$, the right hand side of this inequality tends to 0 as $T \to \infty$ and the assertion follows. □

The following proposition is a straightforward consequence of this result.

Proposition 2.3 *If ω is a faithful state on $\mathcal{F}$ and $\delta \neq 0$, there holds $\omega \cdot \delta \neq 0$.*

Proof: If $\omega \cdot \delta = 0$, it follows from the preceding proposition that for any absolutely integrable function f whose Fourier transform has support in $\mathbb{R}_-$ and any $F \in \mathcal{F}$ there holds $\omega(\alpha_f(F)^* \, \alpha_f(F)) = 0$. Since ω is faithful this implies $\alpha_f(F) = 0$ and $\alpha_{\overline{f}}(F^*) = \alpha_f(F)^* = 0$. As the Fourier transform of the complex conjugate $\overline{f}$ of f has support in $\mathbb{R}_+$ and f, F are arbitrary we arrive at $\alpha_f(F) = 0$ whenever the Fourier transform of f does not contain 0 in its support. Hence $\alpha_t(F) = F$ for $t \in \mathbb{R}$ and consequently $\delta_0 = 0$. Because of relation (2.9) this implies $\omega(\delta(F)^* \delta(F)) = 0$ for all $F \in \mathcal{A}$. But this is incompatible with the assumption that ω is faithful and $\delta \neq 0$. Hence $\omega \cdot \delta \neq 0$. □

More can be said if there acts on $\mathcal{F}$ some group of automorphisms in an asymptotically abelian manner. In order to simplify the discussion we assume that α itself has this property and indicate below which of the subsequent results hold more generally.

Definition: The group α_t, $t \in \mathbb{R}$, is said to act on $\mathcal{F}$ in an asymptotically abelian manner (shortly: it is asymptotically abelian) if

$$
||\alpha_t(F_\pm)\, G_\pm \mp G_\pm \, \alpha_t(F_\pm)|| \to 0 \qquad (2.11)
$$

for $F_\pm, G_\pm \in \mathcal{F}_\pm$ and $|t| \to \infty$.

A well known consequence of asymptotic abelianess is the following result on the asymptotic behaviour of averages of odd operators (Buchholz and Ojima 1997, Narnhofer and Thirring 1994).

Lemma 2.4 *Let α_t, $t \in \mathbb{R}$, be asymptotically abelian. Then*

$$\lim_{T\to\infty} ||T^{-1} \int_0^T dt\, \alpha_t(F_-)|| = 0 \quad \textit{for} \quad F_- \in \mathcal{F}_- \ . \tag{2.12}$$

Proof: Since for $F \in \mathcal{F}$ there holds $||F||^2 = ||F^*F|| \leq ||F^*F + FF^*||$ we obtain the estimate

$$||T^{-1} \int_0^T dt\, \alpha_t(F_-)||^2 \leq ||T^{-2} \int_0^T dt \int_0^T dt'\, (\alpha_t(F_-)^* \alpha_{t'}(F_-) + \alpha_{t'}(F_-)\alpha_t(F_-)^*)||$$
$$\leq T^{-2} \int_0^T dt \int_{-T}^T dt'\, ||F_-{}^*\, \alpha_{t'}(F_-) + \alpha_{t'}(F_-)\, F_-{}^*|| \ . \tag{2.13}$$

According to relation (2.11) the norm under the integral on the right hand side of this inequality tends to 0 if $|t'| \to \infty$, so the statement follows. □

With these preparations we can establish now more detailed information on the representations induced by supersymmetric states. Our result relies on familiar arguments in algebraic quantum field theory (Sakai 1991).

Proposition 2.5 *Let the group $\alpha_t, t \in \mathbb{R}$, be asymptotically abelian. If the state ω is supersymmetric, the GNS–representation π of $\mathcal{F}$ induced by ω is of type I. More specifically, the commutant $\pi(\mathcal{F})'$ of $\pi(\mathcal{F})$ coincides with the center of $\pi(\mathcal{F})''$.*

Proof: According to Proposition 2.2 the automorphisms α are implemented in the GNS–representation $(\pi, \mathcal{H}, \Omega)$ by a continuous unitary group U with positive generator, and Ω is invariant under the action of U. Hence $U(t) \in \pi(\mathcal{F})''$ for $t \in \mathbb{R}$ by a theorem of Araki ((Sakai 1991 Ch. 2.4)). Let E_0 be the projection onto the U–invariant subspace in $\mathcal{H}$. The statement follows if $E_0\pi(\mathcal{F})E_0$ can be shown to be a commutative family ((Sakai 1991 Prop. 2.4.11)). Now by the preceding Lemma there holds for $F_- \in \mathcal{F}_-$

$$E_0\, \pi(F_-)E_0 = E_0\, \pi(\, T^{-1} \int_0^T dt\, \alpha_t(F_-))E_0 \to 0 \tag{2.14}$$

as $T \to \infty$ and consequently $E_0\, \pi(\mathcal{F}_-)E_0 = 0$. Similarly, if $F_+, G_+ \in \mathcal{F}_+$, we obtain from relation (2.11) by standard arguments (mean ergodic theorem)

$$[E_0\, \pi(F_+)E_0, E_0\, \pi(G_+)E_0] = \lim_{T\to\infty} T^{-1} \int_0^T dt\, E_0\, \pi([F_+, \alpha_t(G_+)])E_0 = 0 \ , \tag{2.15}$$

where the convergence is understood in the weak operator topology. □

The preceding result and the second part of Proposition 2.2 imply that any supersymmetric state is a pure supersymmetric ground state or a mixture of such states. Hence supersymmetric states describe the most elementary systems of the theory. Next we analyze the class of states which are not supersymmetric but still admit an implementation of supersymmetry transformation by Hilbert space operators.

Definition: A state ω on $\mathcal{F}$ is said to be super–regular if δ is implementable in its GNS–representation $(\pi, \mathcal{H}, \Omega)$, i.e., if there is some densely defined, closed operator Q on $\mathcal{H}$ such that there holds in the sense of sesquilinear forms

$$\pi(\delta(F_\pm)) = Q\pi(F_\pm) \mp \pi(F_\pm)Q \quad \text{for} \quad F_\pm \in \mathcal{A}_\pm \ . \tag{2.16}$$

Note that it is not required that Ω is contained in the domain of Q.

In order to proceed we need the following technical result which seems of interest in its own right. In its proof we apply similar arguments as in the analysis of even derivations in (Buchholz et al. 1992).

Lemma 2.6 *If ω is a super–regular state on $\mathcal{F}$, there is for any $\varepsilon > 0$ a constant c_ε such that for all $F_\pm \in \mathcal{A}_\pm$*

$$|\,\omega(\delta(F_\pm))| \leq c_\varepsilon(||\pi(F_\pm)\Omega|| + ||\pi(F_\pm{}^*)\Omega||) + \varepsilon(||\delta(F_\pm)|| + ||\overline{\delta}(F_\pm)||) \ . \tag{2.17}$$

Proof: We begin by recalling that if Q is a densely defined, closed operator its adjoint Q^* has the same property and the operators QQ^* and Q^*Q are selfadjoint and positive. From the equality

$$(\Phi, \pi(\overline{\delta}(F_\pm))\Psi) = \mp\, \overline{(\Psi, \pi(\delta(F_\pm{}^*)\Phi))} \ , \tag{2.18}$$

where $F_\pm \in \mathcal{A}_\pm$ and Φ, Ψ are vectors in the domains of Q and Q^*, respectively, it follows that $\overline{\delta}$ is also implementable and

$$\overline{\delta}(F_\pm) = Q^* F_\pm \mp F_\pm Q^* \tag{2.19}$$

in the sense of sesquilinear forms. Putting for $\eta > 0$

$$L_\eta \doteq (1 + i\eta\, QQ^*)^{-1}, \quad R_\eta \doteq (1 + i\eta\, Q^*Q)^{-1} \tag{2.20}$$

it is clear that L_η, R_η as well as $L_\eta Q \subset (Q^* L_\eta^*)^*$ and QR_η are bounded operators. Moreover, by making use of relations (2.16) and (2.19) one finds after a straightforward computation that for $F_\pm \in \mathcal{A}_\pm$

$$L_\eta Q\, \pi(F_\pm) \mp \pi(F_\pm)\, QR_\eta = L_\eta\, \pi(\delta(F_\pm))\, R_\eta \mp i\eta\, L_\eta Q\, \pi(\overline{\delta}(F_\pm))\, QR_\eta \ . \tag{2.21}$$

Now with the help of the spectral theorem one sees that the norms of the vectors $\eta^{1/2}\, QR_\eta\, \Omega$, $\eta^{1/2}\, (L_\eta Q)^*\, \Omega$, $(1 - R_\eta)\, \Omega$ and $(1 - L_\eta)^*\, \Omega$ tend to 0 as $\eta \to 0$. Thus by taking matrix elements of relation (2.21) in the state Ω and applying to the left hand side of the resulting equation the Cauchy–Schwarz inequality one arrives at the statement. □

We can now establish the following result about the asymptotic properties of super–regular states.

Proposition 2.7 *Let α_t, $t \in \mathbb{R}$, be asymptotically abelian. If ω is a super–regular state on $\mathcal{F}$, there holds (point-wise on $\mathcal{A}$)*

$$\lim_{T\to\infty} T^{-1}\int_0^T dt\,\omega\alpha_t\cdot\delta = 0 \ . \tag{2.22}$$

In particular, all limit points of the net of states $\{T^{-1}\int_0^T dt\,\omega\alpha_t\}_{T>0}$ for $T \to \infty$ are supersymmetric.

Proof: Because of Lemma 2.4 we have $T^{-1}\int_0^T dt\,\omega\alpha_t(F_-) \to 0$ for $F_- \in \mathcal{F}_-$ and $T \to \infty$ and since δ is an odd derivation it follows that (2.22) holds on $\mathcal{A}_+$. If $F_- \in \mathcal{A}_-$, we put $F_{-_T} \doteq T^{-1}\int_0^T dt\,\alpha_t(F_-)$ and obtain with the help of the preceding lemma the estimate

$$\begin{aligned}
|\,T^{-1}\int_0^T dt\,\omega\alpha_t\cdot\delta(F_-)| &= |\omega\cdot\delta(F_{-_T})| \\
&\le c_\varepsilon\,(||\pi(F_{-_T})\Omega|| + ||\pi(F_{-_T}{}^*)\Omega||) + \varepsilon\,(||\delta(F_{-_T})|| + ||\overline{\delta}(F_{-_T})||) \\
&\le 2c_\varepsilon\,||F_{-_T}|| + \varepsilon\,(||\delta(F_-)|| + ||\overline{\delta}(F_-)||) \ ,
\end{aligned} \tag{2.23}$$

where, in the final step, we made use of the fact that δ and $\overline{\delta}$ are linear and of the triangle inequality. Applying Lemma 2.4 another time we see that the first term on the right hand side of this inequality vanishes in the limit $T \to \infty$. Since $\varepsilon > 0$ is arbitrary the statement follows. □

This result shows that if there exist in a theory super–regular states there exist also supersymmetric ground states which asymptotically approximate the regular ones. In particular, any α–invariant super–regular state is supersymmetric.

In the proof of these results we made use of the assumption that α is asymptotically abelian, but this condition can be relaxed. It suffices if there is some group of automorphisms β which acts on $\mathcal{F}$ in an asymptotically abelian manner and commutes with δ. If one replaces in the formulations of Lemma 2.4 and Proposition 2.7 the group α by any such β, the resulting statements hold as well.

3 Conclusions

In the preceding analysis we have seen that the implementation of supersymmetry transformations by Hilbert space operators can be accomplished only in a very special class of states. We want to discuss here the physical implications of this observation.

In order to fix ideas let us assume that we are dealing with a theory with an asymptotically abelian time evolution which commutes with the supersymmetry transformation δ (yet it need not necessarily coincide with α). There then emerges the following general picture from our results: According to Proposition 2.3 no thermal state is symmetric with respect to the action of δ, for thermal states are faithful as a consequence of the KMS–condition. This holds therefore *a fortiori* for any mixture of such states. Furthermore, the action of supersymmetry cannot be implemented by operators on the corresponding state spaces (thermal states are not super–regular). This follows from Proposition 2.7, respectively its generalization mentioned at the end of the preceding section, according to which any super–regular state which is invariant under the time evolution is also supersymmetric. We may therefore state:

(a) Thermal states and their mixtures are neither supersymmetric nor do they admit the implementation of supersymmetries by Hilbert space operators.

The fact that one cannot restore supersymmetry in thermal states by proceeding to suitable mixtures was termed spontaneous collapse in (Buchholz and Ojima 1997). Our present results are slightly more general than those in the latter article since they hold without the assumption that the supersymmetry transformation δ is related to the generator of the time evolution.

It is another intriguing consequence of Proposition 2.7 that all super–regular states coincide at asymptotic times with stationary mixtures of supersymmetric states. The latter states in turn are, by Proposition 2.2 and 2.5, mixtures of pure (and hence in their respective representations unique) supersymmetric states which can be distinguished by central observables (macroscopic order parameters). This general result provides evidence to the effect that the asymptotic limits of super–regular states are vacuum states.

In order to substantiate this idea let us consider the pertinent examples of supersymmetry in particle physics (Łopuszański 1991). There the generator δ_0 of the time evolution can be expressed, in any given Lorentz system, by two odd derivations δ_1, δ_2 which commute with space and time translations,

$$\delta_0 = \tfrac{1}{4}(\delta_1 \cdot \overline{\delta_1} + \overline{\delta_1} \cdot \delta_1) + \tfrac{1}{4}(\delta_2 \cdot \overline{\delta_2} + \overline{\delta_2} \cdot \delta_2) \ . \tag{3.1}$$

If the spatial translations act on the field algebra in an asymptotically abelian manner (which is the case if this algebra is generated by local fields), it follows from Proposition 2.7, respectively its generalization, that all states which are super–regular with respect to δ_1 and δ_2 coincide in asymptotic spacelike directions with supersymmetric states. The latter states are, by Proposition 2.2, ground states for the time evolution, and this holds in all Lorentz frames (Buchholz and Ojima 1997). Hence these states are relativistic vacuum states which can be decomposed into pure vacuum states. (The requirement of asymptotic abelianess of the time evolution is not needed here (Sakai 1991 Ch. 2.4)). We can summarize these results as follows:

(b) States admitting the implementation of supersymmetry agree in spacelike asymptotic regions with (mixtures of) pure supersymmetric vacuum states.

Hence in a supersymmetric theory all super–regular states are excitations of vacuum states and therefore describe only elementary systems. More complex systems do not admit an action of supersymmetry.

We are led by these results to the conclusion that supersymmetry is extremely vulnerable to thermal effects and there is no way of restoring the broken symmetry by physical operations on the states. In contrast, such a restoration can in general be accomplished quite easily in the case of broken bosonic symmetries: given a non–isotropic system such as a ferromagnet, say, one can prepare a corresponding rotational invariant (mixed) state by rotating the probe. As we have seen, there is no corresponding symmetry enhancing operation in the case of supersymmetry.

In view of these facts one may wonder how supersymmetry manifests itself in complex physical systems, such as the presumed early supersymmetric stages of the universe, where matter has been in a hot thermal imbroglio. It may well be that the presence of supersymmetry at the microscopic level of fields has no clearly visible consequences for such states. From the theoretical viewpoint this vulnerability of supersymmetry may be a virtue, however. First, it could explain why it is so difficult to establish this symmetry experimentally, should it be present in nature. Second, it might be used to distinguish in the theoretical setting preferred states by imposing supersymmetry as a selection criterion.

Thinking for example of quantum field theory on curved space-time manifolds which do not admit a global time evolution (future directed Killing vector field), the notion of vacuum state becomes meaningless. But there might still exist in such theories some distinguished odd derivation. One could then characterize the preferred states and their corresponding folia by demanding that they be symmetric with respect to its action. The results of the preceding analysis would justify the view that such states describe the most elementary systems of the theory.

This idea suggests the following mathematical question whose solution is known in the case of even derivations (Sakai 1991 Prop. 3.2.18): Under which conditions do there exist for an odd derivation on some graded C^*–algebra states which are annihilated by it? We hope that the present results will stimulate some interest in this problem.

References

Buchholz, D., Doplicher, S., Longo, R., Roberts, J.E., (1992): "A new look at Goldstone's theorem", Rev. Math. Phys. **Special Issue** 47

Buchholz, D., Ojima, I., (1997): "Spontaneous collapse of supersymmetry", Nucl. Phys. **B498**, 228

Haag, R., Łopuszanski, J., Sohnius, M., (1975): "All possible generators of supersymmetries of the S–matrix", Nucl. Phys. **B88**, 257

Narnhofer, H., Thirring,W., (1994): "Clustering for algebraic K systems", Lett. Math. Phys. **30**, 307

Łopuszanski, J., (1991): *An introduction to symmetry and supersymmetry in quantum field theory*, World Scientific

Sakai, S., (1991): *Operator Algebras in Dynamical Systems*, Cambridge University Press 1991

On Generalized Quantum Statistics

W. Marcinek

Institute of Theoretical Physics
University of Wrocław
Poland *

Abstract. A generalization of quantum statistics suitable for the study of one–dimensional models is presented. It is indicated that the essential structure for such generalization is a cross symmetry instead of the braid one. The Fock space representation is discussed. The problem of existence of the well–defined scalar product is also considered.

1 Introduction

Recently the concept of generalized quantum statistics and related topics has been under intensive study (Greenberg 1990, Greenberg 1991, Mohapatra 1990). The generalization of the notion of quantum statistics is motivated by many different applications in quantum field theory and statistical physics (Zee 1995, Jain 1989, 1990, Haldane 1981, Byczuk and Spalek 1995). It is known that there is an approach to particle system with generalized statistics based on the concept of the braid group B_n (Wu 1984, Imbo and March-Russel 1990). In this attempt the configuration space for the system of n–identical particles moving on a manifold $\mathcal{M}$ is $Q_n(\mathcal{M}) = (\mathcal{M}^{\times n} - D)/S_n$, where D is the subset of the Cartesian product $\mathcal{M}^{\times n}$ on which two or more particles occupy the same position and S_n is the symmetric group. The group $\pi_1(Q_n(M)) \equiv B_n(M)$ is known as the n–string braid group on $\mathcal{M}$. There is a group $\Sigma_n(\mathcal{M})$ which is a subgroup of $B_n(\mathcal{M})$ and it is an extension of the symmetric group S_n describing the interchange process of two arbitrary indistinguishable particles. It is obvious that the statistics of the given system of particles is determined by the group Σ_n (Wu 1984, Imbo and March-Russel 1990). This picture breaks up in the one-dimensional case. The difficulty arises with the proper definition of the group $\Sigma_n(\mathcal{M})$, see (Jacak et al. 1995). Hence in this case we need an other approach to the concept of generalized statistics. In this paper we are going to describe a generalization of quantum statistics which works also in the one–dimensional case.

* The work is partially sponsored by Polish Committee for Scientific Research (KBN) under Grant 2P03B130.12

2 Fundamental Assumptions

The starting point for our study of generalized statistics is a system of charged particles interacting with certain external quantum field. The proper physical nature of the system is not essential for our considerations. Our fundamental assumption is that every charged particle is transform under interactions into a system which contains quasiparticles and quasiholes. A quasiparticle is in fact the charged particle dressed with a single quantum of the external field. A particle without quantum is called undressed or a quasihole. It is natural to expect that some new excited states of the system have appear as a result of certain specific interaction. We are going to describe all such excited states as composition of quasiparticles and quasiholes. It is interesting that quasiparticles and quasiholes have also their own statistics. We describe the generalized statistics of a charged particle as the interchange statistics of quasiparticles and quasiholes. We have here the following assumptions:

A0. There is a state $|0>= \mathbf{1}$ called the ground one. There is also the conjugate ground state $<0| \equiv \mathbf{1}^*$.

A1. There is an ordered (finite) set of single quasiparticle states

$$S := \{x^i : i = 1, \ldots, N < \infty\}. \tag{1}$$

These states are said to be elementary, they represent elementary excitations of the system. The set S forms a basis for a linear space E over a field of complex numbers $\mathbb{C}$.

A2. There is also a corresponding set of single quasihole states

$$S^* := \{x^{*i} : i = N, N-1, \ldots, 1\}. \tag{2}$$

These states are said to be conjugated. The set S^* of conjugate states forms a basis for the complex conjugate space E^*.

A3. The pairing $(.|.) : E^* \otimes E \longrightarrow \mathbb{C}$ and the corresponding scalar product is given by

$$g_E(x^{*i} \otimes x^j) \equiv (x^{*i}|x^j) = \langle x^i|x^j\rangle := \delta^{ij}. \tag{3}$$

A4. There is a linear, invertible and Hermitian operator $T : E^* \otimes E \longrightarrow E \otimes E^*$

$$T(x^{*i} \otimes x^j) = \Sigma\, T^{ij}_{kl}\, x^k \otimes x^{*l} \tag{4}$$

called a cross. The operator T describes the interchange statistics of quasiparticles and quasiholes.

A5. There is a pair of algebras $\mathcal{A}$ and $\mathcal{A}^*$ such that

$$\mathcal{A} := \bigoplus_n \mathcal{A}^n,\ \mathcal{A}^* := \bigoplus_n \mathcal{A}^{*n}. \tag{5}$$

and there is an antilinear and involutive isomorphism $(-)^* : \mathcal{A} \longrightarrow \mathcal{A}^*$, i. e. we have the relations

$$m_{\mathcal{A}^*}(b^* \otimes a^*) = (m_{\mathcal{A}}(a \otimes b))^*, \quad (a^*)^* = a, \tag{6}$$

where $a, b \in \mathcal{A}$ and a^*, b^* are their images under the isomorphism $(-)^*$. The algebra $\mathcal{A}^*$ is said to be the conjugated algebra of $\mathcal{A}$.

3 Hermitian Wick Algebras

Let $\mathcal{A}^*$ be the conjugate algebra of $\mathcal{A}$. A linear mapping $\Psi : \mathcal{A}^*\otimes\mathcal{A} \longrightarrow \mathcal{A}\otimes\mathcal{A}^*$ such that

$$\Psi|_{E^*\otimes E} = T + g_E \,, \tag{7}$$

and we have the following relations (Cap et al. 1995)

$$\begin{aligned} \Psi \circ (id_{\mathcal{A}^*} \otimes m_{\mathcal{A}}) &= (m_{\mathcal{A}} \otimes id_{\mathcal{A}^*}) \circ (id_{\mathcal{A}} \otimes \Psi) \circ (\Psi \otimes id_{\mathcal{A}}), \\ \Psi \circ (m_{\mathcal{A}^*} \otimes id_{\mathcal{A}}) &= (id_{\mathcal{A}} \otimes m_{\mathcal{A}^*}) \circ (\Psi \otimes id_{\mathcal{A}^*}) \circ (id_{\mathcal{A}^*} \otimes \Psi) \end{aligned} \tag{8}$$

is said to be a cross symmetry or generalized twist. We use here the notation

$$\Psi(b^* \otimes a) = \Sigma a_{(1)} \otimes b^*_{(2)} \tag{9}$$

for $a \in \mathcal{A}, b^* \in \mathcal{A}^*$. The tensor product $\mathcal{A} \otimes \mathcal{A}^*$ equipped with the multiplication

$$m_\Psi := (m_{\mathcal{A}} \otimes m_{\mathcal{A}^*}) \circ (id_{\mathcal{A}} \otimes \Psi \otimes id_{\mathcal{A}^*}) \tag{10}$$

is an associative algebra called a Hermitian Wick algebra (Jorgensen et al. 1995, Borowiec and Marcinek in preparation) and it is denoted by $\mathcal{W} = \mathcal{W}_\Psi(\mathcal{A}) = \mathcal{A}\bowtie_\Psi\mathcal{A}^*$. Let H be a linear space. We denote by $L(H)$ the algebra of linear operators acting on H.

Let $\mathcal{W} \equiv \mathcal{A}\bowtie_\Psi\mathcal{A}^*$ be a Hermitian Wick algebra. If $\pi_{\mathcal{A}} : \mathcal{A} \longrightarrow L(H)$ is a representation of the algebra $\mathcal{A}$, such that we have the relation

$$\begin{aligned} (\pi_{\mathcal{A}}(b))^*\pi_{\mathcal{A}}(a) &= \Sigma\pi_{\mathcal{A}}(a_{(1)})\pi_{\mathcal{A}^*}(b^*_{(2)}) \,, \\ \pi_{\mathcal{A}^*}(a^*) &:= (\pi_{\mathcal{A}}(a))^* \,, \end{aligned} \tag{11}$$

then there is a representation $\pi_{\mathcal{W}} : \mathcal{W} \longrightarrow L(H)$ of the algebra $\mathcal{W}$ (Borowiec and Marcinek in preparation).

The relations (11) are said to be a commutation relation if there is a positive definite scalar product on $\mathcal{A}$ such that

$$\langle \pi_{\mathcal{A}^*}(x^*) f | g \rangle = \langle f | \pi_{\mathcal{A}}(x) g \rangle \,. \tag{12}$$

Note that if we use the notation

$$\pi_{\mathcal{A}}(x^i) \equiv a^+_{x^i}, \quad \pi_{\mathcal{A}^*}(x^{*i}) \equiv a_{x^{*i}} \,, \tag{13}$$

and the cross T is given by its matrix elements (4), then the commutation relations (11) can be given in the following form

$$a_{x^{*i}} a^+_{x^j} - T^{ij}_{kl} \, a^+_{x^l} a_{x^{*k}} = \delta^{ij} \mathbf{1} \,. \tag{14}$$

4 Fock Space Representation

We have here an interest in the study of Fock representation. In this case the representation act on the algebra $\mathcal{A}$. For the ground state and annihilation operators we assume that

$$\langle 0|0\rangle = 0, \quad a_{s^*}|0\rangle = 0 \quad \text{for} \quad s^* \in \mathcal{A}^* . \tag{15}$$

We define creation operators as multiplication in the algebra $\mathcal{A}$

$$a_s^+ t := m_{\mathcal{A}}(s \otimes t), \quad \text{for} \quad s, t \in \mathcal{A} . \tag{16}$$

The proper definition of the action of annihilation operators on the whole algebra $\mathcal{A}$ is a problem. If the action a of annihilation operators are given in such a way that there is unique, nondegenerate, positive definite scalar product, creation operators are adjoint to annihilation ones and vice versa, then we say that we have the well-defined Fock representation for a system with generalized quantum statistics (Marcinek 1998).

Example 1. We assume here that the algebra of states $\mathcal{A}$ is the full tensor algebra TE over the space E, and the conjugate algebra $\mathcal{A}^*$ is identical with the tensor algebra TE^*. If $T \equiv 0$ then we obtain the most simple example of well-defined system with generalized statistics. The corresponding statistics is the so-called infinite (Bolzman) statistics (Greenberg 1990, Greenberg 1991, Marcinek 1998).

Example 2 Let $T : E^* \otimes E \longrightarrow E \otimes E^*$ be an arbitrary cross. Then there is the cross symmetry $\Psi^T : TE^* \otimes TE \longrightarrow TE \otimes E^*$. It is defined by a set of mappings $\Psi_{k,l} : E^{*\otimes k} \otimes E^{\otimes k} \longrightarrow E^{\otimes l} \otimes E^{*\otimes k}$, where $\Psi_{1,1} \equiv R := T + g_E$, and

$$\begin{aligned} \Psi_{1,l} &:= R_l^{(l)} \circ \ldots \circ R_l^{(1)} , \\ \Psi_{k,l} &:= (\Psi_{1,l})^{(1)} \circ \ldots \circ (\Psi_{1,l})^{(k)} , \end{aligned} \tag{17}$$

here $R_l^{(i)} : E_l^{(i)} \longrightarrow E_l^{(i+1)}$, $E_l^{(i)} := E \otimes \ldots \otimes E^* \otimes E \otimes \ldots \otimes E$ ($l+1$-factors, E^* on the i-th place, $i \leq l$) is given by the relation

$$R_l^{(i)} := \underbrace{id_E \otimes \ldots \otimes R \otimes \ldots \otimes id_E}_{l \ times} ,$$

where R is on the i-th place, $(\Psi_{1,l})^{(i)}$ is defined in similar way like $R^{(i)}$. We also introduce the operator $\tilde{T} : E \otimes E \longrightarrow E \otimes E$ by its matrix elements

$$(\tilde{T})^{ij}_{kl} = T^{ki}_{lj} . \tag{18}$$

If the operator $\tilde{T}$ is a bounded operator acting on some Hilbert space such that we have the following Yang-Baxter equation on $E \otimes E \otimes E$

$$(\tilde{T} \otimes id_E) \circ (id_E \otimes \tilde{T}) \circ (\tilde{T} \otimes id_E) = (id_E \otimes \tilde{T}) \circ (\tilde{T} \otimes id_E) \circ (id_E \otimes \tilde{T}) , \tag{19}$$

and $||\tilde{T}|| \leq 1$, then according to Bożejko and Speicher (Bożejko and Speicher 1994) there is a positive definite scalar product. Note that the existence of nontrivial kernel of operator $P_2 \equiv id_{E\otimes E} + \tilde{T}$ is essential for the nondegeneracy of the scalar product (Jorgensen et al. 1995). One can see that if this kernel is trivial, then we obtain the well–defined system with generalized statistics (Marcinek and Ratowski 1995, Ratowski 1997).

Example 3: If the kernel of P_2 is nontrivial, then the scalar product is degenerate. Hence we must remove this degeneracy by factoring the mentioned above scalar product by the kernel. In this case we have $\mathcal{A} := TE/I, \quad \mathcal{A}^* := TE^*/I^*$, where I is an ideal in TE such that

$$\Psi^T(\mathcal{A}^* \otimes I) \subset I \otimes \mathcal{A}^*\,, \quad P_2(N) = 0, \quad I := gen(N)\,, \quad N \subset E \otimes E \tag{20}$$

and I^* is the corresponding ideal in TE^*. One can see that there is the cross symmetry and the action of annihilation operators can be defined in such a way that we obtain that the system with generalized statistics is well–defined (Marcinek and Ratowski 1995, Ratowski 1997).

Example 4: If a linear and invertible operator $B : E \otimes E \longrightarrow E \otimes E$ defined by its matrix elements $B(x^i \otimes x^j) := B^{ij}_{kl}(x^k \otimes x^l)$ is given such that we have the following conditions

$$\begin{aligned} B^{(1)}B^{(2)}B^{(1)} &= B^{(2)}B^{(1)}B^{(2)}\,, \\ B^{(1)}T^{(2)}T^{(1)} &= T^{(2)}T^{(1)}B^{(2)}\,, \\ (id_{E\otimes E} + \tilde{T})(id_{E\otimes E} - B) &= 0\,, \end{aligned} \tag{21}$$

then $I := gen\{id_{E\otimes E} - B\}$ and one can prove that the corresponding system is well defined (Marcinek and Ratowski 1995, Ratowski 1997).

Acknowledgments

The author would like to thank the PhD students for the help which made his participation in the Max Born Symposium possible.

References

Borowiec, B., Marcinek, W., On twisted product of algebras, in preparation
Bożejko, M., Speicher, R., (1994): Math. Ann. **300**, 97
Byczuk, K., Spalek, J., (1995): Phys. Rev. **B51**, 7934
Cap, A., Schihl, H., Vancura, J., (1995): *On twisted tensor product of algebras*, Comm. Alg. **23**, 4701
Greenberg, O.W., (1990): Phys. Rev. Lett. **64**, 705
Greenberg, O.W., (1991): Phys. Rev. **D 43**, 4111

Haldane, F.D.M., (1981): J. Phys. **C14**, 2585
Imbo, T.D., March-Russel, J., (1990): Phys. Lett. **B252**, 84
Jacak, L., Sitko, P., Wieczorek, K., (1995): Anyons and composite fermions, Scientific Papers of the Institute of Physics of the Technical University of Wroclaw No. 31, Monographs No. 20, Wroclaw, in Polish
Jain, J.K., (1989): Phys. Rev. Lett. **63**, 199; Phys. Rev. **B40**, 8079, **41**, (1990) 7653
Jorgensen, P.E.T., Schmith, L.M., Werner, R.F., (1995): Positive representation of general commutation relations allowing Wick ordering, J. Funct. Anal. **134**, 33
Marcinek, W., Rałowski, R., (1995): Particle operators from braided geometry, in "Quantum Groups, Formalism and Applications" XXX Karpacz Winter School in Theoretical Physics, 1994, Eds. J. Lukierski et al., 149–154 (1995) and J. Math. Phys. **36**, 2803
Marcinek, W., (1998): Remarks on Quantum Statistics, in Proceedings of the Conference "Particles, Fields and Gravitation", April 15 - 19, (1998), Lodz, Poland, ed. by J. Rembieliński, World Scientific, Singapore and math. QA/9806158
Mohapatra, R.N., (1990): Phys. Lett. **B 242**, 407
Rałowski, R., (1997): J. Phys. **A30**, 2633
Wu, Y.S., (1984): J. Math. Phys. **52**, 2103
Zee, A., (1995): Quantum Hall fluids in Field Theory, Topology and Condensed Matter Physics, ed. by H. D. Geyer, Lecture Notes in Physics, Springer

q-Deformed Minkowski Algebra and Its Space-Time Lattice

J. Wess

Sektion Physik der Ludwig-Maximilians-Universität
Theresienstr. 37, D-80333 München
julius.wess@physik.uni-muenchen.de
and
Max-Planck-Institut für Physik
(Werner-Heisenberg-Institut)
Föhringer Ring 6, D-80805 München

Talk given in honour
of Professor Jan Łopuszański's 75th birthday

We have asked how the Heisenberg relations of space and time change if we replace the Lorentz group by a q-deformed Lorentz group (Lorek et al. 1997).

By the Heisenberg relations we mean:

$$
\begin{aligned}
X^a X^b &= X^b X^a \,, \\
P^a P^b &= P^b P^a \,, \\
X^a P^b &= P^b X^a + i\eta^{ab} \,.
\end{aligned}
\tag{1}
$$

The indices a, b run from 0 to 3, 0 being the time component, η^{ab} is the Lorentz metric. This relation is covariant under the Lorentz group, X^a and P^a are four vectors, that is representations or equivalently modules of the Lorentz group.

The relations are compatible with an involution

$$
\overline{X^a} = X^a \,, \qquad \overline{P^b} = P^b \,. \tag{2}
$$

Dividing the free algebra generated by X^a, P^b by the ideal generated by the relations (1) we obtain an algebra.

We can consider this setup as the definition of a relativistic kinematics in the following sense:

To make contact with physics - nature, this should be - we have to produce real numbers. It is only via real numbers that nature talks to us. A natural way to produce real numbers from an algebra is to study its representations in terms of matrices or linear operators in a Hilbert space over the field of complex numbers. In such a representation we shall impose that the involution is represented by the conjugation of the respective linear operators. If we then succeed in realizing the algebraic selfadjointness by the selfadjointness of linear operators in the Hilbert space we know that the spectrum of such operators is real. Thus we can follow the rules of quantum mechanics and identify the eigenvalues of a selfadjoint operator with the possible results of a measurement. In this way we can satisfy the demands of a physicist to relate the algebraic scheme to measurements.

It is desirable, however, also from the point of view of a mathematician to study representations in a Hilbert space - this adds topological properties to the purely algebraic ones.

Following this approach we would find that the spectrum of the operator X^a (or P^a) is continuous, and by identifying X^a with the coordinates of our system we would learn that we live in a fourdimensional Minkowski manifold. Minkowski refers to the Lorentz metric η^{ab}.

We are going to generalize to an algebra where the Lorentz group is replaced by a q-deformed Lorentz group. Why? Just generalizing the algebra without additional assumptions would leave the game too open, we could not do much. Usually we consider the algebra (1) based on a geometrical construction on a differentiable manifold and change the manifold. Here we would like to give the priority to the algebra. Furthermore we try to keep as much algebraic structure as possible - all that can be done explicitly is usually based on some algebra. This is the experience of a physicist who is always trying to do as much as possible explicitly. The q-deformed Lorentz group lends itself quite naturally because we can use an enormous amount of mathematical knowledge that has been gathered within the last twenty years. In addition it has the property that it contains the Lorentz group as a special case. We shall consider real $q \neq 0$ and we know that for $q = 1$ we should obtain the known case.

Let's follow this strategy to find out how physics changes if we change the underlying algebraic relation (1).

We use textbook knowledge to deform the Lorentz group. We also use textbook knowledge for constructing comodules that we identify with coordinates and momenta. On these comodules we impose algebraic relations that ought to be compatible with the comodule property. But it also has to allow an involution. Finally, we impose one more condition concerning the size of the algebra, it should be the same as the size of the undeformed algebra. For a physicist this means that there are no new relations generated. The ordered monomials of the coordinates should remain a basis for the algebra of coordinates. Any new relation would restrict the results of measurements of what we would like to be independent observables. Mathematically we will refer to this condition as the Poincaré-Birkhoff-Witt property. Technically this is closely related to the Yang-Baxter equation, thus we should not be surprised that an $\hat{R}$-Matrix enters the definition of the algebra. This is a matrix with complex entries that satisfies the quantum Yang-Baxter equation.

As a result of these conditions we find a quite unique algebra.

$$X^0 X^c = X^c X^0 , \tag{3}$$

$$\varepsilon_{qCB}{}^{A} X^B X^C = (1-q^2) X^0 X^A .$$

The indices A, B, C run from 1 to 3, X^0 is the time coordinate. As was stated before, q is a real number ($q \neq 0$). The q-deformed ε-symbol is the q-deformed Clebsch-Gordon coefficient that combines two threedimensional representations of the q-deformed rotation group ($SO_q(3)$) to a threedimensional representation again. These q-Clebsch-Gordan coefficients are textbook knowledge. To give an impression of what they look like we write (3) explicitly:

$$\begin{aligned} X^3 X^+ - q^2 X^+ X^3 &= (1-q^2) X^0 X^+ , \\ X^- X^3 - q^2 X^3 X^- &= (1-q^2) X^0 X^- , \\ q X^- X^+ - q X^+ X^- + (1-q^2) X^3 X^3 &= (1-q^2) X^0 X^3 . \end{aligned} \tag{4}$$

As coordinates we use X^+ and X^-, these are deformations of $X^1 \pm X^-$ and they occur naturally if we use textbook knowledge on q-deformed groups.

Equation (4) shows that our space has become noncommutative - something we have the intention to study anyhow.

For the momenta we get the same relations. Coordinates and momenta have the same comodule structure.

The Heisenberg algebra changes as well.

$$\begin{aligned} & P^a X^b - q^{-2} \hat{R}_{II}^{-1}{}^{ab}{}_{cd} X^c P^d \\ = & -\frac{i}{2} \Lambda^{-\frac{1}{2}} \left\{ (1+q^4) \eta^{ab} U + q^2 (1-q^4) V_q^{ab} \right\} . \end{aligned} \tag{5}$$

The indices a, b etc. now run from 0 to 3. As promised there enters an $\hat{R}$ matrix of the q-Lorentz group. It is a 16 by 16 matrix that satisfies the Yang-Baxter equation. The metric η^{ab} is the Clebsch-Gordan coefficient that combines two q-deformed four vectors to a scalar

$$\eta_{ab} X^a Y^b = X^3 Y^3 - q X^+ Y^- + \frac{1}{q} X^- Y^+ - X^0 Y^0 . \tag{6}$$

For real q it is quite generally true that the representations of the q-deformed groups have the same pattern of decomposing the product of two representations into irreducible representations. The respective Clebsch-Gordan coefficients become q-dependent.

At the right hand side of (5) appear new operators, V_q^{ab}, U and $\Lambda^{-\frac{1}{2}}$. V_q^{ab} are the q-deformed generators of the q-Lorentz algebra. They are in the six-dimensional q-antisymmetric representation of the q-Lorentz group. The operator U is related to the Casimir operator of the q-Lorentz group. It becomes one ($U \to 1$) for $q \to 1$.

The additional operator $\Lambda^{-\frac{1}{2}}$ is a scaling operator:

$$\begin{aligned} \Lambda^{-\frac{1}{2}} X^a &= q X^a \Lambda^{-\frac{1}{2}} , \\ \Lambda^{-\frac{1}{2}} P^a &= q^{-1} P^a \Lambda^{-\frac{1}{2}} , \\ \Lambda^{-\frac{1}{2}} U &= U \Lambda^{-\frac{1}{2}} , \\ \Lambda^{-\frac{1}{2}} V_q^{ab} &= V_q^{ab} \Lambda^{-\frac{1}{2}} . \end{aligned} \tag{7}$$

Without this operator an algebra with q-commutators would be very restrictive. An example should illustrate this:

$$XP - qPX = iH . \tag{8}$$

If all the quantities are hermitean it would follow that

$$PX - qXP = -iH , \tag{9}$$

or

$$(1-q^2)XP = i(1-q)H , \qquad PX + XP = 0 . \tag{10}$$

To avoid this the right hand side cannot depend on hermitean operators only. Bearing in mind that operators can usually be written as a product of unitary operators times hermitean operators we have introduced the operator $\Lambda^{-\frac{1}{2}}$.

The algebras (4) and (5) are consistent with the conjugation property:

$$\overline{X^0} = X^0 , \quad \overline{X^3} = X^3 , \quad \overline{X^+} = -qX^- , \quad \overline{X^-} = -\frac{1}{q} X^+ , \tag{11}$$

the same for the momenta and

$$\begin{aligned} \overline{\Lambda^{-\frac{1}{2}}} &= q^{-4} \Lambda^{\frac{1}{2}} , \\ \overline{U} &= U . \end{aligned} \tag{12}$$

The generators V_q^{ab} have the usual conjugation property of the q-Lorentz algebra generators.

This is the legend to the equation (5). It should be noted that the q-Heisenberg algebra does not decompose into an algebraic relation involving

X and P only and a relation defining the generators of the Lorentz algebra in terms of ordered X, P polynomials as we are used to in the case $q = 1$.

Now we would like to study a world that is based on the algebra defined by (3), (5) and (6) as well as by the defining relations of the q-Lorentz group and the comodule structure of X^0 and P^a. The first step is to construct Hilbert space representations where coordinates and momenta are selfadjoint linear operators. This was done in a work with B.L. Cerchiai (Cerchiai and Wess 1998) - let me here state the results.

A complete set of commuting observables is

$$X^0\,, \quad r^2 = g_{AB}X^BX^A = X^3X^3 - qX^+X^- - \frac{1}{q}X^-X^+\,, \quad \mathbf{L}_q^2 \text{ and } L_{q,3}\,. \tag{13}$$

X^0 is the time, r^2 the three-dimensional radius, $L_{q,3}$ the third component of this q-deformed angular momentum and $\mathbf{L}_q^2$ the Casimir operator for this angular momentum. This would be a complete set for $q = 1$ and remains a complete set in the deformed case. Its eigenvalues are sufficient to label all the states in the representation.

As we insist in representations where X^0 and r^2 are selfadjoint we can assume them to be diagonal and compute the spectrum from the algebra.

We find three types of representations. The first are of the timelike nature. This means that the q-Lorentz invariant length

$$s^2 = \eta_{q\,ab}X^aX^b = X^0X^0 - g_{ab}X^BX^A \tag{14}$$

has positive eigenvalues.

There are representations forward timelike, i.e., the eigenvalues of X^0 are positive as well. There is one parameter τ_0, $1 \leq \tau_0 < q$ if q is assumed to be larger than one, that labels inequivalent representations of the timelike forward representations. It is related to a Casimir and carries the dimension time. The eigenvalues of X^0 and r^2 are labeled by two integers n, M where

$$M = -\infty \ldots \infty\,, \qquad n = 0, \ldots \infty$$

$$\begin{aligned} X^0 &: \tau_0 q^M \frac{q^{n+1} + q^{-n-1}}{q + \frac{1}{q}}\,, \\ r^2 &: \tau_0^2 q^{2M} \frac{(q^{n+2} - q^{-n-2})(q^n - q^{-n})}{(q + \frac{1}{q})^2}\,. \end{aligned} \tag{15}$$

The eigenvalue of the angular momentum is restricted by $0 \leq j \leq n$.

The timelike backward (eigenvalue of the time negative) can be obtained from (15) by replacing τ_0 by $-\tau_0$.

There also are the spacelike representations. s^2 has negative eigenvalues. They are also characterized by a parameter l_0, this time carrying the dimension of a length. The eigenvalues are again labeled by two integers n, M but this time

$$M = -\infty \dots \infty\,, \quad n = -\infty \dots \infty$$

and l is not restricted.

We found:

$$\begin{aligned} X^0 : \qquad & l_0 q^M \frac{q^n - q^{-n}}{q + \frac{1}{q}}\,, \\ r^2 : \; & l_0^2 q^{2M} \frac{(q^{n+1} - q^{-n-1})(q^{n-1} - q^{-n+1})}{(q + \frac{1}{q})^2}\,. \end{aligned} \tag{16}$$

What is characteristic for the spectrum is that the eigenvalues are of the form of polynomials in q^M, $q^{\pm n}$.

There is no lightlike representation but the light cone ($s^2 = 0$) is full of limit points of all the representations.

That timelike and spacelike regions should carry independent representations of the algebra appears to be a strange situation. Having a closer look at these representations, that is trying to diagonalize the momenta instead of the coordinates we realize that this is impossible. The momenta have an overcomplete set of eigenvectors with eigenvalues that are complex as well. This clearly shows that the representations of P^a are symmetric but not selfadjoint. Symmetric means that the matrix elements have the property of hermitean operators. Selfadjointness is a question of domain and range of the operators.

We can construct selfadjoint extensions of the individual representations. These extensions will not satisfy the algebra.

It is, however, possible to consider reducible representations composed of a spacelike as well as a timelike forward and timelike backward representation and they will allow a selfadjoint extension of the operators P^a that satisfies the algebra. For a physicist this is somehow similar to the situation that arises with field theoretic anomalies. We have to put together several representations to cancel the anomalies.

The requirement that a selfadjoint operator P^0 should represent the momentum in the algebra tells us that we need a past, a future and spacelike distances. This situation is best illustrated by the Figure 1.

The time t and the three-dimensional radius r are plotted. The points represent the results of an exact measurement of space and time. Each point represents a sphere defined by measuring angular momentum (q-deformed). In the lightlike region the quantum number for angular momentum l is restricted by $l \leq n$ where n counts points on the clearly visible hyperbolas

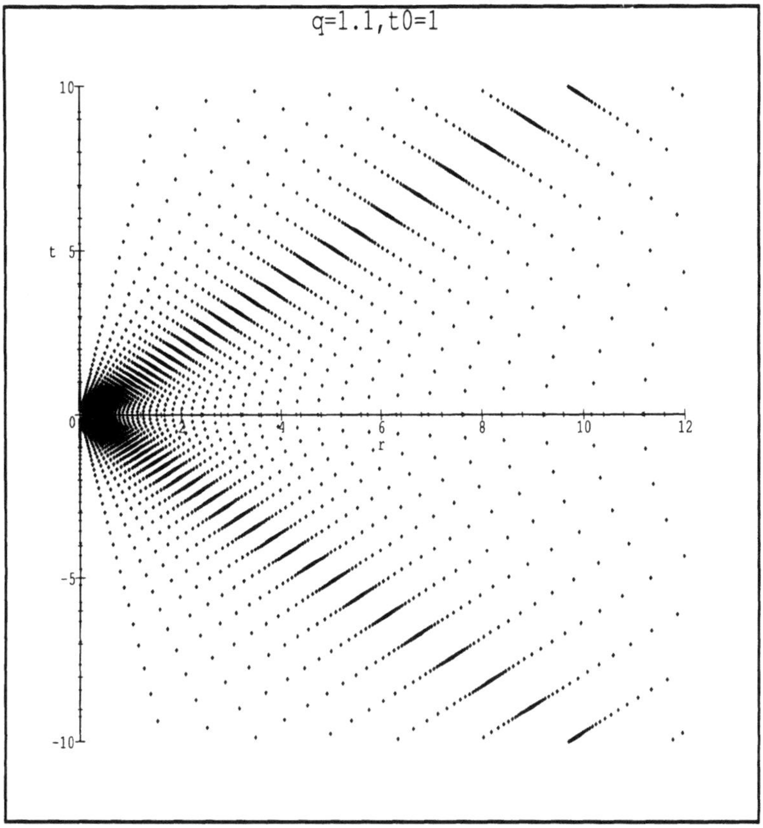

Fig. 1. Admissible values of t versus those of r for $q = 1.1$ and $t_0 = 1$.

starting at $r = 0$. The hyperbolas are labeled by the quantum number M. In the spacelike region there is no limit on l.

Looking at Fig. 1 we realize that this space-time concept is clearly in contradiction with what we know of larger distances in space time, that is larger than 10^{-19} cm. It could, however, be that it approximates a space time structure at very short distances, i.e., at very high energies and for a very short period of time. This could be thought of by assuming that space time exists in different phases - the continuous phase at larger distances and normal (that is quite low) energy density as we know it and a latticized phase at very high energy densities. Describing states of this very high density it might be a better approximation to start from the lattice sphase as represented in Fig. 1, then from the continuous phase.

It should be clear that up to now we have not spoken about any dynamics - all we have done is "quantize" the underlying space time structure.

A dynamics can be introduced by a q-deformed Klein-Gordon equation. That is, we have to solve the eigenvalue problem associated with the operator $g_{ab}P^bP^a = P^2$. This is done by diagonalizing the respective momenta. The eigenfunctions are known and it is not surprising that in a particlelike region of the momenta the eigenvalues of P^2 are discrete and have the form:

$$P^2 : \frac{1}{\tilde{l}_0^2} q^{2\tilde{M}} \qquad \tilde{M} = -\infty \ldots \infty . \tag{17}$$

The eigenvalues at $M \to -\infty$ are clearly not representing a reasonable physical system. It is, however, possible to calculate the size of the corresponding wave packets - it increases very fast for small energies. If the size becomes larger than the size of the system where the lattice is a good approximation the eigenvalues have no physical significance - the system would change its phase. For large energies the wave packets become smaller and smaller - an exponentially growing energy spectrum is not in clear contradiction with what we know about physical systems.

This is just to encourage further research on quantum group guided systems - we might learn not only about mathematically interesting structures but also about physical features that are connected with a noncommutative space time structure.

References

Cerchiai, B.L., Wess, J., (1998): *q-deformed Minkowski Space based on a q-Lorentz Algebra*, E. Phys. J. **C5**, 553–566

Lorek, A., Weich, W., Wess, J., (1997): *Non-commutative Euclidean and Minkowski Structures*, Z. Phys. . **C76**, 375–386

Part IV

Particles, Fields and Geometry

The New Inverse Problem of the Newton Law

M. Gusiew-Czudżak

Uniwersytet Wrocławski, Instytut Fizyki Teoretycznej
Pl. Maksa Borna 9, 50204 Wrocław, Poland
gusiew@ift.uni.wroc.pl

Abstract. The second Newton law is encoded into a completely not integrable Pfaffian system (an ideal) of the differential forms. The aim of this note is to motivate and present a new inverse problem for this Pfaffian system. A new notion of the descendant differential two-form for an exterior differential system is introduced and the set of all descendant forms for the Newton law is determined. The maximal de Rham sub-complex on which the descendant differential form is closed generalize the Hamilton and the Lagrange formalisms.

1 Introduction

An inverse problem reverses an order convenient for physicists: given a Lagrangian (a field of densities) to find it's Euler & Lagrange equation (a vector field). In the inverse problem the order is phenomenological, seems more natural — the equation of the motion (i.e., a vector field) is the primary object, what we try to find is the Lagrangian (Santilli 1978), (Oziewicz 1982), (Oziewicz 1985), (Della Riccia 1982), (Marmo et al. 1990), (Cisło et al. 1995).

In this note we give a sketch of a new inverse problem for the second Newton law. Closely related ideas has been considered by de León & Lacomba (de León and Lacomba 1988), (de León and Lacomba 1989).

By equation of the motion we understand a Pfaffian system or - equivalently - it's annihilator, the Cauchy characteristic distribution. Our considerations are algebraic. What we need to formulate the problem is

[(i)] an associative, unital and commutative $\mathbb{R}$-algebra $\mathcal{F}$, e.g., $\mathbb{R}$-algebra of $\mathbb{R}$-valued smooth functions on $\mathbb{R} \times TT^*Q$. An $\mathbb{R}$-algebra needs to include (or be generated by) time (the proper time), positions, velocities, momenta and forces denoted respectively by the letters ($n = \dim Q \in \mathbb{N}$ is the number of degrees of freedom)

$$i = 1, \ldots, n; \qquad t, q^i, v^i, p_i, f_i \in \mathcal{F}. \tag{1}$$

[(ii)] an $\mathcal{F}$-module M of differential one forms and de Rham $\mathcal{F}$-complex

$$\dim_{\mathcal{F}} M = 4n + 1, \quad M^{\wedge} \equiv \bigoplus_{i \geq 0} M^{\wedge i} \quad \text{with} \quad M^{\wedge 0} \equiv \mathcal{F}, \quad M^{\wedge 1} \equiv M. \tag{2}$$

The time-dependent Hamiltonian formalism is a symplectic form σ (on $\mathbb{R}\times T^*Q$) which is a closed and regular differential two-form. It is the calculus of variations that needs in mechanics a closed differential two-form (Oziewicz 1992). The name 'symplectic' is reserved usually for even-dimensional manifolds, so we should use the name 'pre-symplectic' - however for brevity we will use here the shorter name. The kernel $\ker\sigma$ is a one dimensional distribution of the vector fields, and this distribution is an annihilator of a Pfaffian system (an equation of the motion). In this usual case the Pfaffian system is of a dimension $2n$ and of a codimension 1, due to the fact that we work with the $\mathcal{F}$-module of the differential one-forms generated by $2n+1$ differentials dt, dq^i, dp_i. The velocities and forces are defined in terms of the Hamiltonian function. Such system is completely integrable. For more details see, e.g., (Kocik 1981), (Oziewicz 1985), (Borowiec & Oziewicz 1990).

In the present note we impose no relations between letters (1). The Pfaffian system encoding the second Newton law is again $2n$-dimensional, but its codimension is $2n+1$ and it is completely not integrable. We demand the Cauchy characteristic distribution to be the kernel of a differential two form, a generalized analogon of the symplectic form. Such two-form is called *descendant* for this Pfaffian system. For a Pfaffian system we will find all descendant two-forms (Definition 2.1). The descendant two-form exists only for the even-dimensional Pfaffian systems. The choice of the descendant differential two-form is not unique – this is contrary to the unique choice 'β' (3) given by de León & Lacomba (de León and Lacomba 1988), (de León and Lacomba 1989). One would ask whether within the set of all descendant forms for the Newton law there exists a subset of the closed forms. We prove that within this set of all descendent differential forms there are no closed forms. This is an obstacle if we would like to develop the analogy to symplectic form further on. In this point there arises naturally a need of defining a *formalism* (Definition 4.1), i.e., an epimorphism of the differential graded algebras which maps de Rham complex onto the sub-complex, generated by the smaller number of letters, with the smaller dimension of the module of differential one-forms and with the condition that this is the largest sub-complex on which the image of the descendant form under this epimorphism is a closed differential form. In this way we come back to the known Lagrangian and Hamiltonian formalisms as to the images of the solutions of the equation *on epimorphisms*.

De León and Lacomba (de León and Lacomba 1988), (de León and Lacomba 1989) are starting from the given Lagrangian or Hamiltonian scalar fields, i.e., *functions* and not from the densities, and then they are showing that these functions determine the Lagrangian submanifolds of the symplectic manifold $T^*T(\mathbb{R}\times Q) \simeq TT^*(\mathbb{R}\times Q)$. This approach is exactly the same as the Hamilton & Jacobi theory. Hamilton discovered in 1834 that the integration of a Hamiltonian vector field (a Pfaffian system) can be reduced to the determination of the Hamilton *two-point* 'principal (or characteristic)' function S on the Lagrangian sub-manifold ℓ of the symplectic manifold T^*Q.

Hamilton concluded incorrectly that a function S must satisfy the set of the *two* partial differential equations (of the first order). Jacobi in 1838 pointed out that only one of these equations is needed: for the given differential one form α (e.g., for the Liouville differential one-form) this equation is $dS = \ell\alpha$. This equation (for the Lagrangian submanifold ℓ in fact) is said to be the Hamilton & Jacobi equation. De León and Lacomba define the Liouville - like differential one-form on $T^*T(\mathbb{R} \times Q) \simeq TT^*(\mathbb{R} \times Q)$ (de León and Lacomba 1989 p. 3810),

$$\beta \equiv f_i dq^i - v^i dp_i, \quad \alpha \equiv f_i dq^i + p_i dv^i \quad d\alpha = d\beta, \quad \beta^{\wedge 2n} \neq 0. \tag{3}$$

Then the Hamilton & Jacobi equations for two Lagrangian submanifolds are (de León and Lacomba 1988), (de León and Lacomba 1989),

$$\ell d\beta = 0, \quad -dH = h\beta, \qquad \text{and} \qquad dL = \ell\alpha.$$

In this note, contrary to the canonical approach by de León & Lacomba, we start from *not* symplectic case corresponding to completely not integrable Pfaffian system. Then our phenomenological descendant differential two form Ω is not closed, $d\Omega \neq 0$, $\Omega^{\wedge n} \neq 0$. Instead of the de León & Lacomba equation for *zero* form (for a scalar functions H or L), our Hamilton & Jacobi like equation is an equation on the *formalism*, i.e., on the differential Poincaré & Cartan *one* form α (Definition 4.1),

$$d\alpha = \ell\Omega, \qquad \Omega \simeq dp_i \wedge dq^i + \beta \wedge dt, \qquad d\Omega \simeq d\beta \wedge dt.$$

Therefore our approach can be coined as the higher grade generalization of the Hamilton & Jacobi equation, in our case in mechanics grade $\alpha = 1$, whereas originally grade $S = 0$. The canonical approach of de León & Lacomba seems to show that the Lagrangian formalism and the Hamiltonian formalism are the only possibilities. In our (not canonical) approach there is room for other (mixed) possibilities and we are interested in all other formalisms besides the Lagrangian and Hamiltonian formalisms. However both approaches are strongly related.

Extended version of this approach including proofs of theorems and examples beyond of the known Lagrangian and of the Hamiltonian formalisms, with the mixed formalism among other, will be published elsewhere.

2 Newtonian System and Descendant Differential Form

In this section we give the definition of the special kind of Pfaffian system — the Newtonian system.

For introduction to exterior differential systems we refer to excellent monographs (Cartan 1946), (Ślebodziński 1959, 1970), (Choquet-Bruhat et al. 1977), (Griffiths 1983), (Bryant et al. 1991), (Yang 1992). In the table on the next page we introduce the notation we use throughout this paper.

Some notation

$\mathcal{F}$	is an associative, unital and commutative $\mathbb{R}$-algebra.
M	is a finite dimensional $\mathcal{F}$-module of the differential one-forms.
$M^{\wedge}$	is a differential associative and unital graded $\mathcal{F}$-algebra of the differential forms, *i.e.*, de Rham complex, with $M^{\wedge 0} \equiv \mathcal{F}$ and $M^{\wedge 1} \equiv M$.
M^*	$\equiv \text{mod}_{\mathcal{F}}(M, \mathcal{F}) \equiv \mathcal{F}^M$ is a dual $\mathcal{F}$-module, $M^{**} \simeq M$.
$i\|M^* \in$	$\text{der}_{\mathcal{F}}(M^{\wedge})$ is a graded derivation.
$D < M^*$	is called a distribution (following Chevalley).
$D^{\perp}$	$\equiv \text{ann} D < M$ such that $D^{\perp} D = 0 \in \mathcal{F}$, and $\dim D = \text{codim} D^{\perp}$
$N^{(1)} < M$	is a maximal sub-$\mathcal{F}$-module such that $dN^{(1)} \subset I_N$

Let $N < M$ be sub-$\mathcal{F}$-module of the differential one-forms. N generates a two-sided ideal I_N, called a Pfaffian differential system. We do not assume that it is closed, nor that its integral elements are integrable.

Lemma 2.1 *Let $N < M$ be a Pfaffian system. Then*

$$\text{codim}\, N = 1 \quad \Longrightarrow \quad N^{(1)} = N.$$

Definition 2.2 *A sub-$\mathcal{F}$-module $N < M$ of the differential forms is said to be the Newtonian system if the following two conditions hold*

$$(i) \text{ The first derived system is zero: } \quad N^{(1)} = 0,$$
$$(ii) \text{ codim}\, N = 1 + \dim N = \text{odd}.$$

Let $\dim_{\mathcal{F}} N = 2n$. From *(ii)* it follows that $\dim_{\mathcal{F}} M = 4n + 1$.

The most important fact for us is that for the Newtonian system N (Definition 2.2) the only closed two-form in $N^{\wedge 2}$ is the zero two-form.

Theorem 2.3 *Let $N < M$, $N^{(1)} = 0$ and $\Omega \in N^{\wedge 2}$. Then*

$$d\Omega = 0 \quad \Longleftrightarrow \quad \Omega = 0.$$

Definition 2.4 (Descendant differential form) *A differential two-form $\Omega \in M^{\wedge 2}$ is said to be descendant for the Pfaffian differential system $N < M$ if Ω is such that*

$$\ker \Omega = N^{\perp}.$$

Lemma 2.5 *Let $N < M$, $\dim_{\mathcal{F}} N = 2n$ be a Pfaffian system. The necessary and sufficient conditions that $\Omega \in M^{\wedge 2}$ is descendant for N are that*

$$\Omega \in N^{\wedge 2} \qquad \textit{and} \qquad \Omega^n \neq 0 \quad \in N^{\wedge \dim N}.$$

3 The Second Newton's Law as a Pfaffian System and Its Descendant Differential Form

Let $\mathcal{F}$ be an associative, commutative $\mathbb{R}$-algebra possessing $4n+1$ letters (1) and let the $\mathcal{F}$-module M of the differential one-forms be of $\mathcal{F}$-dimension $4n+1$. The $\mathbb{R}$-algebra $\mathcal{F}$ and M generate the differential graded $\mathcal{F}$-algebra (de Rham $\mathcal{F}$-complex) $M^\wedge$ (2). Following to Newton (Newton 1686, 1990) let

$$\omega_i \equiv dp_i - f_i dt, \quad \vartheta^i \equiv dq^i - v^i dt \quad \in M,$$

$$M > N \equiv \mathrm{span}_\mathcal{F}\{\omega_i, \vartheta^i\}, \quad \dim_\mathcal{F} N = 2n, \quad \mathrm{codim}_\mathcal{F} N = 2n+1. \tag{4}$$

Sub-$\mathcal{F}$-module N is a Pfaffian system and its first derived system is zero. It is therefore, according to Definition 3.2, the Newtonian sub-$\mathcal{F}$-module. The differential one forms ω_i, ϑ^i are formally analogous to the Lie contact forms on a jet manifold $J^1(\mathbb{R}\times T^*Q)$ or better on the manifold (de León and Lacomba 1988), (de León and Lacomba 1989)

$$\mathbb{R}\times TT^*Q \quad \simeq \quad \mathbb{R}\times T^*TQ.$$

We need a descendant differential two-form Ω for the Newtonian system N (4), that is such two-form Ω that $\ker\Omega = N^\perp$.

We will use the following notation.

$$\begin{aligned} K_i^j &\equiv \Omega(\partial_{p_j}\wedge\partial_{q^i}) \in \mathcal{F},\ K\equiv\{K_i^j\},\\ \Gamma_{ij} &\equiv \Omega(\partial_{q^i}\wedge\partial_{q^j}) \in \mathcal{F},\ \Gamma\equiv\{\Gamma_{ij}\},\\ \chi^{ij} &\equiv \Omega(\partial_{p_i}\wedge\partial_{p_j}) \in \mathcal{F},\ \chi\equiv\{\chi^{ij}\}, \end{aligned}$$

Theorem 3.1 *A differential two-form $\Omega \in N^{\wedge 2}$ has the general form,*

$$\Omega = K_j^i\cdot\omega_i\wedge\vartheta^j + \tfrac{1}{2}\Gamma_{ij}\cdot\vartheta^i\wedge\vartheta^j + \tfrac{1}{2}\chi^{ij}\cdot\omega_i\wedge\omega_j, \tag{5}$$

and is descendant differential form for the Newton law (4) iff

$$\mathcal{F} \ni \det\begin{pmatrix} \Gamma & K \\ -K^T & \chi \end{pmatrix} \neq 0. \tag{6}$$

In particular if $\beta \equiv f_i dq^i - v^i dp_i$ is the differential one form on TT^*Q introduced by de León & Lacomba (de León and Lacomba 1989 p. 3810), then

$$\omega_i\wedge\vartheta^j = dp_i\wedge dq^i + \beta\wedge dt \quad \text{and} \quad d(\omega_i\wedge\vartheta^j) = d\beta\wedge dt.$$

4 Formalism

If Ω is a homogeneous differential multi-form (of any grade) and c is a morphism of the differential graded $\mathcal{F}$-algebras then the equations $cd\Omega = 0$ and $c\Omega = d\alpha$ are said to be the Hamilton & Jacobi (like) equation for c (for the given Ω).

The descendant two-forms for the Newton law are, according to the previous sections, not closed. It explains the necessity of defining a formalism. We need a Pfaffian system as, e.g., in (Kocik 1981), (Griffiths 1983), (Oziewicz 1985), (Borowiec & Oziewicz 1990) of even dimension $2n$ and its descendant form σ with $d\sigma = 0$ and $\sigma^n \neq 0$. It means that σ is symplectic. The classical determinism demands that $\dim \ker \sigma = 1$. It follows that the codimension of the Pfaffian system is one, so it must be a sub-$\mathcal{F}$-module of a $\mathcal{F}$-module of dimension $2n+1$, and it also means that it is completely integrable, Lemma 2.1.

In what follows we need a pair of $\mathbb{R}$-algebras, $\mathcal{F}$ and $\mathcal{A}$, and a $\mathcal{F}$-module M and $\mathcal{A}$-module A.

For the given descendant two-form we are looking for an exact split sequence of the differential graded $\mathcal{F}$-algebras, i.e., for the epimorphism of the $\mathbb{R}$-algebras $c \in \operatorname{epi}(\mathcal{F}, \mathcal{A})$ and for de Rham $\mathcal{A}$-complex $A^\wedge$, $\mathcal{A} \equiv A^{\wedge 0}$, and for two morphisms of de Rham complexes

$$0 \longrightarrow \ker c \longrightarrow M^\wedge \stackrel{c}{\longrightarrow} A^\wedge \longrightarrow 0\,,$$

$$\phi \in \operatorname{alg}(A^\wedge, M^\wedge), \quad c \in \operatorname{alg}(M^\wedge, A^\wedge), \quad c \circ \phi = \mathrm{id}_{A^\wedge}\,. \tag{7}$$

We assume that ϕ must be a mono-morphism and that c must be an epimorphism of the differential $\mathbb{Z}$-graded algebras (of de Rham complexes) such that $\dim_{\mathcal{A}} A = 2n+1$ and the image of the descendant form is a symplectic two-form.

Definition 4.1 (Formalism) *Let $\Omega \in M^{\wedge 2}$ be descendant differential form (5) for a Pfaffian system N. An epimorphism $c \in \operatorname{alg}(M^\wedge, A^\wedge)$ is said to be a formalism for Ω (and for N) if $\mathcal{A}$-module $A \equiv \operatorname{im}(c|M)$, $\dim_{\mathcal{A}} A = 2n+1$, is such that $c\Omega \in A^{\wedge 2}$ is symplectic.*

Corollary 4.2 *Every formalism c is defined by the Poincaré & Cartan differential one-form $\{\alpha \in A\}$ modulo one-cocyles Z^1,*

$$c\Omega = d\alpha \in A^{\wedge 2} \quad \textit{and} \quad \dim\ker c\Omega = 1 \quad (\Longleftrightarrow \quad 0 \neq (d\alpha)^n \in A^{\wedge 2n}). \tag{8}$$

For a given descendant differential two-form $\Omega \in N^{\wedge 2}$ the equation (8) is the equation on an epimorphism c and the Poincaré & Cartan differential form α depends on Ω and on c.

References

Borowiec, A., Oziewicz, Z., (1990): Global Pfaff systems in the classical mechanics, in: Mijat Mijatović, editor, *Hadronic Mechanics and Nonpotential Interactions,* Nova Science Publishers, New York - Budapest, pp. 27–36

Bryant, R.L., Shiing-shen Chern, Gardner, R., Goldschmidt, H., Griffiths, P.A., (1991): *Exterior Differential Systems*, Springer-Verlag, New York

Cartan, Élie, (1946): *Les Systèmes Différentieles Extérieurs et leurs Applications Géométriques*, Hermann, Paris

Choquet-Bruhat, Y., de Witt-Morette, C., Dillard-Bleick, M., (1977): *Analysis, Manifold and Physics*, North-Holland, Amsterdam, New York, Oxford

Cisło, J., Łopuszański, J., Stichel, P.C., (1995): On certain class of Lagrange functions with common equations of motions but various Poisson brackets, Fortschritte der Physik (Berlin) **43**, Heft **8**, 745–762

Della Riccia, G., (1982): On the Lagrange representation of a system of Newton equations, in: André Avez, Austin Blaquiére & Angelo Marzollo, editors, *Dynamical Systems and Microphysics, Geometry and Mechanics,* Academic Press, New York, pp. 281–292, ISBN 0-12-068720-8

Griffiths, P.A., (1983): *Exterior Differential Systems and the Calculus of Variations,* Birkhäuser

Hamilton, William R., (1934): On a general method in dynamics by which the study of the motion of all free systems of attracting or repelling points is reduced to the search and differentiation of one central relation or characteristic function, Philosophical Transactions of Royal Society **124**, 247–308

Jacobi, C.G.J., (1838): Über die Reduction der Integration der partiellen Differentialgleichungen erster Ordnung zwischen irgend einer Zahl Variabeln auf die Integration eines einzigen Systems gewölnlicher Differentialgleichungen, Journal für Mathematik, Band **XVII**, 97–162

Kocik, J., (1981): Newtonian versus Lagrangian mechanics, Uniwersytet Wrocławski, Instytut Fizyki Teoretycznej, preprint # 538, 78 pages

de León, M., & Lacomba, E.A., (1988): Les sous-variétés lagrangiennes dans la dynamique lagrangienne d'ordre supérieur, C. R. Acad. Sci. Paris **307**, Série II 1137–1139

de León, M., & Lacomba, E.A., (1989): Lagrangian submanifolds and higher order mechanical systems, Journal of Physics: Math. Gen. **A22**, 3809–3820

Marmo, G., Rubano, C., Morandi, G., Ferrario, C., Lo Vecchio, G., (1990): The inverse problem in the calculus of variations and the geometry of the tangent bundle, Physics Reports **188**, (3 & 4) 147–284

Newton, I., (1686), (1990): *Principes Mathématiques de la Philosophie Naturelle,* 1686. Reprinted from French edition of 1759, by Editions Jacques Gabay , Paris, 1990

Oziewicz, Z., (1982): The meaning of the Lagrangian, in: André Avez, Austin Blaquiére & Angelo Marzollo, editors, *Dynamical Systems and Microphysics, Geometry and Mechanics,* Academic Press, New York, pp. 437–441, ISBN 0-12-068720-8

Oziewicz, Z., (1985): Classical mechanics: inverse problem and symmetries, Reports on Mathematical Physics **22**, (1) 91–111

Oziewicz, Z., (1992): Calculus of variations for multiple-valued functionals, Reports on Mathematical Physics **31**, (1) 85–90

Santilli, R.M., (1978): *Foundations of Theoretical Mechanics I: the Inverse Problem in Newtonian Mechanics,* Springer - Verlag
Ślebodziński, W., (1959): *Formes Extérieures et leur Applications*, PWN Warszawa 1959, 1963; English version (1970): PWN Warszawa
Yang, K., (1992): *Exterior Differential Systems and Equivalence Problems*, Kluwer Academic Publishers, Dordrecht/Boston/London

Particle in Curved 2d Space–Time

G. Jorjadze[1] and W. Piechocki[2]

[1] Razmadze Mathematical Institute,
Tbilisi, Georgia

[2] Sołtan Institute for Nuclear Studies,
Warsaw, Poland

Dedicated to
Professor Jan Łopuszański

Abstract. We investigate classical and quantum dynamics of relativistic particle in 2-dimensional space–time with constant curvature. Singularities of space–time metric, in case of negative curvature, lead to creation and annihilation phenomenon at the classical level. In case of positive curvature, anomaly free quantization leads to discretization of particle mass.

1 Introduction

Consolidation of General Relativity and Quantum Mechanics has not been completed yet. This is an extremely difficult problem. We present some results concerning this problem but at the level of two dimensional theory. In 2d the situation is much simpler, many steps can be done analytically and one can get some insight.

2 Dynamics of Particle in Curved 2d Space-Time

Let us consider a relativistic particle of mass m_0 moving in a gravitational field $g_{\mu\nu}(x^0, x^1)$; $\mu, \nu = 0, 1$. Action describing such a system is proportional to the length of a particle world-line

$$S = \int d\tau L(\tau), \quad L(\tau) := -m_0 \sqrt{g_{\mu\nu}(x^0(\tau), x^1(\tau))\dot{x}^\mu(\tau)\dot{x}^\nu(\tau)}\,, \qquad (1)$$

where τ is an evolution parameter along trajectory $x^\mu(\tau)$ and $\dot{x}^\mu(\tau) := dx^\mu(\tau)/d\tau$.

The action (1) is invariant under reparametrization $\tau \to f(\tau)$, which leads to the constraint

$$\Phi := g^{\mu\nu} p_\mu p_\nu - m_0 = 0\,, \qquad (2)$$

where $p_\mu := \partial L/\partial \dot{x}^\mu$.

In the case of 2d Minkowskian manifold, one can always choose local coordinates in such a way that

$$g_{\mu\nu}(x^0,x^1) = \exp\varphi(x^0,x^1)\begin{pmatrix}1 & 0\\ 0 & -1\end{pmatrix}, \tag{3}$$

where φ is a field.

Since the scalar curvature R for (3) is

$$R(x^0,x^1) = \exp\left(-\varphi(x^0,x^1)\right)\left(\partial_1^2 - \partial_0^2\right)\varphi(x^0,x^1), \tag{4}$$

the Einstein-Hilbert Lagrangian reads

$$\mathcal{L} = -\sqrt{|\,g\,|}\,R = \left(\partial_0^2 - \partial_1^2\right)\varphi(x^0,x^1), \tag{5}$$

and it does not lead to dynamical equation for φ. One can specify the 2d general relativity model (see for example (Jackiw 1984)) assuming that φ is a solution to the Liouville equation (Liouville 1853)

$$(\partial_0^2 - \partial_1^2)\varphi(x^0,x^1) + R_0\exp\varphi(x^0,x^1) = 0, \tag{6}$$

where R_0 is a real constant ($R_0 \in \mathbf{R}$).

For φ satisfying (6) the scalar curvature is $R(x^0,x^1) = R_0$. In what follows we shall consider two cases: $R_0 < 0$ and $R_0 > 0$.

General solution to (6) is

$$\varphi(x^+,x^-) = \log\frac{4A^{+\prime}(x^+)A^{-\prime}(x^-)}{m^2\left[A^+(x^+) - \epsilon A^-(x^-)\right]^2}, \tag{7}$$

where $m^2 := |R_0|/2$, $x^\pm := x^0 \pm x^1$, $A^\pm$ are smooth functions, $A^{\pm\prime} := dA^\pm/dx^\pm$, $\epsilon := |R_0|/R_0$.

For (7) the Lagrangian (1) reads

$$L(\tau) = -2c\sqrt{\frac{A^{+\prime}(x^+(\tau))\,A^{-\prime}(x^-(\tau))}{\left[A^+(x^+(\tau)) - \epsilon\,A^-(x^-(\tau))\right]^2}}, \tag{8}$$

where $c := m_0/m$.

The Lagrangian (8) is formally invariant under the transformations

$$y^+ \longrightarrow \frac{ay^+ + b}{cy^+ + d}, \qquad y^- \longrightarrow \frac{ay^- + \epsilon b}{\epsilon cy^- + d}, \tag{9}$$

where $y^\pm = A^\pm(x^\pm)$. Thus, formally $SL(2,\mathbf{R})/\mathbf{Z_2}$ is the group of symmetry of our system.

The conformal transformation for the metric (3) is defined by

$$x^\pm \longrightarrow y^\pm(x^\pm), \tag{10}$$

$$\varphi(x^+,x^-) \longrightarrow \tilde{\varphi}(x^+,x^-) := \varphi(y^+(x^+),y^-(x^-)) + \log[y^{+\prime}(x^+)y^{-\prime}(x^-)], \tag{11}$$

where $y^{\pm\prime} := dy^\pm/dx^\pm$.

Due to (10) and (11) the general solution (7) is locally equivalent to the following simple solution

$$\varphi(x^+, x^-) = \log \left[\frac{2}{m(x^+ - \epsilon x^-)} \right]^2 . \tag{12}$$

Therefore, dynamics of a particle can be locally described by (12).

2.1 The Case $R_0 < 0$

The solution (12) leads to singular metric

$$g_{\mu\nu}(x^+, x^-) = \left[\frac{2}{m(x^+ + x^-)} \right]^2 \begin{pmatrix} 1 & 0 \\ 0 & -1 \end{pmatrix} . \tag{13}$$

The infinitesimal transformations for (9) are

$$x^\pm \longrightarrow x^\pm \pm \alpha_0, \quad x^\pm \longrightarrow x^\pm + \alpha_1 x^\pm, \quad x^\pm \longrightarrow x^\pm \pm \alpha_2 (x^\pm)^2 \tag{14}$$

and the dynamical integrals corresponding to (14) read

$$P := p_+ - p_-, \quad K := p_+ x^+ + p_- x^-, \quad M := p_+ (x^+)^2 - p_-(x^-)^2 , \tag{15}$$

where $p_\pm := \partial L / \partial \dot{x}^\pm$.

The Poisson brackets on the extended phase space $\Gamma := \{(x^+, x^-, p_+, p_-)\} \subset \mathbf{R}^4$ read

$$\{P, K\} = P, \quad \{K, M\} = M, \quad \{P, M\} = 2K \tag{16}$$

and define Lie algebra isomorphic to the $sl(2, \mathbf{R})$ algebra.

The integrals (15) are not independent. By (2) we get

$$K^2 - PM = c^2 . \tag{17}$$

Equations (15) and (17) define the trajectories of the particle

$$x^1(x^0) = \begin{cases} -(M\eta)/(2c), & \text{for } P = 0 \\ \left(K + \eta \sqrt{(x^0 P)^2 + c^2} \right) / P , & \text{for } P \neq 0 , \end{cases} \tag{18}$$

where $\eta := x^0 / |x^0|$.

From (18) we get that, for $P = 0$, $K = -\eta c$. Therefore the trajectories with $P = 0$, $K = -c$ for $x^0 < 0$ and trajectories with $P = 0$, $K = c$ for $x^0 > 0$ do not exist. This leads to serious problems when we quantize the system.

2.2 The Case $R_0 > 0$

The infinitesimal transformations corresponding to (9) are now defined by

$$x^{\pm} \longrightarrow x^{\pm} + \alpha_0, \quad x^{\pm} \longrightarrow x^{\pm} + \alpha_1 x^{\pm}, \quad x^{\pm} \longrightarrow x^{\pm} + \alpha_2 (x^{\pm})^2 . \tag{19}$$

Making use of the Noether theorem leads to

$$E := -p_+ - p_-, \quad K := p_+ x^+ + p_- x^-, \quad L := -p_+ (x^+)^2 - p_- (x^-)^2 . \tag{20}$$

The Poisson brackets

$$\{E, K\} = E, \quad \{K, L\} = L, \quad \{E, L\} = 2K \tag{21}$$

give again $sl(2, \mathbf{R})$ algebra.

The constraint (2) leads to the relation

$$EL - K^2 = c^2 . \tag{22}$$

Equations (20) and (22) define the trajectories

$$(x^+ + K/E)(x^- + K/E) = -(c/E)^2 , \tag{23}$$

which are hyperbolas with light-cone asymptotics.

3 Quantization

Quantization in our case means finding an appropriate, irreducible self-adjoint representation of $sl(2, \mathbf{R})$ algebra on a Hilbert space.

3.1 The Case $R_0 < 0$

One can find an unique quantum system corresponding to our classical system in the case when we consider the set of all trajectories defined by (18). For anomaly free quantization we have to take into account the following trajectories:

1. For $P \neq 0$ and arbitrary K trajectories are defined by (18) and have discontinuity $2c/P$ at $x^0 = 0$, i.e., particle is 'annihilated' at $(x^0, x^1) = (0, (K - c)/P)$ and then it is 'created' at $(x^0, x^1) = (0, (K + c)/P)$.
2. For $P = 0$, $K = c$ and any M trajectories are defined by $x^1 = M/2c$ for $x^0 < 0$, (18), and there are no trajectories for $x^0 > 0$, i.e., particle is annihilated at $(x^0, x^1) = (0, M/2c)$ and it cannot appear for $x^0 > 0$.
3. For $P = 0$, $K = -c$ and any M there are no trajectories for $x^0 < 0$; for $x^0 > 0$ trajectories are defined by $x^1 = -M/2c$, (18), i.e., particle is created by singularity at $(x^0, x^1) = (0, -M/2c)$.

This set of trajectories is isomorphic to the hyperboloid (17) and has $SL(2,\mathbf{R})/\mathbf{Z_2}$ symmetry (see (Jorjadze and Piechocki 1998a)).

Since P, K and M are gauge invariant and they are constant along the trajectories, we choose them as the observables of the system. Due to (17) only two of them are functionally independent on the constraint surface $\Gamma_c := \{(x^+, x^-, p_+, p_-) \in \Gamma | \;\; \Phi = 0\}$.

To quantize the system we use the following parametrization of the hyperboloid (17):

$$P = J(1 - \cos\beta) - c\sin\beta\,, \tag{24}$$

$$M = J(1 + \cos\beta) + c\sin\beta\,, \tag{25}$$

$$K = -J\sin\beta + c\cos\beta, \tag{26}$$

where $J \in \mathbf{R}$, $\beta \in \mathbf{S^1}$ are the canonically conjugated variables.

The corresponding operators

$$\hat{P} = -i(1 - \cos\beta)\partial/\partial_\beta - (c + i/2)\sin\beta\,, \tag{27}$$

$$\hat{M} = -i(1 + \cos\beta)\partial/\partial_\beta + (c + i/2)\sin\beta\,, \tag{28}$$

$$\hat{K} = i\sin\beta\partial/\partial_\beta + (c + i/2)\cos\beta \tag{29}$$

are self-adjoint on $L^2(S^1)$ and define the unitary irreducible representation of $SL(2,\mathbf{R})/\mathbf{Z_2}$ group.

3.2 The Case $R_0 > 0$

The irreducible self-adjoint representation of $sl(2,\mathbf{R})$ algebra is defined by (see (Jorjadze and Piechocki 1998b))

$$\hat{E} = -(\partial_q + \partial_{q^*}), \quad \hat{K} = -i(q\partial_q + q^*\partial_{q^*})\,, \tag{30}$$

$$\hat{L} = ic(q^* - q) - i(q^2\partial_q + {q^*}^2\partial_{q^*})\,, \tag{31}$$

where

$$q := (K + ic)/E, \quad q^* := (K - ic)/E \tag{32}$$

are the complex coordinates on half plane $Im\; q > 0$.

The corresponding Hilbert space is defined by the set of functions

$$\Psi(q, q^*) := \left(\frac{q - q^*}{2i}\right)^c \psi(q), \quad c = m_0/m > 1/2\,, \tag{33}$$

where

$$\psi(q) := (q+i)^{-2c} \sum_{n\geq 0} b_n \left(\frac{q-i}{q+i}\right)^n, \qquad \sum_{n\geq 0} |b_n|^2 < \infty$$

is a holomorphic function.

The scalar product is given by

$$\langle \Psi_1|\Psi_2\rangle = \int \frac{dqdq^*}{2\pi} \left(\frac{q-q^*}{2i}\right)^{2(c-1)} \psi_1^*(q)\psi_2(q)\,. \tag{34}$$

For $c = m_0/m = 2, 3, 4, \ldots$ the representation (30-34) leads to the unitary irreducible representation of $SL(2,\mathbf{R})/\mathbf{Z_2}$ group. Since m is the parameter of the Liouville equation being well defined for any $m \in \mathbf{R}$, we conclude that the mass of the particle m_0 must be discrete.

4 Summary

The assumptions:

- space–time is curved (we consider the simplest case when the scalar curvature $R =$ const)
- classical and quantum systems should have the same global $SL(2,\mathbf{R})/\mathbf{Z_2}$ symmetry

lead to the results:

- singularity of the metric annihilate and/or creates a particle ($R_0 < 0$ case)
- there is violation of causality (due to discontinuities of trajectories) at the singularity ($R_0 < 0$ case)
- mass of the quantum particle must have discrete value ($R_0 > 0$ case).

Acknowledgments

This work was supported by the grants from: INTAS (96-0482), RFBR (96-01-00344), the Georgian Academy of Sciences, the Polish Academy of Sciences, and the Sołtan Institute for Nuclear Studies. One of us (WP) would like to thank the organizers of the XII Max Born Symposium for a pleasant atmosphere and for financial support.

References

Jackiw, R., (1984); in *Quantum Theory of Gravity*, ed. by S. M. Christensen (Adam Hilger Ltd,. Bristol) p. 403

Jorjadze, G., Piechocki, W., (1998): Class. Quantum Grav. **18**, L41

Jorjadze, G., Piechocki, W., (1998): Theor. Math. Phys. **118**, 183

Liouville, J., (1853): J. Math. Pures Appl. **18**, 71

Are There Two Sterile Neutrinos Cooscillating with ν_e and ν_μ?

W. Królikowski

Institute of Theoretical Physics, Warsaw University
Hoża 69, PL–00–681 Warszawa, Poland

Dedicated to Jan Łopuszański
in honour of his 75th birthday

Abstract. The existence of two sterile neutrinos ν_s and ν_s' (blind to all Standard–Model interactions) is shown to be implied by a model of fermion "texture" that we develop since some time. They may mix nearly maximally with two of three conventional neutrinos, say ν_e and ν_μ, thus leading to neutrino oscillations, say $\nu_e \rightarrow \nu_s$ and $\nu_\mu \rightarrow \nu_s'$, with nearly maximal amplitudes. Then, they can be responsible for the observed deficits of solar ν_e's and atmospheric ν_μ's, respectively, but by themselves do not help to explain the LSND results for $\nu_\mu \rightarrow \nu_e$ oscillations. On the other hand, they are consistent with the CHOOZ negative result. At the moment, the experiment cannot decide, whether the deficit of atmospheric ν_μ's, confirmed by the recent Super–Kamiokande findings, has to be related to the oscillations $\nu_\mu \rightarrow \nu_\tau$ or $\nu_\mu \rightarrow \nu_s'$. In the last Section of the paper, a new notion of "non–Abelian spin–1/2 fermions" is presented in the context of a composite option for fermion families.

PACS numbers: 12.15.Ff , 12.90.+b , 14.60.Gh

1 Introduction

The hypothetic sterile neutrinos, by definition interacting only gravitationally, are blind to all Standard–Model interactions, in contrast to the conventional neutrinos (or, rather, their lefthanded parts) which participate first of all in the weak sector of Standard–Model interactions. Such Standard–Model–inactive fermions are invoked from time to time by theorists, who want to explain (e.g., Smirnow 1997) through neutrino oscillations not only the observed deficits of solar and atmospheric neutrinos, but also the results of LSND experiment. The sterile neutrinos may also form a Standard–Model–inactive fraction of the dark matter.

In the present paper, we demonstrate how two different sterile neutrinos are implied by a model of fermion "texture" (Królikowski 1990, Królikowski 1996) that we develop since some time. As shown previously, this model justifies (Królikowski 1990) the existence of three and only three families of conventional leptons and quarks $(\nu_e\,, e^-\,, u\,, d)$, $(\nu_\mu\,, \mu^-\,, c\,, s)$, $(\nu_\tau\,, \tau^-\,, t\,, b)$ and, moreover, describes reasonably (Królikowski 1996) the masses and mixing parameters of quarks and charged leptons, making also some useful suggestions as to neutrinos. Note that in this model all neutrinos are Dirac particles having both lefthanded and righthanded parts.

In order to make our presentation fairly comprehensible, we will first recapitulate briefly the basic features of the model in its part concerning the existence of fundamental–particle families (Królikowski 1990). Then, we shall discuss the existence of two sterile neutrinos and the related neutrino oscillations.

2 Dirac's Generalized Square Root

The starting point of our model is the conjecture that *all* kinds of matter's fundamental particles existing in Nature can be deduced from Dirac's square–root procedure $\sqrt{p^2} \to \Gamma \cdot p$.

As is easy to observe, this procedure leads in general to the sequence $N = 1, 2, 3, \ldots$ of different (generally reducible) representations

$$\Gamma^\mu \equiv \frac{1}{\sqrt{N}} \sum_{i=1}^{N} \gamma_i^\mu \tag{1}$$

of the Dirac algebra

$$\{\Gamma^\mu , \Gamma^\nu\} = 2g^{\mu\nu} , \tag{2}$$

constructed with the use of the sequence $N = 1, 2, 3, \ldots$ of Clifford algebras

$$\{\gamma_i^\mu , \gamma_j^\nu\} = 2\delta_{ij} g^{\mu\nu} \quad (i, j = 1, 2, \ldots, N) . \tag{3}$$

Then, the sequence $N = 1, 2, 3, \ldots$ of Dirac–type equations follows,

$$\{\Gamma \cdot [p - gA(x)] - M\} \psi(x) = 0 , \tag{4}$$

where $g\Gamma \cdot A(x)$ may symbolize the minimal coupling of $\psi(x)$ to the Standard–Model gauge fields $A_\mu(x)$ including all $SU(3) \times SU_L(2) \times U(1)$ coupling matrices: λ's, τ's, Y and $\Gamma^5 \equiv i\Gamma^0\Gamma^1\Gamma^2\Gamma^3$.

In (4) the matrices (1) can be presented in the reduced forms

$$\Gamma^\mu = \gamma^\mu \otimes \underbrace{\mathbf{1} \otimes \cdots \otimes \mathbf{1}}_{(N-1)\text{times}} \tag{5}$$

with γ^μ and $\mathbf{1}$ denoting the usual 4×4 Dirac matrices. Then, the Dirac–type equations (4) can be rewritten as

$$\{\gamma \cdot [p - gA(x)] - M\}_{\alpha_1\beta_1} \psi_{\beta_1\alpha_2\cdots\alpha_N}(x) = 0 \tag{6}$$

with $\psi(x) = (\psi_{\alpha_1\alpha_2\cdots\alpha_N}(x))$, where $\alpha_1, \alpha_2, \cdots, \alpha_N$ stand for N Dirac bispinor indices: $\alpha_i = 1, 2, 3, 4$ for $i = 1, 2, \ldots, N$. Here, the chiral representations are used to define all α_i $(i = 1, 2, \ldots, N)$. This means that $\alpha_i = 1, 2, 3, 4$ correspond to four different pairs (1,1), (1,-1), (-1,1), (-1,-1) of eigenvalues of the matrices

$$\Gamma_i^5 \equiv i\Gamma_i^0\Gamma_i^1\Gamma_i^2\Gamma_i^3 \,, \qquad \Sigma_i^3 \equiv i\Gamma_i^5\Gamma_i^0\Gamma_i^3 \,, \tag{7}$$

simultaneously diagonal for all i, which choice is allowed because all Γ_i^5 and Σ_i^3 commute both for equal and different i. The Γ_i^μ matrices ($i = 1, 2, \ldots, N$) appearing in (7) are defined as N (properly normalized) Jacobi combinations of γ_i^μ matrices ($i = 1, 2, \ldots, N$), where in particular $\Gamma_1^\mu \equiv \Gamma^\mu$ is given as in (1). Then, $\{\Gamma_i^\mu\,,\,\Gamma_j^\nu\} = 2\delta_{ij}g^{\mu\nu}$ ($i, j = 1, 2, \ldots, N$) due to (3), and also $\{\Gamma_i^\mu\,,\,\Gamma_j^5\} = 0$, but $[\Gamma_i^5\,,\,\Gamma_j^5] = 0$, where particularly $\Gamma_1^5 \equiv \Gamma^5$. Note that in the one–body Dirac–type equations (4) there appear only the "centre-of–mass" Γ_1^μ matrices, while all "relative" matrices $\Gamma_2^\mu, \ldots, \Gamma_N^\mu$ are absent. In spite of this, all $\alpha_1, \alpha_2, \ldots, \alpha_N$ are present in (6): both the "centre–of–mass" Dirac bispinor index α_1 as well as the "relative" Dirac bispinor indices $\alpha_2, \ldots, \alpha_N$, the latter are decoupled, however, even in the presence of Standard–Model coupling $g\Gamma_i \cdot A(x)$.

For $N = 1$ (6) is obviously the usual Dirac equation, for $N = 2$ it is known as the Dirac form (Banks et al. 1982) of the Kähler equation (Kähler 1962; Ivanenko and Landau 1928), whilst for $N = 3, 4, 5, \ldots$ we obtain *new* Dirac–type equations (Królikowski 1990).

If the Standard–Model coupling $g\Gamma \cdot A(x)$ is really present in (6), then the Dirac bispinor index α_1, which is the only α_i affected by the gauge fields $A_\mu(x)$, is *distinguished* by its correlation with the set of all diagonal Standard–Model charges ascribed to any particle of the fields $\psi_{\alpha_1\,\alpha_2\ldots\alpha_N}(x)$ (a label f of this set is here suppressed). The remaining Dirac bispinor indices $\alpha_2, \ldots, \alpha_N$ are all decoupled and so, physically unobservable in the gauge fields $A_\mu(x)$. It is natural to conjecture that they are physically *undistinguishable* and, therefore, are formal objects obeying Fermi statistics along with Pauli principle. This implies that $\psi_{\alpha_1\,\alpha_2\ldots\alpha_N}(x)$ is fully *antisymmetric* with respect to $\alpha_2, \ldots, \alpha_N$.

The above conjecture, together with the probabilistic interpretation of wave functions $\psi_{\alpha_1\,\alpha_2\ldots\alpha_N}(x)$ and the requirement of their relativistic covariance applied to *all* bispinor indices $\alpha_1, \alpha_2, \ldots, \alpha_N$, leads to the conclusion that there are *three* (and only three) families $N = 1, 3, 5$ of leptons and quarks (Królikowski 1990), and *two* (and only two) families $N = 2, 4$ of some, not yet observed, fundamental scalars (Królikowski 1992). They correspond to the wave functions

$$\begin{aligned}
\psi_{\alpha_1}^{(1)} &\equiv \psi_{\alpha_1} \,, \\
\psi_{\alpha_1}^{(3)} &\equiv \frac{1}{4}\left(C^{-1}\gamma^5\right)_{\alpha_2\alpha_3}\psi_{\alpha_1\alpha_2\alpha_3} = \psi_{\alpha_1\,12} = \psi_{\alpha_1\,34} \,, \\
\psi_{\alpha_1}^{(5)} &\equiv \frac{1}{24}\varepsilon_{\alpha_2\alpha_3\alpha_4\alpha_5}\psi_{\alpha_1\alpha_2\alpha_3\alpha_4\alpha_5} = \psi_{\alpha_1\,1234}
\end{aligned} \tag{8}$$

and

$$\phi^{(2)} \equiv \frac{1}{2\sqrt{2}} \left(C^{-1}\gamma^5\right)_{\alpha_1\alpha_2} \psi_{\alpha_1\alpha_2} = \frac{1}{\sqrt{2}} \left(\psi_{12} - \psi_{21}\right) = \frac{1}{\sqrt{2}} \left(\psi_{34} - \psi_{43}\right) ,$$

$$\phi^{(4)} \equiv \frac{1}{6\sqrt{4}} \varepsilon_{\alpha_1\alpha_2\alpha_3\alpha_4} \psi_{\alpha_1\alpha_2\alpha_3\alpha_4} = \frac{1}{\sqrt{4}} \left(\psi_{1234} - \psi_{2134} + \psi_{3412} - \psi_{4312}\right) , \quad (9)$$

respectively. Each of these wave functions carries the (here suppressed) Standard–Model label $f = \nu\, ,\, e\, ,\, u\, ,\, d$ denoting four sorts of fundamental particles corresponding to the signature of conventional neutrinos ν and charged leptons e as well as up quarks u and down quarks d, all four following from the Standard Model (though the existence of three and two fundamental–particle families does not follow from it). In the case of fundamental fermions, the three families are, of course, $(\nu_e\, ,\, e^-\, ,\, u\, ,\, d)$, $(\nu_\mu\, ,\, \mu^-\, ,\, c\, ,\, s)$, $(\nu_\tau\, ,\, \tau^-\, ,\, t\, ,\, b)$, while in the case of fundamental scalars one of (*a priori*) possible options may be that the two families correspond to the first and second fermion family (Królikowski 1992).

Now, in contrast, if the Standard–Model coupling $g\Gamma \cdot A(x)$ is absent from (6), then *only* physically undistinguishable *i.e.*, antisymmetric bispinor indices α_i $(i = 1, 2, \ldots, N)$ can appear at the wave functions $\psi_{\alpha_1\alpha_2\ldots\alpha_N}(x)$. In this case, the argument similar to the used before shows that on the fundamental level there are *two* (and only two) Standard–Model–inactive spin–1/2 fermions $N = 1, 3$ corresponding to the wave functions

$$\psi^{(1)}_{\alpha_1} \equiv \psi_{\alpha_1} ,$$

$$\psi^{(3)}_{\alpha_1} \equiv \frac{1}{6} \left(C^{-1}\gamma^5\right)_{\alpha_1\,\alpha_2} \varepsilon_{\alpha_2\alpha_3\alpha_4\alpha_5} \psi_{\alpha_3\alpha_4\alpha_5} \quad (10)$$

(with no suppressed f label). They can be identified with two *sterile neutrinos* denoted in this paper by ν_s and ν'_s, respectively. Analogically, on the fundamental level there should exist also *two* (and only two) Standard–Model–inactive spin–0 bosons $N = 2, 4$ that may be called *sterile scalars*, $\phi^{(2)}$ and $\phi^{(4)}$ (with no suppressed f label).

3 Neutrino Oscillations Involving ν_s and ν'_s

Let us conjecture tentatively that the sterile neutrinos ν_s and ν'_s are compelled to mix nearly maximally with the conventional neutrinos ν_e and ν_μ, respectively, in order to form four related neutrino mass states ν_1 or ν_4 and ν_2 or ν_5. Other neutrino mixings are assumed not to appear at all or to be negligible. In particular, the third conventional neutrino ν_τ is left not mixed and so, $\nu_3 = \nu_\tau$ is a neutrino mass state. Evidently, the mixings of ν_e with ν_s and ν_μ with ν'_s would be forbidden, if the electroweak $SU_L(2) \times U(1)$ symmetry were not spontaneously broken. Thus, we can say that neutrino oscillations, being a consequence of these mixings, are caused in fact by the spontaneous breaking of electroweak symmetry (if, of course, sterile neutrinos exist).

Under the above conjecture, the unitary transformation $\nu_I = \sum_\alpha V_{I\,\alpha}\nu_\alpha$ between neutrino mass states $\nu_I = \nu_1, \nu_2, \nu_3, \nu_4, \nu_5$ and neutrino flavor states $\nu_\alpha = \nu_e, \nu_\mu, \nu_\tau, \nu_s, \nu_s'$ is given as

$$\begin{aligned} \nu_1 &= V_{11}\nu_e + V_{14}\nu_s \;,\; \nu_4 = V_{41}\nu_e + V_{44}\nu_s \;, \\ \nu_2 &= V_{22}\nu_\mu + V_{25}\nu_s' \;,\; \nu_5 = V_{52}\nu_\mu + V_{55}\nu_s' \;, \\ &\nu_3 = \nu_\tau \;, \end{aligned} \tag{11}$$

where the nonzero coefficients are

$$\begin{aligned} V_{11} = V_{44} &= \frac{1}{\sqrt{1+Y^2}} \;,\; V_{14} = -V_{41}^* = -\frac{Y}{\sqrt{1+Y^2}}e^{i\varphi} \;, \\ V_{22} = V_{55} &= \frac{1}{\sqrt{1+X^2}} \;,\; V_{25} = -V_{52}^* = -\frac{X}{\sqrt{1+X^2}}e^{i\varphi'} \;, \\ V_{33} &= 1 \end{aligned} \tag{12}$$

(in (11) and (12), for notation convenience, we write V_{IJ} in place of $V_{I\,\alpha}$, where $I, J = 1,2,3,4,5$). The magnitudes of these coefficients are determined by the parameters

$$\begin{aligned} Y &= \frac{M_{11} - m_{\nu_1}}{|M_{14}|} = -\frac{M_{44} - m_{\nu_4}}{|M_{14}|} \;, \\ X &= \frac{M_{22} - m_{\nu_2}}{|M_{25}|} = -\frac{M_{55} - m_{\nu_5}}{|M_{25}|} \end{aligned} \tag{13}$$

involving neutrino masses

$$\begin{aligned} m_{\nu_1,\nu_4} &= \frac{M_{11} + M_{44}}{2} \mp \sqrt{\left(\frac{M_{11} - M_{44}}{2}\right)^2 + |M_{14}|^2} \;, \\ m_{\nu_2,\nu_5} &= \frac{M_{22} + M_{55}}{2} \mp \sqrt{\left(\frac{M_{22} - M_{55}}{2}\right)^2 + |M_{25}|^2} \;. \end{aligned} \tag{14}$$

On the other hand $m_{\nu_3} = M_{33}$. Here, (M_{IJ}) $(I, J = 1,2,3,4,5)$ is a 5×5 neutrino mass matrix with $M_{14} = M_{41}^* = |M_{14}| \exp i\varphi$ and $M_{25} = M_{52}^* = |M_{25}| \exp i\varphi'$ as the only off-diagonal elements. Then, (V_{IJ}) $(I, J = 1,2,3,4,5)$ is a 5×5 lepton counterpart of the familiar Cabibbo—Kobayashi—Maskawa matrix for quarks, where now $V_{14} = -V_{41}^*$ and $V_{25} = -V_{52}^*$ are the only nonzero off-diagonal elements.

Some (here neglected) small corrections to the neutrino mixings (11) may be caused by possible small deviations of the charged–lepton mass matrix from a diagonal form (Królikowski 1996). In fact, these deviations produce small deviations of the related diagonalizing unitary matrix from the unit matrix. In turn, such a charged–lepton diagonalizing matrix contributes multiplicatively to the lepton Cabibbo—Kobayashi—Maskawa matrix (Królikowski 1996), changing a little its leading form (12) (in particular, almost all zero elements of its leading form become nonzero but small).

Now, making use of (12), we can calculate the probabilities of neutrino oscillations $\nu_e \to \nu_s$ and $\nu_\mu \to \nu'_s$ (in the vacuum) from the general formula

$$P(\nu_\alpha \to \nu_\beta) = |\langle \nu_\beta | \nu_\alpha(t) \rangle|^2 = \sum_{K\,L} V_{L\,\beta} V^*_{L\,\alpha} V^*_{K\,\beta} V_{K\,\alpha} \exp\left(i \frac{\Delta m^2_{LK}}{2|\mathbf{p}|} t\right) , \tag{15}$$

where $\Delta m^2_{LK} = m^2_{\nu_L} - m^2_{\nu_K}$ (on the rhs of (15), for notation convenience, we will replace α, β by I, $J = 1, 2, 3, 4, 5$). Here, $\nu_\alpha(0) = \nu_\alpha$, $\langle \nu_\beta | = \langle 0 | \nu_\beta$, $\langle \nu_\beta | \nu_\alpha \rangle = \delta_{\beta\,\alpha}$ and, as usual, $t/|\mathbf{p}| = L/E$ $(c = 1 = \hbar)$, what is equal to $4 \times 1.2663 L/E$ if Δm^2_{LK} , L and E are measured in eV^2, m and MeV, respectively (L is, of course, the source–detector distance). Further on, we will denote

$$x_{LK} = 1.2663 \frac{\Delta m^2_{LK} L}{E} \tag{16}$$

and use the identity $\cos 2x_{LK} = 1 - 2\sin^2 x_{LK}$.

In such a way, we derive the following formulae for probabilities of neutrino oscillations $\nu_e \to \nu_s$ and $\nu_\mu \to \nu'_s$ (in the vacuum):

$$P\,(\nu_e \to \nu_s) = 4 \frac{Y^2}{(1+Y^2)^2} \sin^2 x_{41} ,$$
$$P\,(\nu_\mu \to \nu'_s) = 4 \frac{X^2}{(1+X^2)^2} \sin^2 x_{52} , \tag{17}$$

while all other $P\,(\nu_\alpha \to \nu_\beta)$ with $\alpha \neq \beta$ vanish [except, of course, for $P\,(\nu_s \to \nu_e) = P\,(\nu_e \to \nu_s)$ and $P\,(\nu'_s \to \nu_\mu) = P\,(\nu_\mu \to \nu'_s)$]. Thus, the neutrino–oscillation formulae (in the vacuum) for survival probabilities of ν_e and ν_μ are

$$P\,(\nu_e \to \nu_e) = 1 - 4 \frac{Y^2}{(1+Y^2)^2} \sin^2 x_{41} ,$$
$$P\,(\nu_\mu \to \nu_\mu) = 1 - 4 \frac{X^2}{(1+X^2)^2} \sin^2 x_{52} . \tag{18}$$

In the case of solar neutrinos, the observed deficit of ν_e's can be explained through the neutrino oscillations (in the vacuum), when using the two–flavor formula for survival probability of ν_e,

$$P\left(\nu_e \rightarrow \nu_e\right) = 1 - \sin^2 2\theta_{\rm solar} \sin^2 \left(1.27 \frac{\Delta m^2_{\rm solar} L}{E}\right) , \tag{19}$$

with the parameters (Hata and Langacker 1997; Fogli et al. 1997)

$$\sin^2 2\theta_{\rm solar} \sim 0.65 \;\text{ to }\; 1 \;, \quad \Delta m^2_{\rm solar} \sim (5 \;\text{ to }\; 8) \times 10^{-11}\,{\rm eV}^2 \,. \tag{20}$$

These give the so called vacuum fit, in contrast to two other known fits based on the resonant MSW mechanism (Wolfenstein 1978; Mikheyev and Smirnow 1985) in the Sun matter. In our model of neutrino oscillations (where $\nu_e \rightarrow \nu_s$ oscillations are responsible for the deficit of solar ν_e's), this fit leads to

$$\frac{4Y^2}{(1+Y^2)^2} \sim 0.65 \;\text{ to }\; 1 \;, \quad \Delta m^2_{41} \sim (5 \;\text{ to }\; 8) \times 10^{-11}\,{\rm eV}^2 \tag{21}$$

(as $m^2_{\nu_4} > m^2_{\nu_1}$). Hence, $Y \sim 0.507$ to 1 and so, we get a large mixing of ν_e with ν_s: $V_{11} = V_{44} \sim 0.892$ to $1/\sqrt{2}$ and $V_{14} = -V^*_{41} \sim -(0.452$ to $1/\sqrt{2}) \exp i\varphi$ (the phase φ remains not determined).

In the case of atmospheric neutrinos, the recent findings of the Super–Kamiokande experiment (Fukuda et al. 1998) show that the observed deficit of ν_μ's can be explained also through the neutrino oscillations (in the vacuum), when making use of the two–flavor formula for survival probability of ν_μ,

$$P\left(\nu_\mu \rightarrow \nu_\mu\right) = 1 - \sin^2 2\theta_{\rm atm} \sin^2 \left(1.27 \frac{\Delta m^2_{\rm atm} L}{E}\right) , \tag{22}$$

with the parameters

$$\sin^2 2\theta_{\rm atm} \sim 0.82 \;\text{ to }\; 1 \;, \quad \Delta m^2_{\rm atm} \sim (0.5 \;\text{ to }\; 6) \times 10^{-3}\,{\rm eV}^2 \,. \tag{23}$$

In our model of neutrino oscillations (where $\nu_\mu \rightarrow \nu'_s$ oscillations are responsible for the deficit of atmospheric ν_μ's), this implies

$$\frac{4X^2}{(1+X^2)^2} \sim 0.82 \;\text{ to }\; 1 \;, \quad \Delta m^2_{52} \sim (0.5 \;\text{ to }\; 6) \times 10^{-3}\,{\rm eV}^2 \tag{24}$$

(as $m^2_{\nu_5} > m^2_{\nu_2}$). Hence, $X \sim 0.636$ to 1 and thus, we obtain a large mixing of ν_μ with ν'_s: $V_{22} = V_{55} \sim 0.844$ to $1/\sqrt{2}$ and $V_{25} = -V^*_{52} \sim -(0.537$ to $1/\sqrt{2}) \exp i\varphi'$ (the phase φ' remains not determined).

On the other hand, the CHOOZ experiment (Appolonio et al. 1998) found no evidence for neutrino–oscillation modes of $\bar{\nu}_e$ in a parameter region overlapping the range (23) of $\sin^2 2\theta_{\rm atm}$ and $\Delta m^2_{\rm atm}$, what shows that within this parameter range there are no mixings of ν_e with ν_μ, ν_τ, ν_s, ν'_s. In particular for ν_μ, this is consistent with the assumed dominance of mixing between ν_μ and ν'_s over mixing between ν_μ and ν_e within the range (23) of $\sin^2 2\theta_{\rm atm}$ and $\Delta m^2_{\rm atm}$ (at the moment, however, it cannot be decided experimentally

(Fukuda et al. 1998), whether the mixing of ν_μ with ν'_s or the here neglected mixing of ν_μ with ν_τ is responsible for the deficit of atmospheric ν_μ's). For ν_s, this requires that the assumed strong mixing of ν_e with ν_s must correspond to parameters $\sin^2 2\theta_{\rm solar}$ and $\Delta m^2_{\rm solar}$ belonging to a range very different from (23) [in fact, they can lie in the range (20)]. Finally for ν_τ, the lack of mixing between ν_τ and ν_e is one of necessary and sufficient conditions for the assumed identity $\nu_3 = \nu_\tau$ (another is the lack of mixing between ν_τ and ν_μ, if this really is true).

Of course, the sterile neutrinos ν_s and ν'_s by themselves cannot help to explain the results of LSND experiment (Athanassopoulos et al. 1996) which gave evidence for $\bar{\nu}_\mu \to \bar{\nu}_e$ and $\nu_\mu \to \nu_e$ oscillations corresponding to a much larger $\Delta m^2_{\rm LSND}$ than both $\Delta m^2_{\rm solar}$ and $\Delta m^2_{\rm atm}$. These oscillations, if eventually confirmed, would require a considerable mixing of ν_μ with ν_e, corresponding to parameters $\sin^2 2\theta_{\rm LSND}$ and $\Delta m^2_{\rm LSND}$ lying in a range very different from (23). This mixing should be stronger than that induced by the (mentioned before) nondiagonal charged–lepton corrections appearing in our model of fermion "texture" (Królikowski 1996).

The comparison of mass squared differences Δm^2_{41} and Δm^2_{52} as estimated in (21) and (24) suggests that $m^2_{\nu_1}$ and $m^2_{\nu_4}$ are possibly much smaller than $m^2_{\nu_2}$ and $m^2_{\nu_5}$ (alternatively, $m^2_{\nu_1}$ and $m^2_{\nu_4}$ may be much more degenerate than $m^2_{\nu_2}$ and $m^2_{\nu_5}$).

4 Outlook: Non–Abelian Spin–1/2 Fermions

When the Dirac–type equations (4) are considered, one may ask a (perhaps) profound question, as to whether these one–body equations could be understood physically as point–like limits of some N–body equations for tight bound states of N spin–1/2 preons with equal masses. If it was so, the four–positions of such subelementary constituents (called here preons, as usual) should tend practically (within the bound states) to their centre–of–mass four–position,

$$x_i = X + \delta x_i \to X \;,\; X \equiv \frac{1}{N}\sum_{i=1}^{N} x_i \;,\; \sum_{i=1}^{N} \delta x_i \equiv 0 \;, \tag{25}$$

while then δp_i, defined by their four–momenta

$$p_i = P + \delta p_i \;,\; P \equiv \sum_{i=1}^{N} p_i \;,\; \sum_{i=1}^{N} \delta p_i \equiv 0 \;, \tag{26}$$

should vanish in action on the wave functions [here, $x_i = (t_i\,,\,\mathbf{x}_i)$, $\delta x_i = (\delta t_i\,,\,\delta\mathbf{x}_i)$ and $X = (t, \mathbf{X})$].

Of course, the physical mechanism for realization of such practically point–like limits in N–body systems would be provided by an unknown, very strong and shortrange attraction between their N constituents (described, for convenience, in the equal-time formalism, where $\delta t_i \equiv 0$ and δp_i^0 vanish in action on the wave functions). The (necessarily) non-Standard-Model nature of such an attraction would be certainly the most obscure aspect of the compound option for the Dirac-type equations (4).

Let us denote by P_i and X_i $(i = 1, 2, \ldots, N)$, with $P_1 \equiv P$ and $X_1 \equiv X$, the (properly normalized) Jacobi combinations of four-momenta p_i and four-positions x_i $(i = 1, 2, \ldots, N)$, respectively, for N particles. Then,

$$\left[P_i^\mu \, , \, X_j^\nu\right] = i\delta_{ij} g^{\mu\nu} \, . \tag{27}$$

Making use of this notation, we can write the identities

$$\sum_{i=1}^{N} (\gamma_i \cdot p_i - m_i) = \frac{1}{\sqrt{N}} \sum_{i=1}^{N} \left(\Gamma_i \cdot P_i - \sqrt{N} m_i \right) \, , \tag{28}$$

where Γ_i^μ $(i = 1, 2, \ldots, N)$, with $\Gamma_1^\mu \equiv \Gamma^\mu$, stand for the (properly normalized) Jacobi combinations of γ_i^μ matrices $(i = 1, 2, \ldots, N)$ for N particles [the Γ_i^μ matrices were already introduced in (7), though only in reference to the one-body Dirac-type equations(4)]. Then,

$$\left\{\Gamma_i^\mu, \Gamma_j^\nu\right\} = 2\delta_{ij} g^{\mu\nu} \;\; (i, j = 1, 2, \ldots, N) \, , \tag{29}$$

as follows from (3). Here, in particular, $\Gamma_1^\mu \equiv \Gamma^\mu$ is given as in (1).

In this notation, the natural candidates for the hypothetic N-body equations would be

$$\left\{ \Gamma_1 \cdot [P_1 - gA(X_1)] + \sum_{i=2}^{N} \Gamma_i \cdot P_i \right. \\ \left. - \sqrt{N} \left(\sum_{i=1}^{N} m_i + I \right) \right\} \psi(X_1, X_2, \ldots, X_N) = 0 \, , \tag{30}$$

where $I(X_2, \ldots, X_N)$ would symbolize the unknown non-Standard-Model attraction between N constituents. In (30), the Standard-Model gauge fields $A_\mu(x)$ are coupled to the hypothetic N-body systems at four-points describing their centre-of-mass four-positions $X_1 \equiv X$. This is approximately true, when $A_\mu(X + \delta x_i)$ are only weakly dependent on δx_i.

In the point-like limits, where the relative four-positions $X_2, \ldots, X_N$ (*i.e.*, also all δx_i) tend to zero and then the relative four-momenta $P_2, \ldots, P_N$ (*i.e.*, also all δp_i) vanish in action on the wave functions, equations (30) are really reduced to the Dirac-type equations (4) with $p \equiv P \equiv P_1$, $X \equiv X \equiv X_1$, $\Gamma \equiv \Gamma_1$ and $M \equiv \sqrt{N}\,(Nm + I_{X_i \to 0})$ $(m_i \equiv m)$. Note that M grows with N faster than linearly.

In the equal-time formalism, where the relative times $t_2, \ldots, t_N$ (*i.e.*, also all δt_i) are zero and the relative energies $P_2^0, \ldots, P_N^0$ (*i.e.*, also all δp_i^0) vanish in action on the wave functions, equations (30) assume the forms

$$P_1^0\psi(\mathbf{X}_1,\mathbf{X}_2,\ldots,\mathbf{X}_N,t)$$
$$= \left\{\Gamma_1^0\boldsymbol{\Gamma}_1\cdot[\mathbf{P}_1 - g\mathbf{A}(\mathbf{X}_1,t)] + gA^0(\mathbf{X}_1,t) + \sum_{i=2}^{N}\Gamma_1^0\boldsymbol{\Gamma}_i\cdot\mathbf{P}_i\right.$$
$$\left. + \;\Gamma_1^0\left(\sqrt{N}\sum_{i=1}^{N}m_i + I_{X_i^0=0}\right)\right\}\psi(\mathbf{X}_1,\mathbf{X}_2,\ldots,\mathbf{X}_N,t)\,, \qquad (31)$$

where $P_1^0 \equiv P^0 = i\partial/\partial t$ and $I_{X_i^0=0} = I(\mathbf{X}_2,\ldots,\mathbf{X}_N)$.

In the point–like limits, equations (31) are reduced to the equations

$$p^0\psi(\mathbf{x},t) = \left\{\Gamma^0\boldsymbol{\Gamma}\cdot[\mathbf{p} - g\mathbf{A}(\mathbf{x},t)] + gA^0(\mathbf{x},t) + \Gamma^0 M\right\}\psi(\mathbf{x},t) \qquad (32)$$

with $p \equiv P \equiv P_1$, $x \equiv X \equiv X_1$, $\Gamma \equiv \Gamma_1$ and $M \equiv \sqrt{N}(Nm + I_{X_i\to 0})$ $(m_i \equiv m)$. Of course, $p^0 \equiv P^0 = i\partial/\partial t$ and $I_{X_i\to 0}$ stands for a reasonably defined point–like limit of I.

Note that the eigenvalues $(P_1^0)_{\rm kin}$ of the kinetic part of the hamiltonian appearing on the rhs of the state equation (31) get for any N the form $\pm\left[\mathbf{P}_1^2 + \ldots + \mathbf{P}_N^2 + N(Nm)^2\right]^{1/2} = \pm\sqrt{N}\left[\mathbf{p}_1^2 + \ldots + \mathbf{p}_N^2 + N(Nm)^2\right]^{1/2}$, as if our N–body system were a single Dirac particle with the mass Nm in a $(3N+1)$–dimensional spacetime (notice, however, the additional factor $\sqrt{N}$).

A fundamental feature of (30) is that, *via* Γ_i^μ $(i = 1,2,\ldots,N)$, they contain N Dirac nonconventional γ_i^μ matrices $(i = 1,2,\ldots,N)$ which do not commute for different particles, in contrast to Dirac conventional gammas commuting for different particles [in fact, the nonconventional γ_i^μ $(i = 1,2,\ldots,N)$ anticommute for different i, as is seen from (3)]. The spin–1/2 fermions $i = 1,2,\ldots,N$ described within an N–body system with the use of such nonconventional γ_i^μ matrices $(i = 1,2,\ldots,N)$, anticommuting for different particles, might be called *non–Abelian* spin–1/2 fermions (Królikowski 1991). In contrast, in the familiar case of Dirac conventional gammas, commuting for different particles, one could use the term *Abelian* spin–1/2 fermions.

Now, let us observe that the form of spin tensors for spin–1/2 fermions $i = 1,2,\ldots,N$ is identical in the non–Abelian and Abelian case:

$$\sigma_i^{\mu\nu} \equiv \frac{i}{2}\left[\gamma_i^\mu\,,\,\gamma_i^\nu\right] = \begin{cases} i\alpha_i^l & \text{for } \mu = 0\,,\ \nu = l \\ \varepsilon^{klm}\sigma_i^m & \text{for } \mu = k\,,\ \nu = l \end{cases}\,, \qquad (33)$$

where $\alpha_i^l \equiv \gamma_i^0\gamma_i^l$ and $\sigma_i^m \equiv \gamma_i^5\gamma_i^0\gamma_i^m \equiv \gamma_i^5\alpha_i^m$ with $\gamma_i^5 \equiv i\gamma_i^0\gamma_i^1\gamma_i^2\gamma_i^3$. In fact, for each i the components $\frac{1}{2}\sigma_i^{\mu\nu}$ satisfy in both cases the usual Lorentz–group commutation relations, while for different i they commute in both cases as being bilinear in γ_i^μ. Also γ_i^5 commute for different i in both cases. So, the total generators of Lorentz group for a system of N spin–1/2 fermions are given in both cases by the operators

$$J^{\mu} = \sum_{i=1}^{N} \left(x_i^{\mu} p_i^{\nu} - x_i^{\nu} p_i^{\mu} + \frac{1}{2} \sigma_i^{\mu\nu} \right) . \tag{34}$$

Let us note, by the way, the following identity valid for the total spin tensor in both cases

$$\sum_{i=1}^{N} \sigma_i^{\mu\nu} = \sum_{i=1}^{N} \Sigma_i^{\mu\nu} , \tag{35}$$

where

$$\Sigma_i^{\mu\nu} \equiv \frac{i}{2} [\Gamma_i^{\mu}, \Gamma_i^{\nu}] = \begin{cases} i\, A_i^l & \text{for } \mu = 0,\ \nu = l \\ \\ \varepsilon^{klm} \Sigma_i^m & \text{for } \mu = k,\ \nu = l \end{cases} \tag{36}$$

with $A_i^l \equiv \Gamma_i^0 \Gamma_i^l$ and $\Sigma_i^m \equiv \Gamma_i^5 \Gamma_i^0 \Gamma_i^m \equiv \Gamma_i^5 A_i^m$. Evidently, $\Sigma_1^{\mu\nu} \equiv \Sigma^{\mu\nu}$ with $\Sigma^{\mu\nu} \equiv \frac{i}{2} [\Gamma^{\mu}, \Gamma^{\nu}]$ is the centre–of–mass spin tensor for the system of N spin–1/2 fermions, while $\Sigma_2^{\mu\nu}, \ldots, \Sigma_N^{\mu\nu}$ are its relative spin tensors. All spin tensors $\Sigma_i^{\mu\nu}$, being bilinear in Γ_i^{μ}, commute for different i in both cases.

Thus, in this Section, we can draw the important conclusion that for a system of N spin–1/2 fermions the Lorentz–group commutation relations get *two* (and only two) realizations: *either* with the use of Dirac conventional gammas commuting for different particles, *or* with the use of Dirac nonconventional gammas anticommuting for different particles. Such an intriguing statement seems to support the logical consistency and unexpected naturalness of the notion of non–Abelian spin–1/2 fermions. They may provide an unconventional alternative for familiar Abelian spin–1/2 fermions in the potential structure of particle theory. In this Section, their role as hypothetic preons was underlined.

Finally, we should like to emphasize some unconventional features of the quantization procedure which would work in the case of non–Abelian spin–1/2 fermions. It is not difficult to observe that in the case of spin–1/2 fermions, *only* the Fock–space states related to Dirac conventional gammas (commuting for different particles) can be constructed by means of the familiar second–quantization procedure based on Fermi creation and annihilation operators for single particles. This is so, because the repetition of single–particle creation operators can lead to Fock–space states of particles with commuting Dirac gammas *only*. In order to construct the Fock–space states related to Dirac nonconventional gammas (anticommuting for different particles), *new* Fermi or Bose operators creating and annihilating at once *whole* N–particle configurations with odd or even $N = 1, 2, 3, \ldots$, respectively, must be introduced. Of course, these N particles are then non–Abelian spin–1/2 fermions. Such a new procedure might be called the *third quantization* (Królikowski 1991).

References

Appolonio, M., et al. (CHOOZ Collaboration), (1998): Phys. Lett. **B 420**, 397

Athanassopoulos, C., et al. (LSND Collaboration), (1996): Phys. Rev. **C 54**, 2685; Phys. Rev. Lett. **77**, 3082; nucl-ex/9709006

Banks, T., Dothan, Y., Horn, D., (1982): Phys. Lett. **B117**, 413

Fukuda, Y., et al. (1998): Phys. Rev. Lett. **81**, 1562A and references therein

Hata, N., Langacker, P., (1997): hep-ph/9705339; cf. also Fogli, G.L., Lisi, E., Montanino, D., (1997): hep-ph /9709473

Kähler, E., (1962): Rendiconti di Matematica **21**, 425; cf. also Ivanenko, D., and Landau, L., (1928): Z. Phys. **48**, 341

Królikowski, W., (1990): Acta Phys. Pol. **B 21**, 871; Phys. Rev. **D 45**, (1992) 3222; in *Spinors, Twistors, Clifford Algebras and Quantum Deformations (Proc. 2nd Max Born Symposium 1992)*, eds. Z. Oziewicz (1993): et al., Kluwer

Królikowski, W., (1991): Acta Phys. Pol. **B 22**, 613 [E: ibid. **B 23**, (1992) 83]

Królikowski, W., (1992): Phys. Rev. **D 46**, 5188; Acta Phys. Pol. **B 24**, (1993) 1149; **B 26**, (1995) 1217; Nuovo Cimento , **107 A**, (1994) 69

Królikowski, W., (1996): Acta Phys. Pol. **B 27**, 2121; **B 28**, (1997) 1643; **B 29**, (1998) 629; **B 29**, (1998) 755; hep-ph/9803323; hep-ph/9808207

For a review cf. e.g., Smirnow, A.Y., (1997): in *Proc. 28th Inter. Conf. on High Energy Physics, Warsaw 1996*, eds. Z. Ajduk and A.K. Wróblewski, World Scientific

Wolfenstein, L., (1978): Phys. Rev. **D 17**, 2369; Mikheyev, S.P., Smirnow, A.Y., (1985): Sov. J. Nucl. Phys. **42**, 913; Nuovo Cimento, **C 9**, (1986) 17

Magnetic Monopoles and Gravity

D. Maison

Max-Planck-Institut für Physik
— Werner Heisenberg Institut —
Föhringer Ring 6
80805 Munich (Fed. Rep. Germany)
e-mail: dim@mppmu.mpg.de

Dedicated to Prof. Jan Łopuszański
on occasion of his 75th birthday

Abstract. Subject of this talk is an overview of results on self-gravitating non-abelian magnetic monopoles. The coupling to the gravitational field leads to new features absent in flat space: gravitating monopoles have unstable "radial excitations" and they exist only up to a maximal mass (related to a kind of "gravitational confinement" at the Planck scale). In addition to the globally regular monopoles there are "coloured" black holes, i.e., magnetically charged black holes carrying a non-trivial YM field outside their event horizon. The latter give rise to a violation of the "No Hair" Conjecture.

1 Introduction

This talk is an overview of results on self-gravitating magnetic monopoles. It is mainly based on analytical and numerical results obtained in collaboration with P. Breitenlohner and P. Forgács (Breitenlohner et al. 1992, 1995). Many other people, who have contributed in establishing our present understanding of this subject will be mentioned in due course.

As a genuine non-linear structure magnetic monopoles play an important role in the non-perturbative aspects of the Yang-Mills-Higgs (YMH) theory. Originally they were found as solutions of the YMH field equations in flat space, but in a very interesting early paper van Niewenhuizen et.al. (van Nieuwenhuizen et al. 1976) considered also the gravitational self-interaction of monopoles. However these authors made no attempt to actually construct solutions with analytical or numerical methods. Only twenty years later, triggered by the discovery of globally regular solutions of the Einstein-Yang-Mills theory by Bartnik and McKinnon (Bartnik and McKinnon 1988) new interest in the subject arose.

A systematic numerical study of the effects of gravity on magnetic monopoles (Ortiz 1992, Lee et al. 1992, Breitenlohner et al. 1992, 1995) revealed a number of interesting phenomena. In contrast to monopoles in flat space gravitating monopoles allow for "radial excitations", which have some close connection with the solutions discovered by Bartnik and McKinnon. As to be expected gravitating monopoles develop a gravitational instability for sufficiently strong gravitational self-force, manifesting itself in some kind of a "Gravitational Confinement". Systematically increasing the strength of the gravitational self-coupling (resp. letting the monopole mass approach the Planck scale) one reaches a limiting solution, which in its exterior part has the geometry of an extremal black hole.

Obviously one may question the physical relevance of monopoles with a mass close to the Planck mass, since on the one hand even in GUTs the monopole mass would be considerably lower and on the other hand at the Planck scale one would expect quantum gravity effects to come into play.

In addition to the regular monopoles there are also black hole solutions carrying a non-trivial exterior YM field ("Coloured Black Holes"). Taking into account their radial excitations, one finds a rich spectrum of such static black hole solutions. This is to be contrasted with Einstein's theory in vacuum resp. with the Einstein-Maxwell theory, where according to a theorem of Israel (Israel 1967 1968) the Schwarzschild resp. Reissner-Nordstrøm solution are the only static black holes. The co-existence of all these black hole solutions with the same magnetic charge gives rise to an interesting violation of the "No-Hair" Conjecture (Chruściel 1994; Bizon 1993).

2 Magnetic Monopoles and Sphalerons in Flat Space

Let me start with a short reminder on the static, spherically symmetric solutions of the YMH system in flat space. For simplicity I restrict myself to the gauge group $SU(2)$ from now on.

The so-called 't Hooft-ansatz for the static, spherically symmetric YM field in polar coordinates reads

$$W_0^a = 0 \quad W_i^a = \epsilon_{iak}\frac{x^k}{r^2}(W(r) - 1)\ . \tag{1}$$

Inserting it into the standard YM action

$$S_{YM} = -\frac{1}{4\pi}\int d^4\, x\Big[\frac{1}{4g^2}\mathrm{Tr}F^2\Big] \tag{2}$$

yields the reduced action

$$S_{YM,\mathrm{red}} = -\int dr\Big[\frac{1}{g^2}(W'^2 + \frac{(1-W^2)^2}{2r^2})\Big]\ . \tag{3}$$

A rescaling of the radial coordinate $r \to r/\lambda$ leads to a rescaling of the action $S_{YM,\mathrm{red}} \to \lambda S_{YM,\mathrm{red}}$. This property (related to the scale invariance of the 4-dimensional theory) prevents the existence of any non-trivial stationary point of $S_{YM,\mathrm{red}}$ with finite non-zero action (energy), manifestating the general statement, that the flat YM theory has no solitons (Coleman 1975; Deser 1976).

The situation changes with the inclusion of the Higgs field. Through its "vacuum expectation value" v two scales are introduced, the mass $M_W = gv$ of the YM field and the Higgs mass $M_H = \sqrt{\lambda}v$. There are two different cases

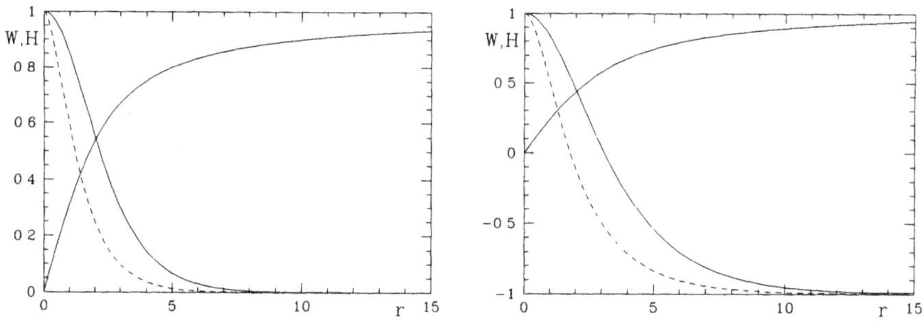

Fig. 1. a) PS-monopole, b) DHN-sphaleron, both for vanishing (solid) and infinite (dashed) Higgs mass

to be considered, leading to rather different types of solutions. The Higgs field can be either in a triplet or in a doublet representation. In either case the action is

$$S_H = \frac{1}{4\pi}\int d^4\,x\Big[\frac{1}{2}|D\phi|^2 - \frac{\lambda}{8}(|\phi|^2 - v^2)\Big]\,. \tag{4}$$

The finite energy solutions in the case of a Higgs triplet are the 't Hooft-Polyakov magnetic monopoles ('t Hooft 1974; Polyakov 1974). They are obtained with the ansatz

$$\phi^a = \frac{x^a}{r}H(r)\,. \tag{5}$$

Inserting this ansatz in the action (4) one gets

$$S_{H,\text{red}} = -\int dr\Big[\frac{r^2}{2}H'^2 + \frac{\lambda r^2}{8}(H^2 - v^2)^2 + W^2H^2\Big]\,. \tag{6}$$

In order to obtain finite total energy the Higgs field has to tend to its vacuum value v for $r \to \infty$, forcing in turn $W \to 0$. Outside a "core" of size $1/M_W$ the solution is essentially equal to the embedded abelian Dirac monopole $W \equiv 0$ avoiding the singular center at $r = 0$ of the latter (compare Fig. 1a).

For large values of M_H and hence of β the function $H(r)$ rises quickly to its asymptotic value v. In the limit $\beta \to \infty$ the Higgs field may be replaced by v for all $r > 0$ and its only role is to give a mass to the YM field. The total energy of the solution stays finite in this limit. In fact, it only varies by a factor ≈ 1.8 as β varies from 0 to ∞. Of particular interest is the exact BPS monopole solution for $\beta = 0$ with the simple exact form (using $h = rH$ for convenience)

$$W(r) = \frac{vr}{\sinh(vr)} \qquad h(r) = vr\coth(vr) - 1 \,, \tag{7}$$

satisfying a system of first order equations (Bogomolny equations)

$$rW' = Wh \,, \tag{8a}$$

$$rh' = h + 1 - W^2 \tag{8b}$$

implying the second order field equations. Considered as a solution of a suitable supersymmetric extension (N=2 SUSY-YM theory) it has the special property of being "half supersymmetric", i.e., it is annihilated by one half of the infinitesimal supersymmetry generators. This implies the relation $E = vP$ between the energy of the solution and its magnetic charge P (equalling the central charge of the $N = 2$ SUSY algebra). If "quantizing" the solution does not destroy supersymmetry, i.e., the above relation is preserved, any quantum corrections to its mass have to vanish (Witten and Olive 1978).

Due to the topological character of the magnetic charge, related to the asymptotic vacuum structure of configurations with finite energy, the monopole is a stable solution.

The second possibility is a Higgs field in the doublet representation. The relevant ansatz of the Higgs field is $\Phi^\alpha = H(r)\xi^\alpha$ with some constant spinor ξ. Although this ansatz is not itself spherically symmetric it leads to a consistent reduction. The corresponding reduced action is

$$S_{H,\mathrm{red}} = -\int dr \left[\frac{r^2}{2}H'^2 + \frac{\lambda r^2}{8}(H^2 - v^2)^2 + \frac{1}{4}(W+1)^2 H^2\right] . \tag{9}$$

The only essential difference of this action to the one for the triplet is the form of the mass term. It destroys the symmetry $W \to -W$ and enforces W to turn to $W = -1$ for $r \to \infty$ in order to have finite total energy (compare Fig. 1b). This asymptotic behaviour implies that the solution has no magnetic charge in contrast to the previous case with $W \to 0$.

Unlike the stable monopole the sphaleron, i.e., the solution minimizing the energy $E = -S$, is unstable. In order to understand this instability it is important to consider the most general spherically symmetric ansatz for the YM field.

$$W_t^a = (0, 0, A_0) \,, \qquad W_\theta^a = (W_1, W_2, 0) \,, \tag{10a}$$

$$W_r^a = (0, 0, A_1) \,, \qquad W_\varphi^a = (-W_2 \sin\theta, W_1 \sin\theta, \cos\theta) \,. \tag{10b}$$

The ansatz used above for the monopole and the sphaleron corresponds to a consistent truncation, putting $A_0 = A_1 = W_2 = 0$ and $W_1 = W$. The sphaleron turns out to be stable under variations staying within the minimal ansatz, but not if $\delta W_2 \neq 0$ and $\delta A_1 \neq 0$. As was discussed by Manton (Manton 1983) this instability may be attributed to the non-trivial topology of the configuration space of the spherically symmetric YM potential, again related to the asymptotic vacuum structure of configurations with finite energy.

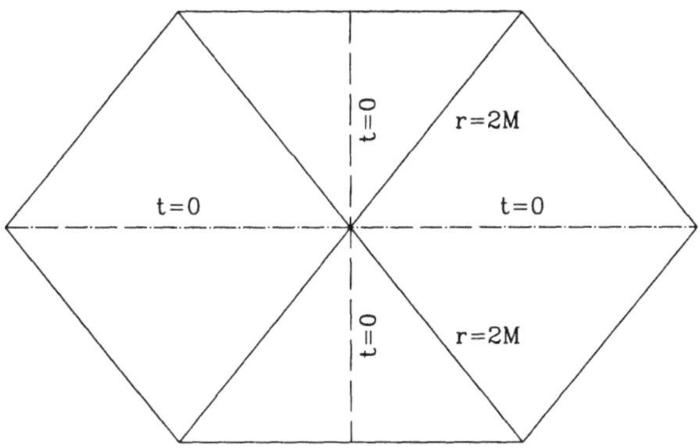

Fig. 2. Conformal diagram for the Schwarzschild solution

3 The Spherically Symmetric Gravitational Field

Spherically symmetric space-times M_4 have the structure of an orthogonal product $M_4 = M_2 \times S^2$ of a 2-dimensional space M_2 with a 2-sphere with a corresponding decomposition of the metric $ds_4^2 = ds_2^2 + r^2 d\Omega^2$, where $d\Omega^2$ is the invariant metric on S^2 and its inverse curvature r is a function on M_2. A convenient parametrization of ds_2^2 is

$$ds_2^2 = A^2 B dt^2 - \frac{dR^2}{B} , \tag{11}$$

with arbitrary time resp. radial coordinates t and R. The standard choice for the latter is the "Schwarzschild" coordinate $R \equiv r$, which is possible as long as $dr/dR \neq 0$. We are only interested in static solutions, with A and B independent of the canonical time coordinate ("Killing-time") t. Insertion of the ansatz into the standard Einstein action then yields

$$S_{G,\mathrm{red}} = \frac{1}{2\sqrt{G}} \int dr \Big[A(B + rB' - 1) \Big] . \tag{12}$$

The dimensionality of G introduces a mass scale $M_{\mathrm{Pl}} = 1/\sqrt{G}$, the Planck mass. Variation with respect to A and B yields (with suitable boundary condition at infinity) the Schwarzschild solution $A = 1$, $B = 1 - 2M/r$. *

* Although $S_{G,\mathrm{red}}$ is also just rescaled under a scaling $r \to \lambda r$ similar to $S_{YM,\mathrm{red}}$, there is now a non-trivial stationary point with vanishing action, because $S_{G,\mathrm{red}}$ is indefinite.

As is well known it describes a static black hole of total mass M and event horizon located at the Schwarzschild radius $r_S = 2M$. The fact that its total mass is finite, although the solution has a real singularity at the origin, illustrates a general difference to flat space, where the finiteness of the total energy of a field configuration in general implies some regularity properties. This remark is not made without hindsight, as it explains the unsuitability of the energy (mass) functional for existence proofs of solutions once the gravitational self-interaction is included.

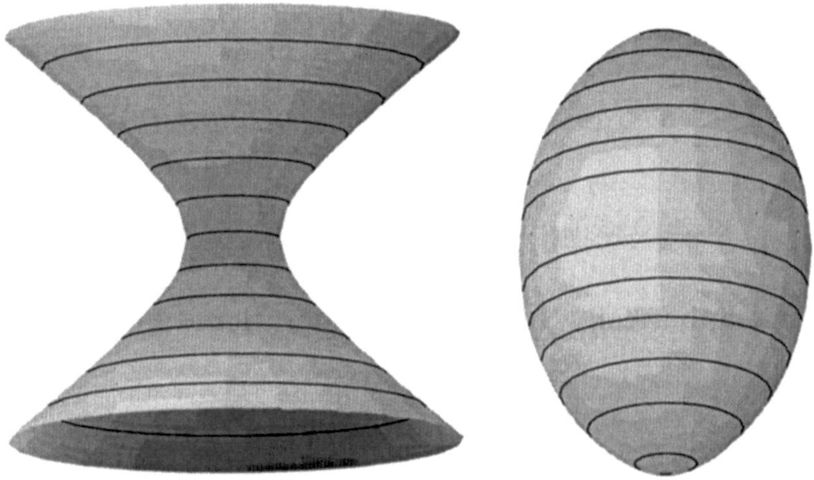

Fig. 3. a) Schwarzschild b.h., $t = 0$ hyperplane for $r > 2M$ (each circle is actually a 2-sphere), b) same for $r < 2M$

The geometry of black hole space-times is best illustrated with their conformal diagram Fig. 2 (Hawking and Ellis 1973). Since we are considering the solutions for fixed t, the hyper-surfaces $t =$ const. are of particular interest. They meet the horizon at the so-called "bifurcation surface" of the horizon. In order to study their geometry it is useful to switch to another radial coordial coordinate, avoiding the apparent singularity of the metric at $r = 2M$. A convenient choice is

$$\rho = \int_{2M}^{r} \frac{dr'}{\sqrt{B}} \,. \tag{13}$$

Due to the square-root ρ is a double-valued function of r, the continuation through the branch point at $r = 2M$ leading to another copy of the original surface and thus giving the surfaces $t =$ const. their famous "wormhole" structure (Fig. 3). A similar construction for the corresponding (time-like)

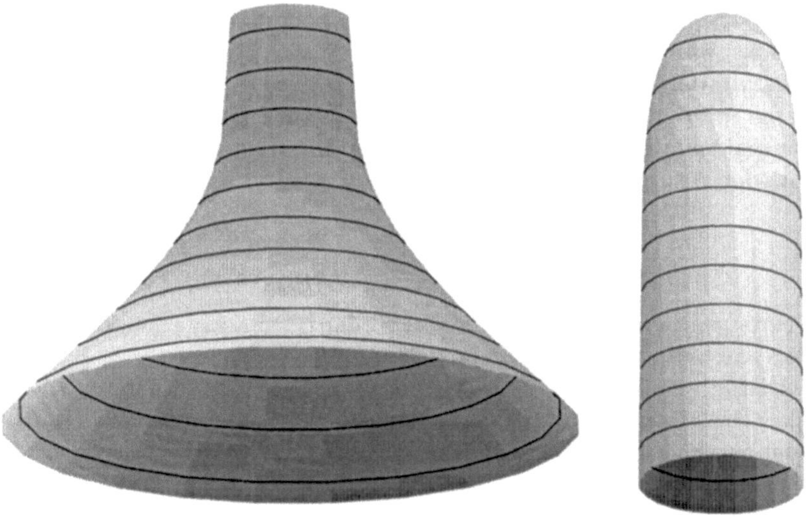

Fig. 4. a) Extremal Reissner-Nordstrøm b.h., part of the t=0 hyperplane for $r > M$, b) same for $r < M$

surfaces inside the horizon (replacing B by $-B$ in (13)) leads to a compact surface with a second singular center ($r = 0$).

Next we consider the static black holes of the Einstein-Maxwell theory, described by the Reissner-Nordstrøm (RN) solution. There are two possibilities for a static, spherically symmetric Maxwell field, the electric monopole with the potential $A_0 = q/r$ or a magnetic monopole with $A_\varphi = q\cos\theta$ with the Dirac string singularity, leading to the same metric given by $A = 1$, $B = 1 - 2M/r + q^2/r^2$. For $M > |q|$ the function B has two zeros, leading to an outer and an inner horizon. Outside the outer horizon the structure of the $t =$ const. surfaces is as before. However, as $|q|$ tends to M the worm-hole develops a long "throat" with $r \approx M$. The limiting case $M = |q|$ represents the "extremal" RN black hole, whose horizon is degenerate, due to the double zero of B at $r = M$. There is no more wormhole, but an infinitely long throat. Also the $t =$ const. (space-like!) surfaces inside the horizon show this infinite throat as $r \to M$ (Fig. 4).

4 Gravitating Monopoles – BPS-Type Solutions

As already mentioned the flat BPS monopole plays a very special role in connection with supersymmetry. Amazingly it is possible to embed the flat solution into certain supergravity theories, satisfying the coupled field equations. The first such embedding is due to Gauntlett et.al. (Gauntlett et al. 1993). The relevant SUGRA theory is the $N = 4$ extended SUGRA coupled to $N = 4$ SUSY-YM, derived from the corresponding $N = 1$ theory in ten dimensions, which itself is a field theory limit of heterotic string theory. Besides the gravitational field the theory contains a dilaton φ and an axion $H_{\lambda\mu\nu}$. For the solution considered, the members of the SUGRA multiplet can be expressed through the YM potential W and the Higgs field h of the flat BPS solution as

$$e^{2\varphi} = \tfrac{2}{r^2}(1 - W^2 + 2h) \qquad g_{\mu\nu} = e^{2\varphi}\eta_{\mu\nu} \tag{14a}$$

$$H_{ij4} = 2\epsilon_{ijk}\tfrac{x^k}{r^4}h(1 - W^2) \qquad H_{ijk} = 0\,. \tag{14b}$$

As in flat space the solution solves first order Bogomolny equations and preserves one half of the supersymmetries.

Another embedding discovered more recently by Chamseddine and Volkov (Chamseddine and Volkov 1977) is even more surprising, since the model contains no Higgs field. The corresponding SUGRA is the $N = 4$ gauged supergravity (Friedman and Schwarz 1978), which may be obtained as a non-trivial Kaluza-Klein reduction from the $N = 1$ SUGRA in ten dimensions (related to type II strings). The YM field results from the non-trivial structure of the internal space $S^3 \times S^3$, on which the compactification is performed. The non-vanishing curvature of the internal space leads to a cosmological constant in four dimensions. After a suitable truncation the 4-dimensional (bosonic) action considered in (Chamseddine and Volkov 1977) is

$$S = -\int d^4x\sqrt{-g}\Big[\frac{R}{2} - \frac{1}{2}(\partial\varphi)^2 + \frac{1}{4g^2}e^{2\varphi}\mathrm{Tr}F^2 - \frac{1}{4}e^{-2\varphi}\Big]\,. \tag{15}$$

The gravitational the gravitational field and the dilaton can be expressed through the flat BPS solution

$$R^2 = 2h - W^2 + 1 \qquad \text{with} \quad R = \frac{r}{\sqrt{2}}e^{-\varphi} \tag{16a}$$

$$B = 1 + \frac{(R^2 + W^2 - 1)^2}{4R^2} \tag{16b}$$

$$A = \frac{r}{\rho} \tag{16c}$$

$$2e^{2\varphi} = A^2B\,, \tag{16d}$$

where the coordinate ρ is chosen such that $ds_2^2 = A^2B(dt^2 - d\rho^2)$. Again the solution satisfies a first order system of Bogomolny equations

$$\rho\frac{d}{d\rho}W = -\frac{W}{2\sqrt{B}}(R^2 + W^2 - 1)\,, \tag{17a}$$

$$\rho\frac{d}{d\rho}\varphi = \frac{1}{4\sqrt{B}}\Big(R^2 - \frac{(W^2-1)^2}{R^2}\Big)\,. \tag{17b}$$

The solution is not asymptotically flat due to the cosmological term in the action (not even asymptotically anti-deSitter) and it preserves 1/4 of the supersymmetries.

5 Gravitating Monopoles without SUSY

Let us now proceed to the case of self-gravitating magnetic monopoles without SUSY. Through gravity another mass scale M_{Pl} has entered and we can form two dimensionless ratios $\alpha = \sqrt{G}v = M_W/gM_{\mathrm{Pl}}$ and $\beta = M_W/M_H$. As long as $M_W << M_{\mathrm{Pl}}$, i.e., $\alpha << 1$ the influence of gravity is small and we expect to find a weakly self-gravitating version of the flat monopole (van Nieuwenhuizen et al. 1976). However, for values $\alpha \approx 1$ the situation changes. The size of the monopole is determined by $R_m = 1/M_W = \sqrt{G}/g\alpha$, whereas its Schwarzschild radius is given by $2GM_{mon} \approx GM_W/g^2 = \alpha\sqrt{G}/g$. For $\alpha \to 1$ both radii approach each other and we expect the monopole to become gravitationally unstable, i.e., we expect regular monopoles to exist only up to some maximal value of α of order one. Since no exact solutions are known (besides the ones involving a dilaton discussed in the previous section) , we have to take recourse to numerical methods for their study.

Combining the flat space ansatz for the YM resp. Higgs field with the one for the static, spherically symmetric gravitational field (11) the reduced Einstein-YM-Higgs (EYMH) action can be expressed as (using Schwarzschild coordinates for simplicity)

$$S = \int drA\Big[\frac{1}{2}(rB' + B - 1) - r^2BV_1 - V_2\Big] , \tag{18}$$

with

$$V_1 = \frac{(W')^2}{r^2} + \frac{1}{2}(H')^2 , \tag{19}$$

and

$$V_2 = \frac{(1-W^2)^2}{2r^2} + W^2H^2 + \frac{\beta^2r^2}{8}(H^2-\alpha^2)^2 . \tag{20}$$

Through a suitable rescaling we have achieved that the action depends only on the dimensionless parameters α and β.

Upon variation we obtain the corresponding field equations

$$(ABW)' = AW\Big(\frac{W^2-1}{r^2} + H^2\Big) \tag{21a}$$

$$(ABr^2H)' = AH\Big(\frac{\beta^2}{2}r^2(H^2-\alpha^2) + 2W^2\Big) \tag{21b}$$

$$rB' = 1 - B - 2(r^2BV_1 + V_2) \tag{21c}$$

$$rA' = 2r^2V_1A . \tag{21d}$$

This system of ODE's has singular points at $r = 0, \infty$ and for $B = 0$. Gravitating monopoles are globally regular solutions of the this singular system. Although it is not too difficult to prove local existence of suitable families of regular solutions the question of global existence is a very difficult problem, still beyond reach (except in some simple cases, e.g., for $\beta = \infty$). Recall that

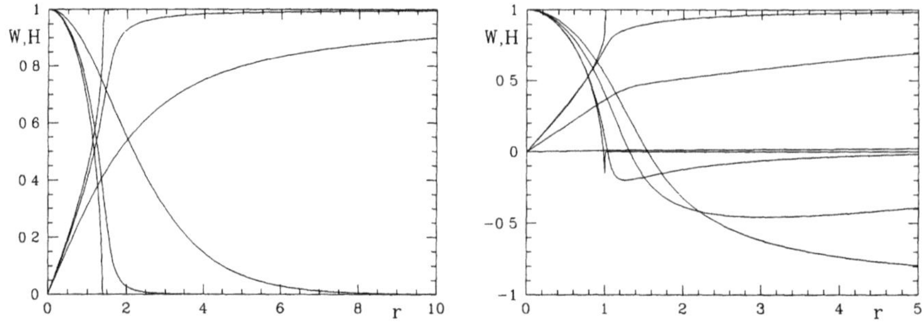

Fig. 5. W and H for ($\beta = 0$) a) the gravitating monopole solutions for $\alpha = 0.05, \alpha_{\rm max} = 1.403$ and $\alpha_c = 1.386$; b) first radial excitation for $\alpha = 0.01, 0.2, 0.5$ and 0.86.

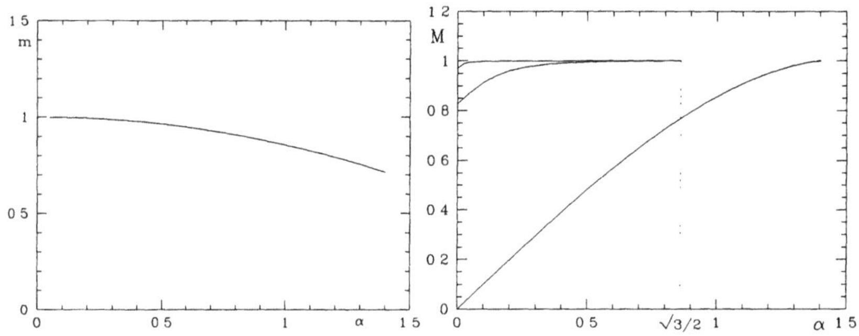

Fig. 6. a) Masses (in units of M_W) of fundamental monopole solutions versus α (for $\beta = 0$); b) Masses (now in units of $M_{\rm Pl}/g$) of fundamental monopole solutions and first and second radial excitations versus α (for $\beta = 0$);

the energy functional has no good functional analytic properties in this case.

Thus our knowledge about the solutions is based to a large extent on numerical computations. There are several methods available for that purpose, one sided ("Shooting and aiming") or two sided methods ("matching") (Breitenlohner et al. 1992, 1995) and for stable solutions also relaxation methods may be applied.

The numerical analysis (Ortiz 1992, Lee et al. 1992, Breitenlohner et al. 1992, 1995) revealed that there are self-gravitating versions of the flat-space non-abelian monopoles for values of α ranging from zero up to some maximal value $\alpha_{\rm max}(\beta)$, fully in accordance with our expectations (Fig. 5a). As α increases the solutions develop a typical limiting behaviour, which may be characterized as "gravitational confinement" of the monopole. As the function B develops a double zero at the finite value $r_l = 1$ (measured in units of $1/M_{\rm Pl}$), the spatial hyper-surface t = const. develops an infinite throat separating an interior region with a smooth origin and non-trivial YM field from an exterior solution with $W \equiv 0$, which is nothing but an extremal RN black hole. Thus there is a "confined" interior essentially non-abelian region and an outer abelian monopole. All this is much like the $t =$ const. surfaces of the extremal RN solution, with the only difference, that the interior part of the throat is in no sense the continuation of the exterior one. In fact the metric function A blows up along the throat coming from the interior, whereas $A \equiv 1$ for the RN solution. Hence the interior solution has no extremal horizon at $r = r_l$, instead it represents a geodesically complete, asymptotically AdS "cosmological" kind of solution.

Actually, what was just said is only true for not too large values of the parameter β (roughly $\beta < 10$), i.e., for not too large Higgs mass. For larger values of β it is still true that $B(r)$ develops a double zero at some finite value $r_l < 1$, thus there is the infinite throat again, but now $W(r)$ does not tend to zero there and $A(r)$ remains bounded. This means that the limiting solution, obtained for some maximal value $\alpha_{\max}(\beta)$, represents an extremal black hole with "non-abelian hair" (E. Weinberg, private communication). Due to the large difference between the mass scales for W and H it seems that this solution can be obtained numerically only with the use of a suitable relaxation method. Amazingly this extremal black hole is completely regular inside its horizon with a regular origin. At the horizon $W(r)$ and $H(r)$ are continuous, but not C^∞, due to some power behaviour of the type $(r - r_h)^{p_i}$ with some real exponents $p_i > 1$.

The observation, that the onset of a gravitational instability as the parameter α becomes too large, manifests itself within the family of static solutions in the formation of an extremal black hole (as far as the outer part of the solution is concerned) seems to be rather general. A similar phenomenon was observed for rigidly rotating dust discs by Neugebauer and Meinel (Meinel 1997). In their case the exterior part of the solution tends to the extremal Kerr solution, whereas the interior part is again some kind of a "cosmological" solution. In contrast to the flat space monopoles gravitating monopoles

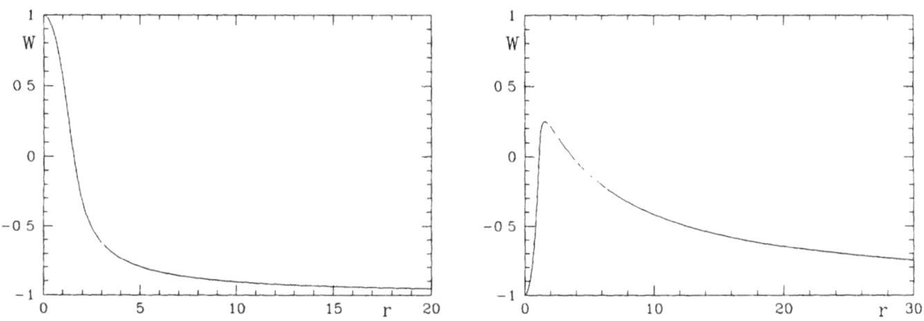

Fig. 7. W for the first two Bartnik-McKinnon solutions

also allow for radial excitations (compare Fig. 5b). As seen from Fig. 6b their mass stays finite (in units of M_{Pl}) as α tends to zero. At this point it is important to observe, that the limit $\alpha \to 0$ can be achieved in two different ways:

i) $G \to 0$, M_W fixed, in which the gravitational field decouples (flat space);
ii) $v = M_W/g \to 0$, G fixed, in which the Higgs field becomes trivial and can be ignored.

Whereas the fundamental monopole tends to its flat version as $\alpha \to 0$, the excited ones have no flat limit, instead tend to a class of solutions without a Higgs field, discovered by Bartnik and McKinnon (Bartnik and McKinnon 1988). There is a countably infinite number of these BM solutions distinguished by the number of zeros of the YM potential W. Their mass is of the order of M_{Pl}, the only scale of the EYM theory. As $r \to \infty$ the function $W(r)$ tends to ± 1 (compare Fig. 7), thus they carry no magnetic charge. In fact, they may be understood as some kind of gravitationally bound sphalerons (Volkov and Gal'tsov 1991; Sudarsky and Wald 1992), in particular as they turn out to be unstable (Straumann and Zhou 1990; Boschung et al. 1994; Volkov 1995). In addition to the "topological" instability of the flat YMH sphalerons however the gravitational BM sphalerons show additional "gravitational" instabilities within the minimal ansatz (Lavrelashvili and Maison 1995).

Turning back to the radially excited monopoles, it appears quite natural (at least for small values of α) to consider them as a Planck scale BM sphaleron sitting inside a $1/M_W$ size monopole. All these radial excitations disappear at the same value of $\alpha = \sqrt{3}/2$ merging in the by now well-known manner with the extremal RN black hole.

A few remarks should be made here about the stability properties of the gravitating monopoles. I shall discuss here only stability against infinitesimal, spherically symmetric perturbations. In view of the time-independence of the solutions this amounts to analyzing the spectrum of perturbations with harmonic time-dependence obeying suitable boundary conditions. Imaginary frequencies correspond to unstable modes of the solution. As to be expected the branch of gravitating monopoles connected to the flat space solution is stable up to α_{max}. All the excited regular monopoles turn out to be unstable (Hollmann 1994).

6 Coloured Black Holes

Apart from the solutions with Minkowskian space-time topology there are non-abelian, "coloured" black holes, parametrized by their radius r_h (the value of r at the event horizon) in addition to α and β (Breitenlohner et al. 1992, 1995, Lee et al. 1992). For $r_h << 1/M_W$ these non-abelian black holes may be interpreted as a tiny Schwarzschild black hole sitting inside a

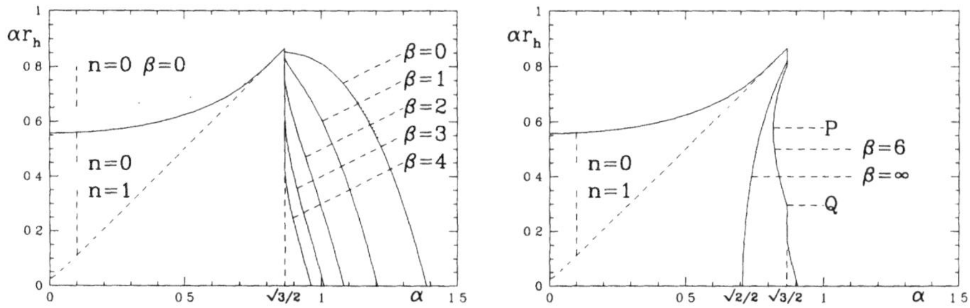

Fig. 8. Domains of existence for non-abelian black holes: a) for $\beta = 0$, 1, 2, 3, and 4; b) for $\beta = 6$ and ∞

monopole (Kastor and Traschen 1992). On the other hand, when r_h becomes bigger than $\approx 1/M_W$ this type of solution disappears and only the abelian RN black holes exist. For $r_h \to 0$ the matter fields tend uniformly to those of the globally regular solutions, whereas for the metrical functions this limit is clearly more delicate.

Detailed numerical analysis reveals that non-abelian black holes exist only in a limited domain of the α-r_h-plane, whose shape undergoes some characteristic changes as β varies from 0 to ∞. Fig. 8 shows some of these domains. Observe that we use αr_h instead of r_h as the abscissa - equivalent to expressing r_h in units of $1/M_W$ - in order to obtain domains remaining bounded for $\alpha \to 0$.

In the following I shall discuss in some more detail the structure of these "Phase Diagrams" and the phenomena happening at their boundaries. Let me start with the case $\beta = 0$. It is appropriate to subdivide the relevant sector $\alpha \geq 0$, $r_h \geq 0$ into the four subregions I-IV (compare Fig. 9).

In regions I and II we find coloured black holes. Above the diagonal, i.e., in regions II and III we have the abelian RN black holes, the extremal RN black holes sitting on the diagonal. Below the diagonal the RN solution has a naked singularity and does not represent a black hole. No black holes, neither abelian nor non-abelian, could be found in region IV. Region I may be subdivided in region Ia, where only the black hole version of the fundamental monopole resides and region Ib, where in addition their radial excitations are found. Thus region Ia contains only one black hole solution ** for given values of α and r_h, whereas in region Ib countably many solutions exist for any given α and r_h.

In region Ib fundamental and radially excited solutions coexist, whereas in region II even abelian and non-abelian black holes coexist. This establishes an obvious violation of the so-called "No-Hair" Conjecture. According to the latter static black holes of a given mass (or size, i.e., given value of r_h) should

** This is not strictly true for small values of β, where two solutions exist in a small interval $\alpha_c(r_h) < \alpha < \alpha_{\max}(r_h)$.

be uniquely determined through their "gauge charges" - their magnetic charge in the present case. However, all these black holes carry the same magnetic charge. Although black holes of the same size differ in general in their mass, degeneracy in mass also occurs. In some cases abelian and non-abelian black holes are even degenerate in mass and size.

As β increases from 0 to $\beta = 4$ the structure of the "Phase Diagram" remains essentially the same, the right boundary curve moving in to the left. However, for $\beta > 4$ this boundary curve develops a second, concave branch (compare Fig. 8b) determined by another mechanism – the formation of a degenerate inner (above P) resp. outer (between P and Q) horizon, leading to extremal black holes with non-abelian hair.

The boundary curve above the diagonal is essentially characterized by the bifurcation of the non-abelian with the abelian RN solution. For a given value of α this happens at some value $r_{c,n}(\alpha)$ depending on the number n of zeros of W (Fig. 8). Approaching this value from below the value W_h of W at the horizon tends to zero, thus abelian and non-abelian black holes merge.

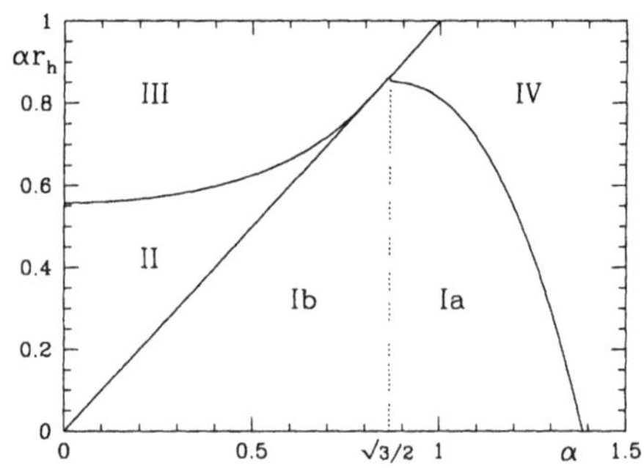

Fig. 9. Domains of existence for abelian and non-abelian black holes, $\beta = 0$

Finally again a remark concerning the stability of the solutions. Similar to the situation with regular monopoles the fundamental coloured black hole solutions are stable, likewise the radially excited ones are unstable. It is, however, interesting to observe that the abelian RN black hole is unstable in the framework of the non-abelian theory for α smaller than some value $\alpha(r_h) < \sqrt{3}/2$ (Bizon and Wald 1991; Lee et al. 1992a, Breitenlohner et al. 1992, 1995). In particular, the extremal RN solution is unstable for $\alpha \leq \sqrt{3}/2$ and stable above this value. At the limiting value $\alpha = \sqrt{3}/2$ the extremal RN solution bifurcates with infinitely many non-abelian solutions and in fact develops infinitely many unstable modes.

7 Additional Remarks

The interpretation of the BM sphalerons as gravitationally bound sphalerons is supported by the fact that similar solutions have been found for a theory where the gravitational field is replaced by a dilaton, serving the same purpose (Lavrelashvili and Maison 1992, 1993; Bizon 1993; Donets and Galtsov 1993). There is also an investigation of magnetic monopoles coupled to gravity and a dilaton (without SUSY) (Forgacs and Gyürüsi 1996). Gravitating monopoles resp.sphalerons for higher gauge groups (*SU(3)* etc.) were studied in (Kleihaus et al. 1995, 1998) with similar results.

Furthermore axially symmetric, static generalizations of the BM solutions were constructed numerically (Kleihaus and Kunz 1997). Similar solutions generalizing multiply charged axially symmetric flat monopoles are expected also with gravity. However more interesting is the question, if there are stationary rotating solutions. It seems that only the neutral BM solutions can rotate (Volkov and Straumann 1997; Brodbeck et al. 1997), whereas rotating magnetic monopoles are excluded (even in flat space) (Brodbeck and Heusler 1997; Heusler et al. 1998).

References

Bartnik, R., McKinnon, J., (1988): Phys. Rev. Lett. **61**, 141

Bizon, P., Wald, R.M. (1991): Phys. Lett. **B 267**, 173;
Lee, K., Nair, V.P., and Weinberg, J. (1992): Phys. Rev. Lett. **68**, 1100

Breitenlohner, P., Forgács, P., Maison, D. (1992): Nucl. Phys. **B 383**, 357;
Nucl. Phys. **B 442**, (1995) 126

Brodbeck, O., Heusler, M. (1997): gr-qc/9706064;
Heusler, M., Straumann,N. and Volkov, M. (1998): gr-qc/9805061

Chamseddine, A.H., Volkov, M.S. (1997): Phys. Rev. Lett. **79**, 3343;
preprint hep-th/9711181

Chruściel, P.T. (1994): preprint, gr-qc/9402032;
Bizon, P. (1993): Acta Phys.Polon. **B24**, 1209

Coleman, S. (1975): *New Phenomena in Subnuclear Physics*, ed. A. Zichichi, Plenum, New York;
Deser, S. (1976): Phys. Lett. **B 64**, 463

Forgacs, P., Gyürüsi, J. (1996): Phys. Lett. **B 366**, 205;
preprint hep-th/9808010

Friedman, D.Z., Schwarz, J.H. (1978): Nucl. Phys. **B 137**, 333

Gauntlett, J.P., Harvey, J.A., Liu, J.T. (1993): Nucl. Phys. **B 409**, 363

Hawking, S.W., Ellis, G.F.R. (1973): *The large scale structure of space-time*, Cambridge University Press

Hollmann, H. (1994): Phys. Lett. **B 338**, 181

't Hooft, G. (1974): NPB 79, 276;
Polyakov, A.M. (1974): JETP Lett. **20**, 194

Israel, W. (1967): Phys. Rev. **164**, 1776;
Commun. Math. Phys. **8**, (1968) 245

Kastor, D., Traschen, J. (1992): Phys. Rev. **D 46**, 5399
Kleihaus, B., Kunz,J. (1997): PRL 79, 1595
Kleihaus, B., Kunz, J., Sood, A. (1995): PLB 354, 240; PLB 418, (1998) 284
Lavrelashvili, G., Maison, D. (1992): Phys. Lett. **B 295**, 67; Nucl. Phys. **B 410**, (1993) 407; Bizon, P. (1993): Phys. Rev. **D 47**, 1656; Donets, E.E. and Galtsov, D.V. (1993): PLB 302, 411
Lavrelashvili, G., Maison, D. (1995): Phys. Lett. **B 343**, 214
Lee, K.-Y., Nair, V.P., Weinberg, E J. (1992): Phys. Rev. **D 45**, 2751
Manton, N.S. (1983): Phys. Rev. **D 28**, 2019
Meinel, R. (1997): in *Proceedings of the Second Mexican School on Gravitation and Mathematical Physics*, ed. Garcia, A., Lämmerzahl, C., Macias, A., Matos, T. and Nuñez, D., Sciencia Verlag Konstanz; gr-qc/9703077
van Nieuwenhuizen, P., Wilkinson, D., Perry, J. (1976): Phys. Rev. **D 13**, 778
Ortiz, M.E. (1992): Phys. Rev. **D 45**, R2586
Straumann N., Zhou Z.-H. (1990): Phys. Lett. **B 237**, 353; Boschung, P., Brodbeck, O., Moser, F., Straumann, N., Volkov, M. (1994): Phys. Rev. **D 50**, 3842; Volkov, M.S., Brodbeck, O., Lavrelashvili, G., Straumann, N. (1995): Phys. Lett. **B 349**, 438
Volkov, M.S., Gal'tsov, D.V. (1991): Phys. Lett. **B 273**, 273; Sudarsky, D and Wald, R.M. (1992): Phys. Rev. **D 46**, 1453
Volkov, M.S., Straumann, N. (1997): PRL 79, 1428; Brodbeck, O., Heusler, M., Straumann,N. and Volkov, M. (1997): PRL 79, 4310
Witten, E., and Olive, D. (1978): Phys. Lett. **B 78**, 97

Critical Behavior of Classical Spin Models and Local Cohomology

A. Patrascioiu[1] and E. Seiler[2]

[1] Physics Department, University of Arizona
Tucson, AZ 85721, U.S.A.
[2] Max-Planck-Institut für Physik
(Werner-Heisenberg-Institut)
Föhringer Ring 6, 80805 Munich, Germany

Dedicated to Jan Łopuszański

Abstract. Using reflection positivity as the main tool, we establish a connection between the existence of a critical point in classical spin models and the triviality of a certain local cohomology class related to the Noether current of the model in the continuum limit. Furthermore we find a relation between the location of the critical point and the momentum space autocorrelation function of the Noether current.

1 Introduction: Lattice and Continuum

It is well known that one possible approach to the construction of a Quantum Field Theory (QFT) goes by way of taking the continuum limit of a system of Classical Statistical Mechanics on a lattice, such as the Ising model, the classical Heisenberg model or more generally a classical spin model. Taking the continuum limit means in this context that one has to drive the system to a critical point, that is a point at which the dynamically produced scale(s) become infinite in terms of the lattice spacing; the continuum limit is then obtained by an infinite rescaling of the lattice model (see below; a rather detailed discussion of how this is done is contained in (Patrascioiu and Seiler 1997)). A bonus of this construction is that the continuum limit inherits certain properties of the lattice model, such as Reflection Positivity (RP) which leads to positivity of the state space metric and the spectrum condition of the QFT.

More precisely we have to distinguish between two kinds of continuum limits:

- The massive continuum limit: one chooses the dynamically generated length (correlation length) ξ of the system as the standard of length, considers the system at length scales that are fixed multiples of that standard, and sends $\xi \to \infty$ by driving the system to criticality.
- The massless continuum limit: the lattice system is put right on a critical point; one then chooses an arbitrary length scale that becomes infinite in lattice units and rescales the system accordingly.

The first option will produce a (Euclidean) QFT with unit mass, the second one a massless QFT, which according to standard lore will also be conformally invariant. In $2D$ it is believed that this allows to classify the critical behaviors according to the well-studied (rational) Conformal QFTs (Belavin et al. 1984, Friedan et al. 1984).

In this talk we want to discuss this connection, and actually close some gaps. In the course of the argument it turns out that one has to prove the triviality of a certain 'local cohomology class' related to the Noether current. This is possible with the use of lattice Ward identities (WI) and RP.

The same ingredients lead at the same time to an interesting and maybe unexpected relation between the location of the critical point of the lattice model and the Noether current 2-point function of the continuum model. This leads to a new criterion that allows to discriminate between the 'conventional wisdom' about nonabelian spin models in $2D$, which posits that they do not become critical at any temperature and the scenario long advocated by us (Patrascioiu and Seiler 1993, 1995) that they do have a transition to a massless spin wave phase, just as the plane rotator model.

2 Local Cohomology

It has been noted long ago (Strocchi 1967, Pohlmeyer 1972, Roberts 1978) that the imposition of locality (local commutativity, Einstain causality) may make the cohomology of Minkowski space nontrivial.

The problem of local cohomolgy may be stated as follows: assume that an antisymmetric tensor field $\Phi_{\mu_1,\ldots,\mu_k}(x)$ is given, which satisfies Wightman's axioms and is closed, i.e., satisfies

$$d\Phi \equiv d(\sum \Phi_{\mu_1,\ldots,\mu_k} dx^{\mu_1}...dx^{\mu_k}) = 0 \tag{1}$$

(in the notation of alternating differential forms).

The question is then under which conditions the field Φ is exact, i.e., there exists a local antisymmetric tensor field Ψ such that $\Phi = d\Psi$.

There are some well-known examples where the answer is 'no', even though Minkowski space is topologically trivial:

(1) the free Maxwell field F in dimension $D \geq 2$ (Strocchi 1967);

(2) the gradient of the massless free scalar field ϕ in $2D$, because the field ϕ does not exist as a local (Wightman) field.

There is also a simple $2D$ example on which we hit in our analysis of $2D$ classical spin models: let

$$\Phi = \phi_c dx^1 dx^2 , \tag{2}$$

where ϕ_c has the Euclidean two-point function

$$\langle \phi_c(0)\phi_c(x)\rangle = \frac{1}{(x^2)^2} . \tag{3}$$

Then Φ is trivially closed in $2D$, but it is not exact, i.e., there is no local vector field j_μ such that

$$\phi_c = \epsilon_{\mu\nu}\partial_\mu j_\nu \,. \tag{4}$$

This example can be made more explicit by requiring ϕ_c to be a generalized free, i.e., Gaussian field, with its two-point function given by (3). If we solve the differential equations that the two-point function of j_μ has to fulfill in order to satisfy (4) and impose euclidean covariance, we find that there is no scale invariant solution. The covariant solutions are

$$G_{\mu\nu}(x) = -\delta_{\mu\nu}\frac{\ln x^2 + \lambda}{8x^2} + x_\mu x_\nu \frac{\ln x^2 + 1 + \lambda}{4x^2} \,. \tag{5}$$

This is not the two point function of a local vector field, continued to euclidean times: it violates the so-called reflection positivity (Osterwalder and Schrader 1973, 1975), because the logarithm changes sign. Similarly it also does not obey the positivity required for a random field.

3 What Is the Massles Continuum Limit of a Critical Classical Spin Model?

There is an old argument (Affleck 1985) that a classical spin model with a continuous symmetry group G will have a massless contiuuum limit that has an enhanced $G \times G$ symmetry; this is supposed to come about due to the splitting of the model into two independent 'chiral' theories. Affleck (Affleck 1985) gave this argument in the the framework of Quantum Field Theory in Minkowski space, but it can be easily rephrased in the euclidean setting. In (Patrascioiu and Seiler 1998a, Patrascioiu and Seiler 1998b) we pointed out two possible gaps in those arguments coming from hidden assumptions whose validity ahs to be checked. But in those papers we also showed that these gaps can be closed, using properties like reflection positivity.

The core of the euclidean version of Affleck's argument is the following: assume that we have a scale invariant continuum theory with a conserved current $j_\mu(x)$. Euclidean covariance requires that the two-point function $G_{\mu\nu}$ of j_μ is of the form

$$G_{\mu\nu} \equiv \langle j_\mu(0) j_\nu(x)\rangle = \delta_{\mu\nu}\frac{b}{x^2} + \frac{a x_\mu x_\nu}{(x^2)^2} \quad (x \neq 0) \,. \tag{6}$$

Imposing current conservation means

$$\partial_\mu G_{\mu\nu} = 0 \,, \tag{7}$$

for $x \neq 0$, which implies

$$a = -2b \,, \tag{8}$$

$$G_{\mu\nu}(x) = b\left(\frac{\delta_{\mu\nu}}{x^2} - \frac{2x_\mu x_\nu}{(x^2)^2}\right) \,. \tag{9}$$

This is, up to the factor b, equal to the two point function of $\partial_\mu\phi$ where ϕ is the massless free scalar field (it is irrelevant here that the massless scalar field does not exist as a Wightman field). If we look at the two-point function of the dual current $\epsilon_{\mu\nu}j_\nu$, it turns out to be

$$\tilde{G}_{\mu\nu} \equiv \epsilon_{\mu\lambda}\epsilon_{\rho\nu}G_{\lambda\rho} = -G_{\mu\nu}\,, \tag{10}$$

so the dual current two point function satisfies automatically the conservation law. Conservation of the two currents j and $\tilde{j}$ is equivalent to conservation of the two chiral currents $j_\pm = j_0 \pm j_1$ in Minkowski space.

By general properties of local quantum field theory (Reeh-Schlieder theorem, see (Reeh and Schlieder 1961, Streater and Wightman 1978)) it follows that the dual current is conserved as a quantum field. So the two conservation laws together imply that

$$j_\mu = \sqrt{b}\partial_\mu\phi\,, \tag{11}$$

where ϕ is the massless scalar free field, and also that

$$j_\mu = \sqrt{b}\epsilon_{\mu\nu}\partial_\nu\psi\,, \tag{12}$$

where ψ is another 'copy' of the massless scalar free field.

As presented, this argument is certainly correct. But it depends on the assumption that the Noether currents *exist* as Wightman fields, and this assumption is in fact nontrivial and could a priori fail in the critical spin models. A simple example of a Quantum Field Theory with a continuous symmetry in which the Noether current does not exist as a Wightman field is given by the two-component free field in $2D$ in the massless limit. It is simply given by a pair of independent Gaussian fields $\Phi^{(1)}, \Phi^{(2)}$, both with covariance

$$C(x) = \frac{1}{(2\pi)^2}\int d^2p \frac{e^{ipx}}{p^2+m^2}\,, \tag{13}$$

where we are interested in the limit $m \to 0$. This system has a global $O(2)$ invariance rotating the two fields into each other. It is well known that the massless limit only makes sense for functions of the gradients of the fields. But the Noether current of the $O(2)$ symmetry is given by

$$j_\mu(x) = \Phi^{(1)}(x)\partial_\mu\Phi^{(2)}(x) - \Phi^{(2)}(x)\partial_\mu\Phi^{(1)}(x) \tag{14}$$

and it cannot be written as a function of the gradients. It is also easy to see directly that its correlation functions do not have a limit as $m \to 0$ (see (Patrascioiu and Seiler 1998b)). The Noether current itself makes sense as a quantum field only if it is smeared with test functions f_μ satisfying

$$\int d^2x f_\mu(x) = 0\,. \tag{15}$$

On the other hand, it is not hard to see that $\phi_c(x) = curl(j)$ can be written as a function of the gradients:

$$\phi_c(x) = 2\big((\partial_2\Phi^{(1)}(x))(\partial_1\Phi^{(2)}(x)) - (\partial_1\Phi^{(1)}(x))(\partial_2\Phi^{(2)}(x))\big) \qquad (16)$$

and its two-point function is of the form

$$\langle\phi_c(0)\phi_c(x)\rangle \propto \frac{1}{(x^2)^2}\,. \qquad (17)$$

In other words, in this model we have found exactly the nontrivial local cohomology class described in the previous section. The problem in the massless contiuum limits of classical spin models is then the following: it is conceivable that both *curl* j and *div* j have bona fide continuum limits, but the current itself does not. In other words, it could happen that there is a nontrivial second 'local cohomology class' just as in the example discussed above. But it turns out that reflection positivity can be used to rule out such a possibility, provided we are dealing with a model that becomes critical at a finite value of the inverse temperature β (this is, however, a prerequisite for constructing a massless continuum limit anyway). Our arguments show that both *curl* j and *div* j have correlations that are pure contact terms in the continuum limit; this means that in Minkowski space both the current and its dual are conserved, thereby justifying Affleck's claim.

For completeness, let us mention an even more exotic possible way in which the conformal classification of the critical behavior of the classical spin models could fail: one could be imagine that the current itself has correlations that are pure contact terms in the continuum, which would mean that the Noether current simply vanishes as a quantum field. Of course this would also imply vanishing of the corresponding charge, and since the commutator of the charge with the (renormalized) spin field should be a component of the (renormalized) spin field, those fields themselves would have to vanish, leading to a totally trivial theory containing only the vacuum. There is a huge body of numerical results that makes this inconceivable, and we also did some numerical simulations to eliminate this possibility directly in the case of the $O(2)$ model (Patrascioiu and Seiler 1998a, Patrascioiu and Seiler 1998b).

4 The Noether Current: Some Generalities

The $O(N)$ model is determined by its standard Hamiltonian (action)

$$H = -\sum_{\langle ij\rangle} s(i)\cdot s(j)\,, \qquad (18)$$

where the sum is over nearest neighbour pairs on a square lattice and the spins $s(.)$ are unit vectors in $\mathbb{R}^N$. As usual Gibbs states are defined by using the Boltzmann factor $\exp(-\beta H)$ together with the standard a priori measure on the spins first in a finite volume, and then taking the thermodynamic limit.

It is rigorously known (Froehlich and Spencer 1981) that for $N = 2$ the model has a transition to a massless spin wave phase at a certain $\beta = \beta_{KT} \approx 1.12$, the so-called Kosterlitz-Thouless transition (Kosterlitz and Thouless 1975). This transition separates a high temperature phase with exponential clustering from a low temperature one with only algebraic decay of correlations. For $N > 2$ the standard wisdom is that there is no such transition and the model does not become critical at any finite β, but is asymptotically free. For many years, however, we have been criticizing the arguments on which this standard wisdom is based and gave arguments for an alternative scenario according to which ALL the $O(N)$ models have a transition to a spin wave phase (Patrascioiu and Seiler 1993, 1995).

Here we do not want to enter into this discussion, but we will produce a criterion that distinguishes between these two scenarios.

But at first let us assume that our model has a finite critical point and study the consequences. We are in particular interested in the correlations of the Noether currents, given by

$$j_\mu^{ab}(i) = \beta\Big(s^a(i)s^b(i+\hat{\mu}) - s^a(i)s^b(i+\hat{\mu})\Big)\,. \tag{19}$$

Typically we will restrict ourselves to the case $a = 1, b = 2$ and omit the flavor indices on the current.

On a torus the current can be decomposed into 3 pieces, a longitudinal, a transverse and a constant (harmonic) piece. This decomposition is easiest in momentum space, and effected by the projections

$$P^T_{\mu\nu} = \left(\delta_{\mu\nu} - \frac{(e^{ip_\mu}-1)(e^{-ip_\nu}-1)}{\sum_\alpha(2-2\cos p_\alpha)}\right)(1-\delta_{p0})\,, \tag{20}$$

$$P^L_{\mu\nu=} = \frac{(e^{ip_\mu}-1)(e^{-ip_\nu}-1)}{\sum_\alpha(2-2\cos p_\alpha)}(1-\delta_{p0}) \tag{21}$$

and

$$P^h_{\mu\nu} = \delta_{\mu\nu}\delta_{p0} \tag{22}$$

with $p_\mu = 2\pi n_\mu/L$, $n_\mu = 0, 1, 2, ..., L-1$.

In the following we will mostly discuss these correlations in momentum space. In particular we study the tranverse momentum space 2-point function

$$\hat{F}^T(p, L) \equiv \hat{G}(0, p; L) = \langle|\hat{j}_1(0,p)|^2\rangle \tag{23}$$

(for $p \neq 0$; the hat denotes the Fourier transform)
and the longitudinal two-point function

$$\hat{F}^L(p, L) \equiv \hat{G}(p, 0; L) = \langle|\hat{j}_1(p,0)|^2\rangle \tag{24}$$

(for $p \neq 0$).

Because the current is conserved, its divergence in the Euclidean world should be a pure contact term, and for dimensional reasons the two-point function should be proportional to a δ function, i.e.,

$$\hat{F}^L(p,L) = const. \tag{25}$$

The constant is in fact determined by a Ward identity in terms of $E = \langle s(0)\cdot s(\hat{\mu})\rangle$: consider (for a suitable finite volume) the partition function

$$Z = \int \prod_i ds(i) \prod_{\langle ij\rangle} \exp(\beta s(i)\cdot s(j))\,, \tag{26}$$

where ds denotes the standard invariant measure on the $(N-1)$-sphere. Replacing under the integral $s(i)$ by $\exp(\alpha L_{12})$, where L_{12} is an infinitesimal rotation in the 12 plane, does not change the integral. So expanding in powers of α, all terms except the one of order α^0 vanish indentically in $\alpha(i)$. This leads in a well-known fashion to Ward identities expressing the conservation of the current. Looking specifically at the second order term in α and Fourier transforming, we obtain for all $p \neq 0$

$$\langle |j_1(p,0)|^2\rangle = \hat{F}^L(p,L) = \frac{2}{N}\beta E\,. \tag{27}$$

This is confirmed impressively by the Monte Carlo simulations (Patrascioiu and Seiler 1998b).

The thermodynamic limit is obtained by sending $L\to\infty$ for fixed $p = 2\pi n/L$, so that in the limit p becomes a continuous variable ranging over the interval $[-\pi,\pi)$. The $O(N)$ models do not show spontaneous symmetry breaking according to the Mermin-Wagner theorem, and presumably have a unique infinite volume limit at any temperature.

The massive continuum limit is contructed as follows: First one takes the thermodynamic limit of the the model in its high temperature phase. There is a dynamically generated length scale ξ, the correlation length regulating the exponential decay of the correlations. This is now taken as the standard of length, and the fields are rescaled accordingly. In particular the Noether current is rescaled as follows:

$$j_\mu^{ren}(x) = \xi j_\mu(i) \tag{28}$$

with $x = i/\xi$. After that, the system is driven to the critical point, where $\xi\to\infty$.

The massless continuum limit, on the other hand, is obtained as follows: we take the thermodynamic limit of the model right at its critical point. Since there is no dynamically generated scale, we take an arbitrary sequence l_n going to infinity as our standard of length. The currents are then rescaled as

$$j_\mu^{ren}(x) = l_n j_\mu(i) \tag{29}$$

with $x = i/l_n$ and we take the limit $n\to\infty$.

5 The Noether Current: Bounds and Inequalities

The Gibbs measure formed with the standard action on the periodic lattice has the property of reflection positivity (see for instance (Osterwalder and Seiler 1978)). Reflection positivity means that expectation values of the form

$$\langle A\theta(A)\rangle \tag{30}$$

are nonnegative, where A is an observable depending on the spins in the 'upper half' of the lattice ($\{x|x_1 > 0\}$, and $\theta(A)$ is the complex conjugate of the same function of the spins at the sites with x_1 replaced by $-x_1$. Applied to the current two-point functions this yields:

$$F^L(x_1, L) = \sum_{x_2} \langle j_1(x_1, x_2) j_1(0,0)\rangle \leq 0\,, \tag{31}$$

for $x_1 \neq 0$ and

$$F^T(x_1, L) = \sum_{x_2} \langle j_2(x_1, x_2) j_2(0,0)\rangle \geq 0\,, \tag{32}$$

for all x_1. From these two equations and (27) it follows directly that

$$0 \leq \hat{F}^T(p, L) \leq \hat{F}^T(0, L) = \hat{F}^L(0, L) \leq \hat{F}^L(p, L) = \frac{2}{N}\beta E\,. \tag{33}$$

These inequalities remain of course true in the thermodynamic limit, but we have to be careful with the order of the limits. If we define $\hat{F}^T(p, \infty)$ and $\hat{F}^L(p, \infty)$ as the Fourier transforms of $\lim_{L\to\infty} F^T(x, L)$ and $\lim_{L\to\infty} F^L(x, L)$, respectively, one conclusion can be drawn immediately:

Proposition: $\hat{F}^T(p, \infty)$ and $\hat{F}^L(p, \infty)$ are continuous functions of $p \in [-\pi, \pi)$.

The proof is straightforward, because due to the inequalities (32) (33) and (34) together with the finiteness of β_{crt} the limiting functions F^L and F^T in x-space are absolutely summable. But it is not assured that the limits $L \to \infty$ and $p \to 0$ can be interchanged, nor that the thermodynamic limit and Fourier transformation can be interchanged. On the contrary, by the numerics presented in (Patrascioiu and Seiler 1998a, Patrascioiu and Seiler 1998b), as well as finite size scaling arguments, it is suggested that

$$\lim_{p\to 0}\lim_{L\to\infty} \hat{F}^L(p, L) > \lim_{L\to\infty} \hat{F}^L(0, L) \tag{34}$$

and therefore also

$$\lim_{p\to 0}\lim_{L\to\infty} \hat{F}^L(p, L) > \lim_{p\to 0}\lim_{L\to\infty} \hat{F}^T(p, L)\,. \tag{35}$$

The fact that these are strict inequalities plays an important role in the justification of Affleck's claim, as will be seen below.

To continue, let us describe how the two types of continuum limit are taken in Fourier space, concretely for our functions $\hat{F}^T(p,\infty), \hat{F}^L(p,\infty)$.

The massive continuum limit means considering $\hat{F}^T(p)$ etc. for a sequence of ξ values diverging to ∞ as functions of $q \equiv p/m = p\xi$, i.e., taking

$$\lim_{\xi\to\infty} \hat{T}(q) \equiv \hat{F}^T(qm). \tag{36}$$

In this context it is important to note that the functions $\hat{F}^T(p)$ depend explicitly on β which is sent to β_{crt}, and through this on the correlation length ξ, which is sent to ∞.

The massless continuum limit on the other hand is obtained by going to the critical point and considering $\hat{F}^T(p)$ etc. as a function of $q \equiv p/l_n$, i.e., taking

$$\lim_{n\to\infty} \hat{T}(q) \equiv \hat{F}^T(q/l_n). \tag{37}$$

In this case we are always dealing with only one function $\hat{F}^T(p)$, not depending on n, because β is fixed to its critical value.

6 Consequences

For the massless continuum limit the inequalities (33) lead to an important consequence, which closes the main gap in Affleck's argument by showing the triviality of the second local cohomology class defined by the *curl* of the Noether current:

Proposition: In the massless continuum limit both $\hat{F}^L(p,\infty)$ and $\hat{F}^T(p,\infty)$ converge to constants for $p \neq 0$.

Corollary: The local cohomology class defined by $curl(j)$ is trivial.

Proof: Let $\hat{F}(p)$ be the Fourier transform of either $\hat{F}^T(p,\infty)$ or $\hat{F}^L(p,\infty)$. We consider $\hat{F}(p)$ as a distribution on $[-\pi,\pi)$. We extend $\hat{F}(p)$ to a periodic distribution on the whole real line. The continuuum limit of $F(n)$ (the corresponding function in x space) also has to be considered in the sense of distributions. If we change our standard of length to $l_M = M$, the lattice spacing will be $a = 1/M$, respectively. For an arbitrary test function f (infinitely differentiable and of compact support) on the real axis we then have to consider the limit $M \to \infty$ of

$$(F,f)_M \equiv \sum_n f(n/M)F(n). \tag{38}$$

We claim that the right hand side of this is equal to

$$\frac{1}{2\pi}\int_{\infty}^{\infty} dq\hat{F}(q/M)\hat{f}(q). \tag{39}$$

Proof: Insert in (38)

$$F(n) = \frac{1}{2\pi} \int_{-\pi}^{\pi} dp \hat{F}(p) e^{ipn} \tag{40}$$

and

$$f(n/M) = \frac{1}{2\pi} \int_{-\pi}^{\pi} dp \hat{f}(p) e^{ipn/M} \tag{41}$$

and use the identity

$$\sum_n e^{ipn+iqnb} = 2\pi \sum_r \delta(p+qb+2\pi r) . \tag{42}$$

This produces, after carrying out the trivial integral over q using the δ distribution,

$$\begin{aligned} &\frac{M}{2\pi} \int_{-\pi}^{\pi} dp \sum_{r=-\infty}^{\infty} \hat{F}(-p)\hat{f}((p+2\pi r)M) \\ &= \frac{1}{2\pi} \sum_{r=-\infty}^{\infty} \int_{-M\pi}^{M\pi} dq \hat{F}(-q/M)\hat{f}(q+2\pi Mr) . \end{aligned} \tag{43}$$

Finally, using the periodic extension of $\hat{F}(p)$, this becomes what is claimed in (39).

From (39) one sees that what is relevant for the continuum limit is the small momentum behavior of $\hat{F}(p)$. In particular, if $\lim_{p\to 0} \hat{F}(p) \equiv \hat{F}(0)$ exists, we obtain

$$\lim_{M\to\infty} (F, f)_M = \frac{1}{2\pi} \hat{F}(0) \int dq \hat{f}(q) = \frac{1}{2\pi} f(0) \hat{F}(0) \tag{44}$$

expressing the fact that in this case the limit of F is a pure contact term. This finishes the proof of the proposition.

In spite of this result, Affleck's claim could still fail in a different way if $\hat{F}^T(p,\infty)$ and $\hat{F}^L(p,\infty)$ converged to the same constant in the continuum limit. Let us denote the continuum limit of $\hat{F}^T(p,\infty)$ by g. Then the current-current correlation in this limit is

$$\langle j_\mu j_\nu \rangle\hat{}(p) = \beta E P^L_{\mu\nu} + g P^T_{\mu\nu} = g\delta_{\mu\nu} + (\beta E - g)\frac{p_\mu p_\nu}{p^2} . \tag{45}$$

So we see that if $g = \beta E$, the current-current correlation reduces to a pure contact term and vanishes in Minkowski space. Above we proved only that

$$g \le \beta E . \tag{46}$$

But if the current-current correlation were a pure contact term, it would be unavoidable to conclude that also the spin field becomes ultralocal. This can be seen as follows: if the current is ultralocal in the euclidean world, by the Osterwalder-Schrader reconstruction (Osterwalder and Schrader 1973, 1975) the current field operator in Minkowski space has to vanish, and so does the charge operator $Q_{12} = \int dx j_o(x,t)$. But if the charge operator generates a global $O(N)$ symmetry, it has to have the following commutation relation with the (renormalized, Minkowskian) spin field $s(x)$:

$$[Q_{12}, s_a(x)] = s_a(x), \quad a > 2\,, \tag{47}$$

$$[Q_{12}, s_1(x)] = s_2(x)\,, \tag{48}$$

$$[Q_{12}, s_2(x)] = -s_2(x)\,, \tag{49}$$

which would then imply that $s(x)$ vanishes identically. This argument is not fully rigorous, because it assumes (47) as well as the validity of the Osterwalder-Schrader axioms; both have not been proven in full rigor for the continuum limit of the $O(N)$ models. Also there is only numerical evidence, but no rigorous proof, that the continuum limit of the spin field is not ultralocal. For these reasons we presented in (Patrascioiu and Seiler 1998b) numerical data which (together with finite size scaling arguments) rule out directly ultralocality of the current.

Let us now turn to the massive continuum limit. For this the inequalities (33) yield the announced bound on the transition temperature in terms of the tranverse Noether current in momentum space:

Proposition: For the $O(N)$ models the critical inverse temperature satisfies

$$\beta_{crt} \geq \frac{N}{2} \sup_p \hat{F}^T(p)\,. \tag{50}$$

The quantity $J(p) = \hat{F}^T(p) - \hat{F}^T(0)$ satisfies

$$J(p) \leq \frac{2}{N}\beta_{crt}\,. \tag{51}$$

Proof: Both statements follow directly by taking first the thermodynamic and then the massive continuum limit of (33), using also the trivial fact $E \leq 1$.

This result is the announced criterion distinguishing between $\beta_{crt} < \infty$ and $\beta_{crt} = \infty$ by the boundedness or unboundedness of $J(p)$ or $\hat{F}^T(p)$. Of course it is a highly nontrivial matter to verify this criterion. Balog and Niedermaier (Balog and Niedermaier 1997) gave arguments that $J(p)$ is unbounded in their form factor construction of the $O(3)$ model, which seems to suggest $\beta_{crt} = \infty$. But we found by very precise numerical simulations evidence (Patrascioiu and Seiler 1998) that the form factor construction disagrees with the (massive) continuum limit of the lattice $O(3)$ model, leaving open the possibility that indeed $\beta_{crt} < \infty$ as long advocated by us.

Acknowledgment

A.P. is grateful to the Max-Planck-Institut for its hospitality and financial support; E.S. wishes to thank the University of Arizona for its hospitality and financial support.

References

Affleck, I., (1985): Phys. Rev. Lett. **55**, 1355
Balog, J., Niedermaier, M., (1997): Phys. Rev. Lett. **78**, 4151
Belavin, A., Polyakov, A., Zamolodchikov, A., (1984): Nucl. Phys. **B241**, 333
Friedan, D., Qiu, Z., Shenker, S., (1984): Phys. Rev. Lett. **52**, 1575
Froehlich, J., Spencer, T., (1981): Comm. Math. Phys **81**, 527
Kosterlitz, J.M., Thouless, D.J., (1975): J. Phys. (Paris) **32**, 581
Osterwalder, K., Schrader, R., (1973): Comm. Math. Phys. **31**, 83; Comm. Math. Phys. **42**, (1975) 281
Osterwalder, K., Seiler, E., (1978): Ann. Phys. (N.Y.), **110**, 440–471
Patrascioiu, A., Seiler, E., (1997): J. Stat. Phys. **89**, 947
Patrascioiu, A., Seiler, E., (1998): Phys. Lett. B **417**, 123
Patrascioiu, A., Seiler, E., (1998): Phys. Rev. E **57**, 111
Pohlmeyer, K., (1972): Comm. Math. Phys. **25**, 73
Patrascioiu, A., Seiler, E., (1993): Nucl. Phys. B (Proc. Suppl.) **30**, 184; Phys. Rev. Lett. **74**, (1995) 1920
Patrascioiu, A., Seiler, E., (1998): *Nonlinear σ-model, form factors and universality*, preprint MPI-PhT/98–40, AZPH–TH–98–07, hep–th/9809154
Reeh, H., Schlieder, S., (1961): Nuovo Cim. **22**, 1051
Roberts, J., (1978): *Mathematical Problems in Theoretical Physics* (G. Dell'Antonio, S. Doplicher and G. Jona–Lasinio, eds.), Springer-Verlag, Berlin
Streater, R.F., Wightman, A.S., (1978): *PCT, Spin and Statistics and All That*, Benjamin/Cummings, Reading, Mass.,
Strocchi, F., (1967): Phys. Rev **162**, 1429

Bose–Einstein Correlations in High Energy Multiple Particle Production Processes*

K. Zalewski

Jagellonian University
and
Institute of Nuclear Physics, Kraków, Poland

Abstract. Correlations among identical bosons, which are familiar from statistical physics, play an increasingly important role in high energy multiple particle production processes. They provide information about the region, where the particles are produced and, if Einstein's condensation can be reached, they can lead to spectacular new phenomena.

1 Introduction

In this paper we will consider Bose-Einstein correlations in high-energy particle production processes, i.e., the correlations among identical bosons in the final state, which follow from Bose-Einstein statistics. When hundreds of identical bosons are being produced in a single scattering act, as happens e.g., in heavy ion collisions at high energy, such correlations can lead to spectacular phenomena. They are also, most probably, the best way of getting information about the space-time structure of the region, where the final state particles are produced. Let us begin with a very simple example.

Consider the elastic scattering of two alpha particles with initial momenta equal in magnitude, opposite and parallel to a horizontal axis, say the x-axis. Suppose that the detectors register the final state particles if and only if the scattering is at 90° and one of the final particles goes up and hits the upper detector (U), while the other goes down and hits the lower detector (L). There are two possibilities. Either particle 1, say the particle coming from the left, hits detector U and particle 2 hits detector L, or particle 1 hits detector L and particle 2 hits detector U. Let us denote the probability amplitudes for these two processes by A and B respectively. Since a rotation around the x-axis can convert these two processes into each other, $|A| = |B|$. Since the alpha particles are identical bosons and an exchange of the two final state particles converts the two processes into each other, $A = B$. The detection probability is $|A+B|^2 = 4|A|^2$. If the particles were distinguishable, the probability would be $|A|^2+|B|^2 = 2|A|^2$. Thus, the fact that the particles are indistinguishable increases the probability by a factor of two.

* Supported in part by the KBN grant 2P03B 086 14

Let us make a few comments about this simple result.

- In the example the two amplitudes interfere constructively, because they are coherent. This is sometimes called first order interference. We will see in the following that the Bose-Einstein correlations of interest for us are due to the incoherence of the production process, and are a manifestation of the so called second order interference.
- The statement that the scattering probability for identical particles is twice the corresponding probability for distinguishable particles is not possible to check experimentally, because non identical alpha particles are not available. The best one can do is to compare the experimental result for the identical alpha particles with the calculation for the non identical ones. In the present example, where the calculation is simple and non controversial, this is not much of a problem, but in multiple production processes a calculation from first principles is not possible and the definition of the distribution for distinguishable particles, which should be modified by the Bose-Einstein correlations to yield the distribution which can be compared with experiment, is a difficulty.
- The final state can be represented by the density operator

$$\hat{\rho} = \frac{1}{2}|U_1L_2 + U_2L_1\rangle\langle U_1L_2 + U_2L_1| \,, \tag{1}$$

where U_iL_k is the state, where particle i is registered by the detector U and particle k by detector L. Expanding the left-had-side one obtains four terms, if, however, the density operator is to be used only for calculating averages of operators symmetric with respect to exchanges of the identical particles, which is sufficient for all practical applications, one can use the simpler form

$$\hat{\rho} = |U_1L_2\rangle\langle U_1L_2| + |U_1L_2\rangle\langle U_2L_1| \,. \tag{2}$$

It is useful to rewrite this formula in the form

$$\hat{\rho} = \sum_P |U_1L_2\rangle\langle U_{P1}L_{P2}| \,, \tag{3}$$

where the summation is over all the permutations P of the indices 1 and 2 and Pi denotes the index obtained from index i under permutation P. For our simple example this formula is ridiculously complicated, for more difficult cases, however, its analogues are very convenient.

2 HBT Contribution

An interesting application of the Bose-Einstein interference to find the sizes of the emitting objects was discovered in the fifties by two astronomers, R. Hanbury Brown and R.Q. Twiss (Hanbury Brown and Twiss 1956). By studying the second order Bose- Einstein interference of photons, they were able to measure the radii of some stars. The idea may seem obvious today, but it was not so at the time it was put forward. In the seventies Hanbury Brown wrote (Hanbury Brown and Twiss 1974) (quoted after (Silverman 1995))

> *Now to a surprising number of people this idea seemed not only heretical but patently absurd and they told us so in person, by letter, in publications, and by actually doing experiments which claimed to show that we were wrong. At the most basic level they asked how, if photons are emitted at random in a thermal source can they appear in pairs at the two detectors. At a more sophisticated level the enraged physicist would brandish some sacred text, usually by Heitler, and point out that ... our analysis was invalidated by the uncertainty relation ...*

Since the distances to many stars are known, their radii could be determined, if the opening angles between the light rays coming from the stars could be measured. These angles, however, are in most cases too small for a direct measurement. Hanbury Brown and Twiss suggested the following procedure. Consider two light rays coming from two points on the surface of the star – ray a from point a and ray b from point b. The problem is to measure the angle Θ between the two rays. Each of the rays falls on two photodetectors denoted 1 and 2. The distance between the photodetectors is d. Elementary trigonometry yields to first order in Θ the relation

$$\Theta = \frac{\Delta_a - \Delta_b}{d \sin \alpha}, \tag{4}$$

where Δ_i is the difference of distances between point i on the star and the two photodetectors, while α is the angle between the line connecting the two photodetectors and the direction of the two rays. Thus the problem of measuring the opening angle Θ reduces to the problem of measuring $\Delta_a - \Delta_b$. Of course in practice, in order to find the radius of the star a suitable averaging over the possible emission points is necessary, this, however, is rather simple and we shall not discuss it any further.

The current generated in the photodetector is proportional to the intensity of the incident light. Thus for counter 1 it is

$$i_{1u} = K_1 \left[E_a \sin(\omega_a t + \phi_a) + E_b \sin(\omega_b t + \phi_b)\right]^2 , \tag{5}$$

where ϕ_i, E_i, ω_i denote respectively the phase at the star surface, the amplitude and the frequency for ray i, and K_1 is a proportionality coefficient dependent on the working of the photodetector 1. For simplicity, the polarization effects have been ignored and the time necessary to reach photodetector 1 has been put equal t for both point a and point b. In the apparatus the current i_{1u} is further filtered so that only frequencies from 1 Hz to 100 Hz survive. Thus finally, the current from the first photodetector is

$$i_1 = K_1 E_a E_b \cos[(\omega_a - \omega_b)t + (\phi_a - \phi_b)]\,. \tag{6}$$

This current is zero on the average and does not look particularly interesting. The analysis for the second photodetector is similar except that the time necessary to reach the detector for the ray from point i on the star is increased by Δ_i/c . One obtains

$$i_2 = K_2 E_a E_b \cos\left[(\omega_a - \omega_b)t + \frac{\omega}{c}(\Delta_a - \Delta_b) + (\phi_a - \phi_b)\right]\,, \tag{7}$$

where $\omega \approx \omega_a \approx \omega_b$. This is again a rather uninteresting current, but the average of the product of the filtered currents from the two photodetectors

$$\langle i_1 i_2 \rangle = K_1 K_2 E_a E_b \cos\left[\frac{\omega}{c}(\Delta_b - \Delta_a)\right]\,, \tag{8}$$

which is measurable, yields $\Delta_a - \Delta_b$ and consequently the necessary opening angle Θ.

Note that the result is obtained in spite of the fact that the presence of the random phases ϕ_i means that light from a is incoherent with respect to light from b. Because of these phases the product of two amplitudes, one for the ray a and one for the ray b averages to zero, The product of four amplitudes, two from a and two from b, however, can survive. For this reason one calls this effect second order interference or intensity interferometry.

3 The GGLP Contribution

The first application of intensity interferometry in particle physics was made by the Goldhabers, Lee and Pais (Goldhaber et al. 1960). Their problem was somewhat different from that that in the HBT case. The interfering particles were like sign pion produced at two points within the interaction region of a hadron–hadron collision. The interference of interest was not between the measurements at two points in space, but between momentum measurements. Assuming that at the production space-time point x_k the pion wave function has phase ϕ_k and that the momentum of the pion p_k is well-defined, one expects at the registration point x an amplitude proportional to $\exp[ip_k(x_k - x) + i\phi_k]$. The probability of finding the two pions produced at points x_1 and x_2 with momenta p_1 and p_2, after proper symmetrization of the wave function, is proportional to

$$\frac{1}{2}\left|e^{i(p_1x_1+p_2x_2)} + e^{i(p_1x_2+p_2x_1)}\right|^2 = 1 + \cos[(p_1 - p_2)(x_1 - x_2)]\,. \tag{9}$$

We assume now that the production process is incoherent, so that the averaging over the times and positions x_1, x_2 should be made at the level of probabilities and not of amplitudes. Then the distribution of the difference in momenta should be approximately given by the formula

$$C(p_1 - p_2) = 1 + \langle \cos[(p_1 - p_2)(x_1 - x_2)] \rangle , \qquad (10)$$

where the averaging is over x_1 and x_2. Qualitatively, the result does not depend much on the actual prescription being used for the averaging. For $p_1 \approx p_2$ the argument of the cosine is close to zero and consequently $C \approx 2$. For large momentum differences, the argument of the cosine is a rapidly oscillating function of $x_1 - x_2$, which is strongly suppressed by the averaging process, and $C \approx 1$. If the weight function used for the averaging contains just one parameter with the dimension of length, let us denote it R, the width of region in $p_1 - p_2$, where C is significantly bigger than one, must be of the order of R^{-1}. There are many specific recipes how to perform the averaging. The results obtained for the correlations of momenta and for the sizes and shapes of the interaction regions are reasonable. For reviews see (Boal et al. 1990) and (Haywood 1995). In spite of this success many difficulties remain.

- Since the energy of a pion is determined by its momentum, one has data only on the three-dimensional distribution of the differences of spatial momenta. This is not enough to derive the four-dimensional distribution of the sources in space-time. Therefore, the results are strongly model dependent.
- The averaging over the square of the wave function corresponds to the assumption that the density matrix of the final pions in coordinate representation is diagonal. This in turn implies that the momentum distribution should be flat, which contradicts experiment. A closely related question is, how the pion can be initially localized at the production point and then represented by a plane wave corresponding to well-defined momentum.
- Information about the production region is, in practice, obtained only from pairs of pions with similar momenta. Consequently, what is being measured is not the whole interaction region, but the region, where the pions with similar momenta are produced. This, incidentally, explains the fact that the interaction regions usually come out roughly spherical, while one believes that the full interaction region is more string like.
- There are many corrections, which probably should be applied, but it is controversial how. Here belong the corrections for Coulomb repulsion between the charged like sign pions, the corrections for the final state interactions due to strong coupling, the corrections due to resonance production, the corrections due to partial coherence of the source etc.

4 Density Matrix Approach

In order to obtain a more general formulation for the GGLP problem it is convenient to use the formalism of density matrices. This has been described by a number of people, here we are the closest to the formulation used by Bialas and Krzywicki (Bialas and Krzywicki 1995). We introduce an auxiliary, unphysical process, where all the particles produced are distinguishable. We assume that for this process simple, intuitive ideas work. Then we correct for the Bose-Einstein correlations in order to obtain results comparable with experiment. This approach has its defects as discussed in the Introduction (see also (Boal et al. 1990), (Haywood 1995)), but for lack of a better idea it is widely used. As our starting point for the distinguishable particles we use an independent production model (cf (Bialas and Zalewski 1998a), (Bialas and Zalewski 1998b), (Bialas and Zalewski 1998c) and references contained there). In this model the multiplicity distribution for the particles is Poissonian

$$P_N^{(0)} = \frac{\nu^N}{N!}e^{-\nu} \tag{11}$$

and for each multiplicity the density matrix is a product of single particle density matrices

$$\rho_N^{(0)}(q,q') = \prod_{i=1}^{N} \rho_1^{(0)}(q_i,q_i')\,. \tag{12}$$

The momentum distribution is given, as usual, by the diagonal elements of the density matrix in the momentum representation

$$\Omega_{0N}(q) = \rho_N^{(0)}(q,q)\,. \tag{13}$$

It is convenient to normalize it to unity

$$\sigma_N^{(0)} = \int dq\, \Omega_{0N}(q) = 1\,. \tag{14}$$

For identical particles the density matrix should be symmetrized as explained in the Introduction

$$\rho_N(q,q') = \sum_P \rho_N^{(0)}(q,q_P')\,. \tag{15}$$

The corresponding momentum distribution for a given multiplicity is

$$\Omega_N = \rho_N(q,q)\,. \tag{16}$$

This, however, is no more normalized to unity, because

$$\sigma_N = \int dq\, \Omega_N(q) = 1 + \ldots\,. \tag{17}$$

The first term in the last expression corresponds to the identity permutation, but there are $(N-1)!$ further terms. This yields the multiplicity distribution

$$P_N = \mathcal{N} P_N^{(0)} \sigma_N\,, \tag{18}$$

where $\mathcal{N}$ is an N-independent normalizing factor, which ensures that $\sum P_N = 1$.

5 Simple Case: Pure Final State

In order to present simply the qualitative features of the result, let us consider first the case, when for each multiplicity the final state is pure. This is a grossly oversimplified model, but we will find that it contains some features of the much more realistic approach presented in the following section. For the pure state model

$$\hat{\rho}_N^{(0)} = |\psi_N^{(0)}\rangle\langle\psi_N^{(0)}| \,. \tag{19}$$

We assume that each of the state vectors $|\psi_N^{(0)}\rangle$ is symmetric with respect to exchanges of particles. Thus the effect of the summation over the permutations P is simply to multiply the operator $\hat{\rho}_N^{(0)}$ by $N!$. As a result the probability of producing exactly N particles is also multiplied by the factor $N!$. The Poisson distribution goes over into a geometrical distribution and after evaluating the normalization factor we get

$$P_N = (1-\nu)\nu^N \,. \tag{20}$$

This formula makes sense only if $\nu < 1$, because otherwise the sum of the probabilities P_N diverges. For the average number of particles one finds

$$\overline{N} = \frac{\nu}{1-\nu} \tag{21}$$

with a singularity at $\nu = 1$. From the model presented in the following section it will be seen that this singularity corresponds to Einstein's condensation.

In order to avoid the repeated summation of series it is convenient to introduce the generating functions. The generating function for the multiplicity distribution is

$$\Phi(z) = \sum_{N=0}^{\infty} P_N z^N = \frac{1-\nu}{1-z\nu} \,. \tag{22}$$

The logarithmic derivative of this function with respect to z at $z = 1$ yields the average multiplicity. The second derivative of the logarithm with respect to z at $z = 1$ is the dispersion and in general the p-th cumulant of the multiplicity distribution is given by

$$K_p = (p-1)! \left(\frac{\partial^p Log\Phi}{\partial z^p}\right)_{z=1} = (p-1)! \left(\frac{\nu}{1-\nu}\right)^p \,. \tag{23}$$

Inclusive and exclusive momentum distributions, as well as all the correlation functions in momentum space, can be calculated by functional differentiation from the generating functional

$$\Phi[u] = \sum_{N=0}^{\infty} \mathcal{N} P_N^{(0)} \int dq\, \Omega_N(q) \prod_{i=1}^{N} u(q_i) = \frac{1-\nu}{1-\nu \int dq_i \Omega_{01}(q_i) u(q_i)} \,. \tag{24}$$

For instance, the single particle distribution is

$$\left(\frac{\delta\Phi[u]}{\delta u}\right)_{u=1} = \frac{\nu}{1-\nu}\Omega_0(q)\,. \tag{25}$$

Thus symmetrization (Bose-Einstein statistics) introduces in this simple model only a change of normalization.

6 Independent Production

Let us consider now the full independent production model. In order to find the modification of the multiplicity distribution due to Bose-Einsten statistics it is necessary to calculate the correction factors

$$\sigma_N = \sum_P \int dq \prod_{i=1}^{N} \rho_1^{(0)}(q_i, q_{Pi})\,. \tag{26}$$

Since each permutation can be decomposed into cycles, this integrals can be expressed in terms of the cycle integrals

$$C_{k>1} = \int d^{3k}q\, \rho_1^{(0)}(q_1,q_2)\rho_1^{(0)}(q_2,q_3)\ldots\rho_1^{(0)}(q_k,q_1)\,. \tag{27}$$

It is convenient to add the definition

$$C_1 = 1\,. \tag{28}$$

Similarly the integrals necessary to calculate the generating functional for the momentum distributions can be expressed in terms of the cycle integrals

$$C_k[u] = \int d^{3k}\, qu(q_1)\rho_1^{(0)}(q_1,q_2)u(q_2)\rho_1^{(0)}(q_2,q_3)\ldots u(q_k)\rho_1^{(0)}(q_k,q_1)\,. \tag{29}$$

After some combinatorics, very similar to that used when deriving the linked clusters expansion familiar from quantum field theory and many body theory, one finds the generating functional

$$\Phi[u] = \exp\left[\sum_{k=1}^{\infty} \nu^k \frac{C_k[u] - C_k[1]}{k}\right]\,. \tag{30}$$

Substituting in this functional z for u one obtains the generating function for the multiplicity distribution. Without exhibiting the actual calculations we will now present some general results, obtained for this model.

- The single particle momentum distribution and all the momentum correlation functions can be expressed in terms of one function depending on two single particle momenta

$$L(q_1, q_1') = \sum_{k=1}^{\infty} \nu_k \int d^3q_2 \ldots d^3q_k \, \rho_1^{(0)}(q_1, q_2)\rho_1^{(0)}(q_2, q_3) \ldots \rho_1^{(0)}(q_k, q_1') \,. \tag{31}$$

For instance the momentum distribution is

$$\Omega(q) = L(q, q) \,. \tag{32}$$

The two particle cumulant is

$$K_2(q_1, q_2) = L(q_1, q_2)L(q_2, q_1) \,. \tag{33}$$

In general the p-th correlation function is

$$\begin{aligned} K_p(q_1, q_2, \ldots, q_p) &= L(q_1, q_2)L(q_2, q_3) \ldots L(q_p, q_1) \\ &+ (\text{permutations of the indices } 2, \ldots, p) \,. \end{aligned} \tag{34}$$

- For typical density matrices the average square of the difference of momenta between two particles $\langle q^2 \rangle$ decreases due to symmetrization.
- For typical density matrices the average difference between the production points of pairs of particles decreases due to symmetrization.
- For typical density matrices the size of the interaction region as evaluated from the width of the two-particle momentum correlation function decreases due to symmetrization.

We will discuss these predictions in a further section, where we will rederive them in a more intuitive way. The references to "typical density matrices" mean that the statement is true for most density matrices, but not for all. We have not been able to find a condition defining the relevant class of density matrices.

Probably the most interesting implication is the possibility of Einstein's condensation, but this will be discussed in the following section.

7 Einstein's Condensation

Using matrix notation one can rewrite the definition of the function L given in the previous section in the form

$$L(q, q') = \sum_{k=1}^{\infty} \nu^k \langle q| \left(\hat{\rho}_1^{(0)}\right)^k |q'\rangle \,. \tag{35}$$

Expanding the single particle density operator in terms of its eigenvectors and eigenvalues, we find

$$\hat{\rho}_1^{(0)} = \sum_n |n\rangle \lambda_n \langle n| \tag{36}$$

and for its k-th power

$$\left(\hat{\rho}_1^{(0)}\right)^k = \sum_n |n\rangle \lambda_n^k \langle n| . \tag{37}$$

Thus in the momentum representation

$$L(q,q') = \sum_n \langle q|n\rangle \langle n|q'\rangle \sum_{k=1}^{\infty} \nu^k \lambda_n^k . \tag{38}$$

This expression makes sense only if for all n there is $\lambda_n \nu < 1$. Denoting the largest eigenvalue of the density operator by λ_0, we expect problems when $\nu\lambda_0 \to 1$.

Performing the summations of the geometric series, we can rewrite the expression for $L(q,q')$ in the form

$$L(q,q') = \sum_n \frac{\psi_n(q)\psi_n^*(q')\nu\lambda_n}{1-\nu\lambda_n} . \tag{39}$$

For $\nu\lambda_0 \to 1$ it is convenient to use the equivalent formula

$$L(q,q') = \frac{\psi_0(q)\psi_0^*(q')}{1-\nu\lambda_0} + \tilde{L}(q,q') , \tag{40}$$

where $\tilde{L}$ remains bounded in the limit. Putting $q = q'$ and integrating over q we get the corresponding formula for the average multiplicity

$$\overline{N} = \frac{1}{1-\nu\lambda_n} + \text{bounded term} . \tag{41}$$

From these formulae it is clear that when $\nu\lambda_0$ tends to one, Einstein's condensation occurs. Increasing ν corresponds to the increasing of the number of particles in the system. In all the states with indices $n \neq 0$ there is place only for a limited number of particles, while all the surplus, which can be arbitrarily large, gets located in the state $|0\rangle$. In this sense, when the number of particles becomes very large, we recover the model with the pure state discussed previously. A very interesting question is, whether experimentally it is possible to create condition, where the Einstein condensate would dominate.

Let us conclude this section with two remarks. For the Gaussian single particle density matrix

$$\rho_1^{(0)}(q,q') = \frac{1}{\sqrt{2\pi\Delta^2}} \exp\left[-\frac{q_+^2}{2\Delta^2} - \frac{R^2}{2}q_-^2\right] , \tag{42}$$

where $q_+ = (q + q')/2$ and $q_- = q - q'$, the eigenvalues and eigenfunction are known (Bialas and Zalewski 1998a). Thus all the calculations can be easily performed. In fact they have been performed by various methods (Pratt 1993), (Pratt and Zelevinsky 1994), (Pratt 1994), (Csörgö and Zimanyi 1998).

The theory can be reformulated in the second quantization formalism. Then the function $L(q, q')$ appears as the Green function $\langle \hat{a}_q^\dagger \hat{a}_{q'} \rangle$ and the possibility of expressing all the correlation functions in terms of $L(q, q')$ is the Wick theorem with L as the only non zero contraction.

8 Statistical Physics Interpretation

Many results from the previous sections can be simply reinterpreted and rederived using standard statistical physics. Consider the single particle unsymmetrized density operator

$$\hat{\rho}_1^0 = \sum_n |n\rangle \lambda_n \langle n| \tag{43}$$

with the condition $\sum_n \lambda_n = 1$. This can be reinterpreted as the density operator corresponding to the canonical ensemble, if we put

$$\lambda_n = \frac{1}{Z} e^{-\beta\varepsilon_n} , \tag{44}$$

where, as usual, ε_n is the energy of state $|n\rangle$, β is the inverse temperature in energy units and $Z = \sum_n \exp[-\beta\varepsilon_n]$ is the canonical partition function. The corresponding (single particle) Hamiltonian is

$$\hat{H} = \sum_n |n\rangle \varepsilon_n \langle n| . \tag{45}$$

This Hamiltonian, when written in the coordinate representation, may look quite unusual, but some cases are simple. For instance, the Gaussian density matrix corresponds to the Hamiltonian of a harmonic oscillator.

For indistinguishable particles a single particle is not a convenient subsystem and, as suggested by Pauli long ago, it is better to choose as subsystem the open system consisting of all the particles in state $|n\rangle$. The state of this subsystem is defined by the number of particles (N) in it. The probability of state N of the subsystem is

$$P_n(N) = \frac{1}{\mathcal{Z}_n} \overline{\nu}^N e^{-\beta N \varepsilon_n} . \tag{46}$$

The grand partition function

$$\mathcal{Z}_N = \frac{1}{1 - \overline{\nu} e^{-\beta\varepsilon}} \tag{47}$$

is chosen so that $\sum_{N=0}^{\infty} P_n(N) = 1$, the parameter $\overline{\nu}$ is known in statistical physics as the fugacity and is connected to the chemical potential μ by the formula

$$\overline{\nu} = e^{\beta\mu} . \tag{48}$$

In order to reproduce the formulae from the previous sections, one puts $Z\overline{\nu} = \nu$. The grand partition function can be used to find the moments of the multiplicity distribution very much like the multiplicity generating function. For instance, for the average occupation of state n we find

$$\langle N_n \rangle = -\frac{1}{\beta}\frac{\partial Log \mathcal{Z}_n}{\partial \mu} = \frac{1}{e^{\beta(\varepsilon_n - \mu)} - 1} = \frac{\nu\lambda_n}{1 - \nu\lambda_n} . \tag{49}$$

The probability of no particles in state n is

$$P_n(0) = \frac{1}{\mathcal{Z}_n} = 1 - \overline{\nu}e^{-\beta\varepsilon_n} = \frac{1}{\langle N_n \rangle + 1} . \tag{50}$$

The probability of no particle in the whole system is

$$P(0) = \prod_n \frac{1}{\langle N_n \rangle + 1} . \tag{51}$$

Let us consider two limiting cases. When all the occupation numbers are very small, the product equals approximately $\exp[-\langle N \rangle]$ and for large multiplicities it is very small. Very large fluctuations of the multiplicity are very unlikely to occur. When most particles are in the state $n = 0$, the product is approximately $(\langle N \rangle + 1)^{-1}$, which is much bigger than in the previous case. Thus, when there is much Einstein condensate, large multiplicity fluctuations become much more probable. Cosmic ray physicists have been reporting (Lates et al. 1980) observations of centauro and anticentauro events. This are high multiplicity events, where respectively either the neutral pions or the charged pions are missing. One could speculate that this phenomena are related to Einstein's condensation.

Statistical physics gives also a simple interpretation for the function $L(q, q')$. One finds

$$L(q, q') = \sum_n \psi_n(q)\psi_n^*(q') \frac{1}{e^{\beta(\varepsilon_n - \mu)} - 1} . \tag{52}$$

This is the canonical density matrix with the Maxwell-Boltzmann weights replaced by the Bose-Einstein weights. The fact that the Bose-Einstein weights fall with increasing energy ε_n faster than the Maxwell-Botzmann weights explains qualitatively most of the observations reported previously. For most Hamiltonians the wave function spreads in ordinary space and in momentum space, when energy is increased. Since the Bose-Einstein weights enhance the low energies, they reduce the average momenta and radii. Also the reduction of the effective radius of the interaction region, as determined from the width of the correlation function in momentum space, can be easily understood. If in the previous formula all the terms had equal weights, we would obtain $L(q,q') = \delta^3(q-q')$. The stronger the cut on the sum, the broader the peak in $q-q'$ becomes. Since the Bose-Einstein weights are more peaked at low energies than the Maxwell-Boltzmann ones, they correspond to a broader peak in the correlation function. Since the width of this peak is inversely proportional to the radius of the production region, symmetrization reduces the radius of this region. All these qualitative arguments are usually true. It is, however, easy to show examples of Hamiltonians, where e.g., with increasing energy the wave function shrinks either in ordinary space, or in momentum space. Additional assumptions necessary to convert these qualitative arguments into rigorous theorems are, therefore, necessary, but not yet known.

References

Bialas, A., Krzywicki, A., (1995): Phys. Letters **B354**, 134

Bialas, A., Zalewski, K., (1998a): hep-ph/9803408 and Eur. J. Phys. C in print

Bialas, A., Zalewski, K., (1998b): hep-ph/9806435 and Phys. Letters B in print

Bialas, A., Zalewski, K., (1998c): hep-ph/9807382 and Slovak J. Phys in print

Boal, D.H., Gelbke, C.-K., Jennings, B.K., (1990): Rev. Mod. Phys. **62**, 553

Csörgö, T., Zimanyi, J., (1998): Phys. Rev. Letters **80**, 916

Goldhaber, G., Goldhaber, S., Lee, W., Pais, A., (1960): Phys. Rev. **120**, 300

Hanbury Brown, R., Twiss, R.Q., (1956): Nature **177**, 27

Hanbury Brown, R., (1974): *The Intensity Interferometer*, Taylor and Francis New York p. 7.

Haywood, R., (1995): Rutherford lab. report RAL94-07

Lates, C.M.G., Fujimoto, Y., Hasegawa, S., (1980): Phys. Rep. **65**, 151

Pratt, S., (1993): Phys. Letters **B301**, 159

Pratt, S., Zelevinsky, V., (1994): Phys. Rev. Letters **72**, 816

Pratt, S., (1994): Phys. Rev. **C50**, 469

Silverman, M.P., (1995): *More than one mystery* Springer-Verlag New York

Reduction of Couplings in Massive Models of Quantum Field Theory

W. Zimmermann

Max-Planck-Institut für Physik
Munich, Germany

Dedicated to Prof. Jan Łopuszański on the occasion of his 75th birthday

Abstract. The method of reducing the number of couplings is reviewed for massive models of quantum field theory. It is shown that the principle of reduction is independent of the scheme of renormalization used. Finally the possibility of eliminating the mass parameters is discussed.

First I would like to thank the organizers for this invitation which gives me the opportunity to congratulate Jan Łopuszański personally to the coming event of his 75th birthday. Between us this has been a long friendship. The other day we recollected that we met for the first time in New York in 1960, this is almost four decades ago. Among many other things we both share a certain conservative attitude towards particle physics, in particular our dedication to local quantum field theory. Most of Łopuszański's work is concerned with this subject as is mine. In this context it should be mentioned that local quantum field theory is just completing its seventh decade. It is still alive and well, having passed all experimental and mathematical tests - at least so far. Quite appropriate for this occasion I will discuss issues of local quantum field theory in four dimensions.

The purpose of the reduction method is to find relations among coupling constants which are compatible with the renormalization group (Zimmermann 1985a). This is a generalization of coupling relations which follow from symmetry properties. Such relations can be used to express some couplings of a system in terms of other parameters (see refs. (Zimmermann 1985b, Oehme 1986, Sibold 1988) for a review). In a paper with Kubo and Sibold an application was made to the standard model (Kubo et al. 1985). The main result was a prediction of the top mass. In lowest order we computed a value of about 90 GeV, including two-loop corrections Kubo obtained appr. 100 GeV (Kubo 1991). At the time, when we wrote the paper, this was in 1985, this was considered much too high. But now the experimental value is around 175 GeV. There is no chance of improving or correcting our calculations which would be substantial enough to bring the value up close to the experimental mass. So it has to be accepted that the application to the standard model as such failed. On the other hand the method itself is certainly correct, therefore we think that the deviation is due to the influence of heavier particles beyond the standard model. Moreover there is the important aspect of asymptotic freedom (Gross and Wilczek 1973, 1974, 1985, Politzer 1973). The principle of reduction is equivalent to having asymptotic freedom simultaneously for several couplings in the ultraviolet or infrared region - apart from the case that all β functions vanish identically (Oehme and Zimmermann 1985, Oehme et al. 1985) (see refs. (Sibold 1985, Zimmermann 1986) for a review). Obviously the standard model as a whole cannot be asymptotically free due to the opposite signs of the β functions for the gauge couplings. Therefore, we applied the reduction method only to QCD extended by the Higgs and Yukawa couplings. The remaining electroweak couplings were treated as perturbations of the system.

This deficiency of the standard model - I mean the violation of asymptotic freedom in the gauge sectors - is removed by unifying the gauge couplings. Then asymptotic freedom becomes possible. Applying the reduction method to supersymmetric grand unified theories Kubo, Mondragón and Zoupanos indeed found asympotically free solutions. In this way they were able to obtain acceptable values of the top mass (Kubo et al. 1994).

In this lecture I want to talk about problems with formulating the reduction method in massive models of quantum field theory. Originally the reduction method was developed only for massless models. We applied it to massive models nevertheless, since the β functions on which all calculations are based are massless, if computed by dimensional renormalization (Weinberg 1973), (Collins and Mac Farlane 1974). So the question arises, whether or not the reduction principle is scheme independent. In the first part of my talk I will set up the reduction method in massive models. Then the simplifications occurring for massless β functions will be discussed. Next the scheme independence of the reduction method is sketched. In the final part it will be shown - following a suggestion of Maison - how the mass parameters can be eliminated in the beta functions without referring to dimensional renormalization.

1 Reduction in Massive Models

We consider a model with m fields

$$\phi_1, ..., \phi_m ,$$

$n + 1$ dimensionless coupling parameters

$$\lambda_0, \lambda_1, ..., \lambda_n$$

and pole masses

$$m_1, ..., m_c$$

with normalization mass κ $(\kappa^2 < 0)$. Starting point are the differential equations of the renormalization group. In a model with dimensionless couplings they have the form (Osviannikov 1956, Callan 1970; Symanzik 1970)

$$\left(\kappa^2 \frac{\partial}{\partial \kappa^2} + \sum \beta_j \frac{\partial}{\partial \lambda_j} + \sum \gamma_l \right) \tau = 0$$

for the Fourier transforms

$$\tau = \tau(k_1, \ldots, k_s; \lambda_0, \ldots, \lambda_n; m_1^2, \ldots, m_c^2, \kappa^2)$$

of the time ordered functions

$$\langle T\phi_{j_1}(x_1) \ldots \phi_{j_s}(x_s) \rangle$$

of field operators. The coefficients β and γ depend on couplings and dimensionless mass ratios

$$\beta_j = \beta_j \left(\lambda_0, \ldots, \lambda_n, \frac{m_1}{|\kappa|}, \ldots \frac{m_c}{|\kappa|} \right) .$$

These differential equations are based on Stueckelberg's concept of the renormalization group first formulated in 1953 within perturbation theory and further developed by Bogoliubov and Shirkov (Stueckelberg and Petermann 1953, Bogoliubov and Shirkov 1955). Renormalization group invariance actually concerns the exact theory as well, and the consequences are sometimes in contradiction to perturbation theory. The phenomenon of asymptotic freedom, for instance, is a rigorous consequence of renormalization group invariance, but is not valid in a given order of perturbation theory. Stueckelberg's renormalization group is simply defined as the group of transformations

$$\phi_1'(x) = z_j^{1/2} \phi_j(x) \qquad \text{with} \qquad z_j > 0 ,$$

which multiply each field operator by a positive number. So from the mathematical point of view this group is very trivial. Invariance under this group reflects the arbitrariness which one has in normalizing a field operator. The differential equations follow from the requirement that the normalization of the field operators and the couplings be uniquely determined by the normalization conditions

$$\lambda_j = \Gamma_j , \qquad (m_j^2 - k^2) G_j = 1 \qquad \text{at} \qquad k^2 = \kappa^2 .$$

Γ_j is a suitable vertex function with a certain configuration of momenta so that Γ_j is a function of a momentum square only. G_j is a suitable structure function of a propagator. The differential equation describes the variation of the correlation functions under an infinitesimal change of the normalization mass. Stueckelberg's concept seems to be too general, but the differential equations become non-trivial by the form of the β and γ functions which involve the dynamics of the particular model. They are derived as asymptotic series in the couplings whose coefficients are computed in perturbation theory. In particular, the lowest order coefficients are relevant for the asymptotic behavior of the correlation functions.

Let me now introduce the concept of the reduction principle. It generalizes a certain aspect of symmetries. An important consequence of symmetries is that the number of independent parameters of a system is reduced. So masses in a multiplet become equal and simple relations among coupling constants follow. Suppose a symmetry of a model involving several dimensionless coupling parameters is strong enough to constrain the couplings such that only one, say λ_0, remains independent, then all other couplings become functions of λ_0,

$$\lambda_j = \lambda_j(\lambda_0)\,.$$

Of course, this is only correct, if the symmetry can be implemented to all orders of perturbation theory, that means that no anomalies occur which spoil the symmetry of the classical theory.

Mostly one has simple relations like

$$\lambda_j = \rho_j \lambda_0 \qquad \text{or} \qquad \lambda_j = \rho_j \lambda_0^2\,,$$

where the ρ_j are certain numerical coefficients given by the structure of the group. The relations hold to all orders, usually, provided the normalization conditions defining the couplings can be chosen in a way which respects the symmetry to all orders. For more general normalization conditions unrelated to the symmetry one obtains power series expansions instead,

$$\lambda_j = \rho_{j1}\lambda_0 + \rho_{j2}\lambda_0^2 + \dots\,,$$

where the higher order coefficients are uniquely determined and may depend on the masses of the system.

It is this aspect of symmetry which is generalized by the reduction principle using the renormalization group concept. From now on we will not assume any symmetry properties. Instead we ask ourselves, whether it is possible to express all couplings as functions of a single one, say λ_0:

$$\lambda_j = \lambda_j\left(\lambda_0, \frac{m_1^2}{|\kappa^2|}, \dots, \frac{m_c^2}{|\kappa^2|}\right)\,.$$

As requirements we impose

1. $\lambda_j \to 0$ simultaneously with $\lambda_0 \to 0$,
2. λ_j power series in λ_0,
3. renormalization group invariance for the reduced model.

Taking a more general point of view one might drop the power series requirement. In that case one has only the first condition that all couplings simultaneously approach zero. This could be interpreted as as generalization of broken symmetry constraints. For appropriate signs of the β functions this represents the case of asymptotic freedom in several variables.

An interesting possibility which is not discussed in this talk should be mentioned: One can apply the reduction principle as well to the behavior of couplings near a non-trivial fixed point instead of the origin.

Next we discuss the problem of finding functions

$$\lambda_j = \lambda_j \left(\lambda_0, \frac{m_1^2}{|\kappa^2|}, \dots \right) ,$$

which are compatible with the renormalization group invariance for both, the original and the reduced model. To this end we compare the original differential equation

$$\left(\kappa^2 \frac{\partial}{\partial \kappa^2} + \sum \beta_j \frac{\partial}{\partial \lambda_j} + \sum \gamma_l \right) \tau = 0$$

with the corresponding equations for the reduced model

$$\left(\kappa^2 \frac{\partial}{\partial \kappa^2} + \beta_0' \frac{\partial}{\partial \lambda_0} + \beta_0' \right) \tau' = 0 ,$$

τ' is the time ordered function with

$$\lambda_j \left(\lambda_j (\lambda_0, \frac{m_1^2}{|\kappa^2|}, \dots \right)$$

substituted for λ_j. Inserting

$$\frac{\partial \tau'}{\partial \lambda_0} = \frac{\partial \tau}{\partial \lambda_0} + \sum \frac{\partial \tau}{\partial \lambda_j} \frac{\partial \lambda_j}{\partial \lambda_0} ,$$

$$\frac{\partial \tau'}{\partial \kappa^2} = \frac{\partial \tau}{\partial \kappa^2} + \sum \frac{\partial \tau}{\partial \lambda_j} \frac{\partial \lambda_j}{\partial \kappa^2}$$

and comparing the coefficients we obtain

$$\beta_0' = \beta_0, \qquad \kappa^2 \frac{\partial \lambda_j}{\partial \kappa^2} + \beta_0' \frac{\partial \lambda_j}{\partial \lambda_0} = \beta_j .$$

So the result is a system of partial differential equations

$$\kappa^2 \frac{\partial \lambda_j}{\partial \kappa^2} + \beta_0 \frac{\partial \lambda_j}{\partial \lambda_0} = \beta_j$$

for the functions λ_j. This system must be solved under the condition

$$\lambda_j \to 0 \qquad \text{for} \qquad \lambda_0 \to 0$$

and the power series condition, if so desired. These are the reduction equations. In this form - with the mass dependence - they were set up by Piguet and Sibold (Piguet and Sibold 1989). They further derived reduction equations from the Callan-Symanzik equation and related partial differential equations of the system. For those reduction solutions which are uniquely determined power series in the primary coupling Piguet and Sibold proved that the reduction of couplings and dependence on parameters (like masses) are consistent. Due to the partial with respect to κ^2 it is hard to study reduction equations in the general case. Fortunately, a systematic treatment of finding all solutions is possible by eliminating the normalization mass κ along with the other masses. This will be the subject of my talk. The issue is connected to the question of scheme independence. Before we come to that, let me first review the simplifications in the massless case.

2 Massless β Functions

For a massless model the dependence on κ^2 drops out in the β functions. This is also the case for a massive model in the scheme of dimensional renormalization provided pole masses are used as mass parameters. Then we have a system of ordinary differential equations

$$\beta_0 \frac{d\lambda_j}{d\lambda_0} = \beta_j \,.$$

These equations, of course, are much easier to treat than the partial differential equations involving κ. It is obvious that these equations can always be solved. We may take any point $\lambda_0, \lambda_1, \ldots, \lambda_n$, where the β functions are sufficiently regular, so that a Lipschitz condition holds. Then exactly one solution passes through this point. But the conditions

1. $\lambda_j \to 0$ for $\lambda_0 \to 0$,
2. λ_j power series (optional)

have to be imposed. Already the first condition is very restrictive. For the point $\lambda_0, \lambda_j = 0$ is not regular, since β functions in 4-dimensional models vanish quadratically at the origin

$$\frac{d\lambda_j}{d\lambda_0} = \frac{\beta_j}{\beta_0} \to \frac{0}{0} \qquad \text{for} \qquad \lambda_j, \lambda_0 \to 0 \,.$$

If the power series condition is included, there are only a finite number of solutions possible in most cases and sometimes none at all.

As a simple example I mention the pseudoscalar interaction. The interaction term

$$ig\bar{\psi}\gamma_5 A\psi$$

must be supplemented by a direct scalar interaction

$$-\frac{\lambda}{4!}A^4 \,.$$

The reason is that diagrams like the box diagram lead to divergent contributions which cannot be compensated by a counter term, since there is no available renormalization constant. Therefore a self-coupling of the scalar field must be introduced in order to make the renormalization program work. But then it is natural to require that all Green's functions should only depend on the original coupling instead of having a model with two independent parameters. Then also λ, being defined through Green's functions, will be a power series in g. This is the principle of reduction to demand that λ be a function of g,

$$\lambda = \lambda(g^2)$$

consistent with the renormalization group and

$$\lambda \to 0 \qquad \text{for} \qquad g \to 0\,,$$

moreover

$$\lambda = \rho g^2 + \rho_2 g^4 + \dots\,.$$

The reduction equation becomes

$$\beta_{g^2}\frac{d\lambda}{dg^2} = \beta_\lambda$$

with

$$\beta_{g^2}\frac{d\lambda}{dg^2} = bg^4 + \dots\,, \qquad b = \frac{5}{16\pi^2}$$

and

$$\beta_\lambda = c_1\lambda^2 + c_2\lambda g^2 + c_3 g^4 + \dots = \frac{1}{16\pi^2}\left(\frac{3}{2}\lambda^2 + 4\lambda g^2 - 24 g^4\right) + \dots\,.$$

For solving we make the ansatz

$$\lambda = \rho g^2 + \rho_2 g^4 + \rho_3 g^6 + \dots\,.$$

ρ satisfies the quadratic equation

$$c_1\rho^2 + (c_2 - b)\rho + c_3 = 0$$

with the roots

$$\rho = \frac{1}{3} \pm \frac{1}{3}\sqrt{145}\,.$$

Since $\lambda > 0$ we choose the positive root so that

$$\lambda = \frac{1}{3}\left(1 + \sqrt{145}\right) g^2 + \rho_1 g^4 + \dots\,.$$

All higher order coefficients are uniquely determined recursively. By a reparametrization, i.e., a simple redefinition of the coupling λ,

$$\lambda' = \lambda + a_2\lambda^2 + \dots\,,$$

it can always be arranged that the lowest order approximation

$$\lambda' = \frac{1}{3}\left(1 + \sqrt{145}\right) g^2$$

becomes exact like in case of a symmetry. This is an example of a reduction which does not seem to be related to a symmetry because of the root $\sqrt{145}$.

3 Scheme Independence

In this section we turn to the proof of the scheme independence. This will indicate a natural way of eliminating κ and the masses. We return to the general case of mass dependent β functions. For simplicity a model is chosen with only two couplings, λ_0 and λ_1. In this case the differential equations of the renormalization group are

$$\kappa^2 \frac{\partial \tau}{\partial \kappa^2} + \beta_0 \frac{\partial \tau}{\partial \lambda_0} + \beta_1 \frac{\partial \tau}{\partial \lambda_1} + \sum \gamma_l \tau = 0\,, \tag{1}$$

$$\beta_j = \beta_j \left(\lambda_0, \lambda_1, \frac{m_1}{|\kappa|}, \dots\right)\,, \qquad \gamma_j = \gamma_j \left(\lambda_0, \lambda_1, \frac{m_1}{|\kappa|}, \dots\right)\,. \tag{2}$$

Next the scheme will be changed, for instance by using other vertex functions or momentum configurations in defining the couplings. Then the new couplings are given by the following transformations

$$\Lambda_j = \Lambda_j(\lambda_0, \lambda_1, m_1^2, \dots, \kappa^2); \qquad j = 0, 1\,. \tag{3}$$

The dependence of the Green's functions on the couplings of the new scheme is given by

$$\tau(k_1, \dots; \lambda_0, \lambda_1, m_1^2, \dots, \kappa^2) = \hat{\tau}(k_1, \dots, \kappa^2)\,. \tag{4}$$

From this follow the differential equations of the renormalization group in the new scheme

$$\kappa^2 \frac{\partial \hat{\tau}}{\partial \kappa^2} + \hat{\beta}_0 \frac{\partial \hat{\tau}}{\partial \Lambda_0} + \hat{\beta}_1 \frac{\partial \hat{\tau}}{\partial \Lambda_1} + \sum \hat{\gamma}_l \hat{\tau} = 0 \tag{5}$$

with

$$\kappa^2 \frac{\partial \Lambda_j}{\partial \kappa^2} + \beta_0 \frac{\partial \Lambda_j}{\partial \lambda_0} + \beta_1 \frac{\partial \Lambda_j}{\partial \lambda_1} = \hat{\beta}_j, \qquad \gamma_j = \hat{\gamma}_j\,. \tag{6}$$

In the old renormalization scheme the differential equations of the renormalization group for a reduced system are

$$\kappa^2 \frac{\partial \tau'}{\partial \kappa^2} + \beta_0 \frac{\partial \tau'}{\partial \lambda_0} + \sum \gamma_l \tau' = 0\,. \tag{7}$$

As a consequence of eqs. (1) and (7) the reduction equations

$$\kappa^2 \frac{\partial \lambda_1}{\partial \kappa^2} + \beta_0 \frac{\partial \lambda_1}{\partial \lambda_0} = \beta_1 \tag{8}$$

follow.

In the new scheme the corresponding equations are

$$\kappa^2 \frac{\partial \hat{\tau}}{\partial \kappa^2} + \hat{\beta}_0 \frac{\partial \hat{\tau}}{\partial \Lambda_0} + \hat{\beta}_1 \frac{\partial \hat{\tau}}{\partial \Lambda_1} + \sum \hat{\gamma}_l \hat{\tau} = 0 \tag{9}$$

for the original system and

$$\kappa^2 \frac{\partial \hat{\tau}'}{\partial \kappa^2} + \hat{\beta}_0 \frac{\hat{\tau}'}{\partial \Lambda_0} + \sum \hat{\gamma}_l \hat{\tau}' = 0 \tag{10}$$

for the reduced system. The reduction equations in the new system are

$$\kappa^2 \frac{\partial \Lambda_1}{\partial \kappa^2} + \hat{\beta}_0 \frac{\partial \Lambda_1}{\partial \Lambda_0} = \hat{\beta}_1 \, . \tag{11}$$

In order to establish the scheme independence we have to show that the reduction equations (8) and (11) in the old and new system are equivalent. That means that each solution $\lambda_1(\lambda_0)$ expressed as function $\Lambda_1(\Lambda_0)$ in terms of the new variables Λ_0 and Λ_1 is also a solution of (11). Vice versa, each solution $\Lambda_1(\Lambda_0)$ should provide a solution of (8) by change of variables. A direct proof of this statement is possible, but quite lengthy. It is much easier to prove the equivalence of (8) and (11) in an indirect manner by first showing the equivalence of the renormalization group equations (original and reduced) in both schemes, i.e., the equivalence of (1) and (7) to (5) and (10). The equivalence of the reduction equations (8) and (11) is then an obvious consequence.

4 Elimination of Mass Parameters

With the result that the reduction principle is scheme independent one might believe that the problem of mass dependence is already resolved, since dimensional renormalization is just another scheme of renormalization. Accordingly, it seems justified to drop the mass dependence in the reduction equations, because the β functions are massless by dimensionless renormalization at least for models with dimensionless couplings only as considered in this talk. However, the connection between dimensional renormalization and other methods of renormalization is not well understood. Therefore, instead of relying on consequences of dimensional renormalization, we will try to remove the mass dependence from the β functions. For the model of the ϕ^4 -coupling Maison has shown that the mass can be eliminated in the β function by a transformation of the coupling combining the renormalization group with the Callan-Symanzik equation (Maison private communication). I modified Maison's approach by using the renormalization group equations alone in order to eliminate the masses in the β functions. This method applies to general systems provided the massless limits of the β functions exist and are approached smoothly for vanishing masses. As example we take again a model involving two dimensionless couplings.

In the last section we obtained relations (6) which provide the form of the new β functions after transforming the couplings. We now use the same relations but with a completely different meaning: They will be interpreted as the defining relations for the transforming functions Λ_j yet to be determined with the functions $\hat{\beta}_j$ chosen to be the massless limits of the original β functions.

There are, of course, many solutions of (6). But it turns out that only one is reasonable. One could, for instance, impose the condition that the old and new couplings be equal at a certain value of the renormalization mass

$$\Lambda_j = \lambda_j \qquad \text{at} \qquad \kappa^2 = \kappa_0^2 \,.$$

It can be shown that such solution exists uniquely. But the disadvantage would be that the Green's functions now involve a new dimensional parameter, the mass κ_0, where the couplings are adjusted. There is an ideal way out, namely to adjust the two couplings at infinite normalization mass. For formulating this in a precise manner one replaces κ^2 by

$$\zeta = \frac{1}{|\kappa|} \,.$$

Then the equations to be solved are

$$\beta_0 \frac{\partial \Lambda_j}{\partial \lambda_0} \partial \lambda_0 + \beta_1 \frac{\partial \Lambda_j}{\partial \lambda_j} - \frac{1}{2} \frac{\partial \Lambda_j}{\partial \zeta} = \hat{\beta}_j \,,$$

with the initial condition

$$\Lambda_j = \lambda_j \qquad \text{at} \qquad \zeta = 0$$

to be imposed. The dependence of the β functions on the coupling and mass ratios is of the form

$$\beta_j \left(\lambda_0, \lambda_1, \frac{m_1}{|\kappa|}, \dots, \frac{m_c}{|\kappa|} \right) \quad = \quad \beta_j \left(\lambda_0, \lambda_1, m_1 \zeta, \dots, m_c \zeta \right) \,,$$

$$\hat{\beta}_j = \beta_j \left(\lambda_0, \lambda_1, 0, \dots, 0 \right) \,.$$

Infinite normalization mass may be a dangerous limit considering the evolution of a system. But it is harmless when couplings of different schemes are adjusted. First of all, this is a very natural choice. For the new β functions represent the massless limit. On the other hand, for the β functions the massless limit is equivalent to the limit $\zeta = 0$. Moreover, it can be shown that there is a unique solution by expanding with respect to powers of the couplings provided the β functions have a sufficiently smooth behavior in the massless limit.

As an example we construct the coefficient c_{00} in the ansatz

$$\Lambda_0 = \lambda_0 + c_{00} \lambda_0^2 + c_{01} \lambda_0 \lambda_1 + \dots \,.$$

For the expansions of the β functions we use the notation

$$\beta_0 = a_{00} \lambda_0^2 + a_{01} \lambda_0 \lambda_1 + a_{11} \lambda_1^2 + \dots \,,$$

$$\hat{\beta}_0 = \hat{a}_{00} \Lambda_0^2 + \hat{a}_{01} \Lambda_0 \Lambda_1 + \hat{a}_{11} \Lambda_1^2 + \dots \,.$$

c_{00} satisfies the ordinary differential equation

$$a_{00} - \frac{1}{2}\frac{\partial c_{00}}{\partial \zeta} = \hat{a}_{00} \, .$$

With the initial condition this is solved uniquely by

$$c_{00} = 2 \int_0^\zeta \frac{a_{00} - \hat{a}_{00}}{\zeta'} d\zeta' \, .$$

In conclusion it can be said that the method of reduction works independently of the renormalization scheme used. Moreover, the masses can be eliminated from the reduction equations provided certain conditions for the massless limit are satisfied.

References

Bogoliubov, N.N., Shirkov, D.V., (1955): Dokl. Akad. Nauk SSSR **103**, 391

Callan, C., (1970): Phys. Rev. **D2**, 1541; Symanzik, K., (1970): Comm. Math. Phys. **18**, 227

Collins, J., Mac Farlane, A., (1974): Phys. Rev. **D10**, 1201

Gross, D., Wilczek, F., (1973): Phys. Rev. Lett. **3O**, 134; Phys. Rev. **D8**, (1974) 3633 and Phys. Rev. **D9**, (1985) 980

Kubo, J., Sibold, K., Zimmermann, W., (1985): Nucl. Phys. **B259**, 331

Kubo, J., (1991): Phys. Lett. **B262**, 472

Kubo, J., Mondragón, M., Zoupanos, G., (1994): Nucl. Phys. **B424**, 29

Maison, D., private communication

Oehme, R., Zimmermann, W., (1985): Comm. of Math. Phys. **97**, 569

Oehme, R., Sibold, K., Zimmermann, W., (1985): Phys. Lett. **B153**, 147 and unpublished

Oehme, R., (1986): Progr. Theor. Phys. Suppl. **86**, 2125

Osviannikov, L.V., (1956): Dokl. Akad. Nauk SSSR **109**, 1112

Piguet, O., Sibold, K., (1989): Phys. Lett. **B229**, 83

Politzer, H., (1973): Phys. Rev. Lett. **30**, 1346

Sibold, K., (1985): Proc. of the Intern. Europhysics Conf. on High-Energy Physics, Bari, Italy, p. 197

Sibold, K., (1988): Acta Phys. Polon. **Bl9**, 295

Stueckelberg, E., Petermann, A., (1953): Helv. Phys. Acta **26**, 499

Weinberg, S., (1973): Phys. Rev. **D8**, 3497

Zimmermann, W., (1985a): Comm. Math. Phys. **97**, 211

Zimmermann, W., (1985b): in XIV. Intern. Coll. on Group Theor. Methods in Physics, Seoul, Korea, p. 145

Zimmermann, W., (1986): in Renormalization Group 1986, Dubna, USSR, p. 51

Abstracts

Particle–Hole Asymmetry in the BCS Thermodynamics*

J. Czerwonko

Institute of Physics,
Wrocław University of Technology
Wybrzeże Wyspiańskiego 27, 50-370 Wrocław, Poland

It has been shown that the particle-hole asymmetry (PHA) of DOS leads to the first-order phase transition, a small deviation from the Luttinger theorem, and to very strange behaviour of subcritical specific heat. Because of the accuracy of the BCS thermodynamics in the thermodynamic limit (Bogolubov) it is strange that in trying to strengthen the theory while taking into account the tendency of DOS, we are in fact causing the deterioration of the theory. The answer lies in the retardation of the electron- phonon interaction for low temperature superconductors. Hence, if some elements of the BCS theory are applied for HTSC, it becomes necessary to be very careful in the question of thermodynamic properties. Moreover, the criteria of stability of the superconducting state has been formulated, at constant p and V as well, for one-component superconductors and isotropic Fermi superfluids. These criteria are free of the strong connection with the BCS model, they are purely thermodynamical. It is also shown that for the superconducting/superfluid Fermi systems the specific heat at constant p and V differ substantially, in contrast to any other low-temperature systems.

* Presented at the XXXVIII Cracow School of Theoretical Physics, Zakopane, Poland, June 1-10, 1998.

On Generalizations of the Gravitational Interaction

L. Halpern

Department of Physics, Florida State University
Tallahassee, Florida 32306–3016

Views of Einstein and Schrödinger on the limitations of the validity of the general theory of relativity are compared with the mainstream view held today.

A modernized relativistic version of the principle if inertia serves as guideline for the formulation of a theory which describes elementary particle spin as the analog of a gravitational charge generalizing the character of a gauge theory and removing some of the isolation of gravitation from the rest of physics.

The relativistic version of the principle of inertia is formulated on the manifold of the Anti De Sitter group $G = SO(3,2)$. It prescribes the orbits of structureless and spinning test particles as the natural projection $\pi : G \to B$ of orbits of one–dimensional subgroups on the Anti De Sitter universe B which is the space of right cosets $B = G/H$ with $H = SO(3,1)$ the Lorentz subgroup. Einstein's equations with a cosmological member are fulfilled on the group manifold for the Cartan–Killing metric γ. They project with π on Einstein's equations on B with the corresponding projected metric g.

$P(G, H, \pi, B)$ forms a principal fibre bundle with typical fibre H. A connection is chosen by defining four tangent vector fields as horizontal and the tangent vectors of H everywhere as vertical; it is a metric connection if horizontal and vertical vector fields are mutually perpendicular with respect to a generalized metric γ for which the vertical vectors retain the Killing property and the commutation relations of the group H. Only the commutation relations of the horizontal vectors are generalized; they determine the curvature two–form of a gauge formalism. Restricting such geometries to solutions of Einstein's equations results in a Kaluza–Klein formalism, the only one in which the metric g and the curvature two–form are truly unified and determine the geometry on B. The curvature produces the correct force on a spinning particle's orbit (an orbit which includes vertical components). The spin precession is however not taken into account because the Kaluza–Klein formalism does not consider charges with space–time properties and is thus only an approximation. To obtain the spin precession we note that the commutation relations of the ten tangent vectors are at every point those of a Lie algebra. The modified connection is then constructed on the group manifold G as a linear connection formed out of left invariant vectors only; it is left invariant and no more bi–invariant as in conventional K–K theories, but it is a metric connection. The connection has thus contortion terms on G (not necessarily also on B). These terms are functions of the curvature on B. The Einstein equations projected on B now result there in an Einstein term with a source term which is bilinear in the curvature and apart from this in a Maxwell–Yang term which is formed out of covariant derivatives of the curvature (Halpern 1996; Yang 1974). The term bilinear in the curvature conteracts gravitational collapse and leads to violations of equivalence. Such effects become significant only in domains where curvature is excessively large. The forces resulting from the dependence of the center of gravity on the system of reference are not taken into account here and will be dealt with in a following publication. The introduction of a connection which is not bi–invariant with contortion on G is a new feature, necessary to modify the K–K formalism to the present case.

References

Halpern, L., (1996): Astrophys. & Space Science **224**, 263
Yang, C.N., (1974): Phys. Rev. Lett. **33**, 44

MIX
Papier aus verantwortungsvollen Quellen
Paper from responsible sources
FSC® C105338

If you have any concerns about our products,
you can contact us on
ProductSafety@springernature.com

In case Publisher is established outside the EU,
the EU authorized representative is:
Springer Nature Customer Service Center GmbH
Europaplatz 3, 69115 Heidelberg, Germany

Printed by Libri Plureos GmbH
in Hamburg, Germany